Fundamentals of Protein Structure and Function

Fundamentals of Protein Structure and Function

Engelbert Buxbaum
Portsmouth, Dominica

Engelbert Buxbaum
Department of Biochemistry
Ross University School of Medicine
Box 266, Portsmouth Campus, Roseau
Commonwealth of Dominica (West Indies)
engelbert_buxbaum@web.de

Please note that glasses must be included with all returned books. Springer will not accept any returns if the glasses have been separated from the book.

Library of Congress Control Number: 2005928847

ISBN 978-0-387-26352-6 ISBN 978-0-387-68480-2 (eBook)

9 8 7 6 5 4 3 2 1

springer.com

Preface

... to describe the things that are, as they are.
(FRIEDRICH II. (1164–1250): On the Art to Hunt with Birds)

In those simple words described FRIEDRICH, emperor of the Holy Roman Empire of German Nation, around 1240 the job of a scientist. He was an avid hunter who particularly loved hunting with birds of prey. He realised that to really enjoy that sport one had to know as much as possible about animals, a knowledge that can be derived only by careful, unbiased observation. Thus he broke with the antique tradition of mixing observation with philosophy, religion and superstition, and wrote the first truly scientific book in history.

Today science is all-pervading in our life. We live about twice as long as people did in the days of FRIEDRICH and our lives are much more comfortable, as a result of scientific research and its technical application.

Proteins are the main actors in living organisms. While nucleic acids store the information required to make functional proteins, proteins do almost all the real work. So it is not surprising that after several years of emphasis in nucleic acid biochemistry, culminating in the human genome project, attention is now returning to proteins. Modern methods, in particular protein crystallography, have confirmed that "nature is greatest in the smallest things". So apart from it's utility, which makes training in science a necessity, understanding nature is, today just as much as in FRIEDRICHs days, also a source of joy and satisfaction. I wish every reader of this book a share in it.

How to read this book

I did not intend to write another biochemistry or cell biology text book, as there are many on the market already (for example [48, 56]). Rather, I wanted to focus on the fascinating world of proteins.

The four parts of this book follow largely the curriculum for biochemistry training of undergraduate medical and science students. The parts on protein structure and enzymology can be covered at the beginning of the course, the special proteins and membrane transport at the end, once molecular biology has been studied.

I have added some more advanced material for science students (placed into boxes), which could form the basis for seminars, term papers and the like. This material is not exam-relevant for medical students, but may broaden your view on the more basic topics.

Each chapter contains some example questions, which resemble those that might be asked in exams. The answers are provided in the appendix. Special emphasis was put on multiple-choice questions. One might question the didactic value of such questions, but fact is that they are now in common use. A short explanation on how to answer such question is also found in the appendix.

Acknowledgements

I wish to thank all my students, friends and colleagues who have given me their support and suggestions for this text, and who have gone through the arduous task of proof-reading. All remaining errors are, of course, mine. Corrections of errors, additions and updates will be published on http://ebuxbaum.fortunecity.de/corrections.html

This text was created with LaTeX using the MikTeX system (`www.dante.de`). Chemical structures were created with ISIS-Draw (`www.mdli.com`), drawings with XFig (`www.xfig.org`) under Linux or with TurboCAD (`www.turbocad.com`) under Windows. Gnuplot (`www.gnuplot.org`) was used for plotting mathematical functions. For graphic handling the Gnu Image Manipulation Package (Gimp, `www.xcf.berkely.edu/~gimp/gimp.html`) was used. Three-dimensional graphics were rendered with the Persistence Of Vision Ray-tracer (POV-Ray, `www.povray.org`). Molecular models were calculated from PDB coordinates obtained from OCA (`bip.weizmann.ac.il/oca-bin/ocamain`) using DeepView (formerly SwissPDB, `www.expasy.ch/spdbv/mainpage.html`).

A big "thank you" to all who made these programs freely available, or who maintain information services on the internet. Without your generosity this book would not be.

Ports Mouth (Dominica)
January 2007

Engelbert Buxbaum

Contents

Part I Protein structure

Part IV Membrane transport

Part I

Protein structure

1

Amino acids

1.1 Basic structure of amino acids

Amino acids are compounds which have a carboxy group at one end and an amino group at the carbon atom next to the carboxy group, the so called α-carbon (see fig. 1.1). Several amino acids contain additional acidic or basic groups.

The carboxy group will donate a proton to the amino group, so that an amino acid (in the absence of other acids or bases) will carry both a negative and a positive charge, making the whole molecule appear uncharged (zwitter-ion).

The simplest amino acid is glycine, where R is a hydrogen atom. Since the α-carbon carries only 3 different ligands (carboxy group, amino group and hydrogen), it is not enantiomeric. Thus glycine is not chiral, unlike all other amino acids which carry 4 different ligands on the α-carbon. We thus distinguish D- and L-amino acids, only L-amino acids occur in proteins and in mammalian metabolism. D-amino acids do occur in bacterial cell walls (murein) however.

1.2 The isoelectric point

From your chemistry lessons you know how to determine the $p\mathrm{K}_\mathrm{a}$of an acid or base, the pH at which half of the molecules are charged. A compound which can act as both acid and base (like an amino acid) has another important property: The isoelectric point pI, which is the pH at which the number of positive charges on the molecule is the same as the number of negative charges. At the pI the molecule would therefore appear uncharged. At this pH the molecules ability to interact with water is lowest, and therefore its solubility.

O OH
C
H_2N—C—H
R

$\rightleftharpoons$

O O⁻
C
$H_3\overset{+}{N}$—C—H
R

NH_2
H_2N—C^ε—C^δ—C^γ—C^β—C^α—C'(=O)OH
H_2 H_2 H_2 H_2 H

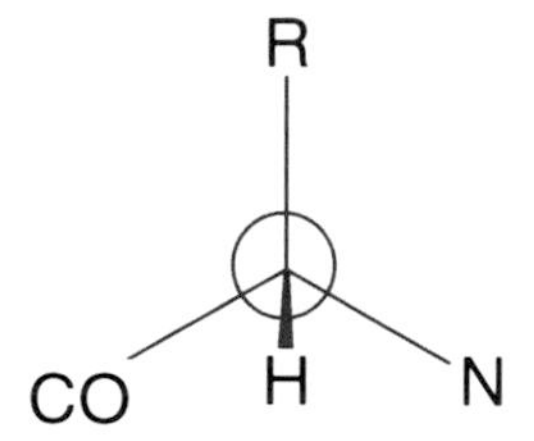

Figure 1.1. Top: Basic structure of an amino acid. Amino acids can form zwitter-ions. **Middle**: Nomenclature of carbon atoms, using lysine as example. The Carboxy-carbon is designated C', the following carbon atoms are labelled with the letters of the Greek alphabet. Sometimes the last C-atom is called ω, irrespective of the chain length. **Bottom**: In l-amino acids if the α-carbon is placed on the paper plane, with the hydrogen facing you, the remaining substituents read "CORN".

How can we get the isoelectric point? Looking at the titration curve of glycine (see fig. 1.3, top) we see that below $p\mathrm{K}_1$ most of the molecules bear one positive charge at the amino group, the carboxy group is uncharged.

Above $p\mathrm{K}_2$ it is the other way round, the carboxy group bears a (negative) charge and the amino group is uncharged. Right in the middle between $p\mathrm{K}_1$ and $p\mathrm{K}_2$ there is an inflection point in the titration curve, this is the $p\mathrm{I}$.

Thus we remember: In a chemical which has one acidic and one basic group, $p\mathrm{I}$ is the average of the two $p\mathrm{K}$ values:

$$p\mathrm{I} = \frac{1}{2} * (p\mathrm{K}_1 + p\mathrm{K}_2) \tag{1.1}$$

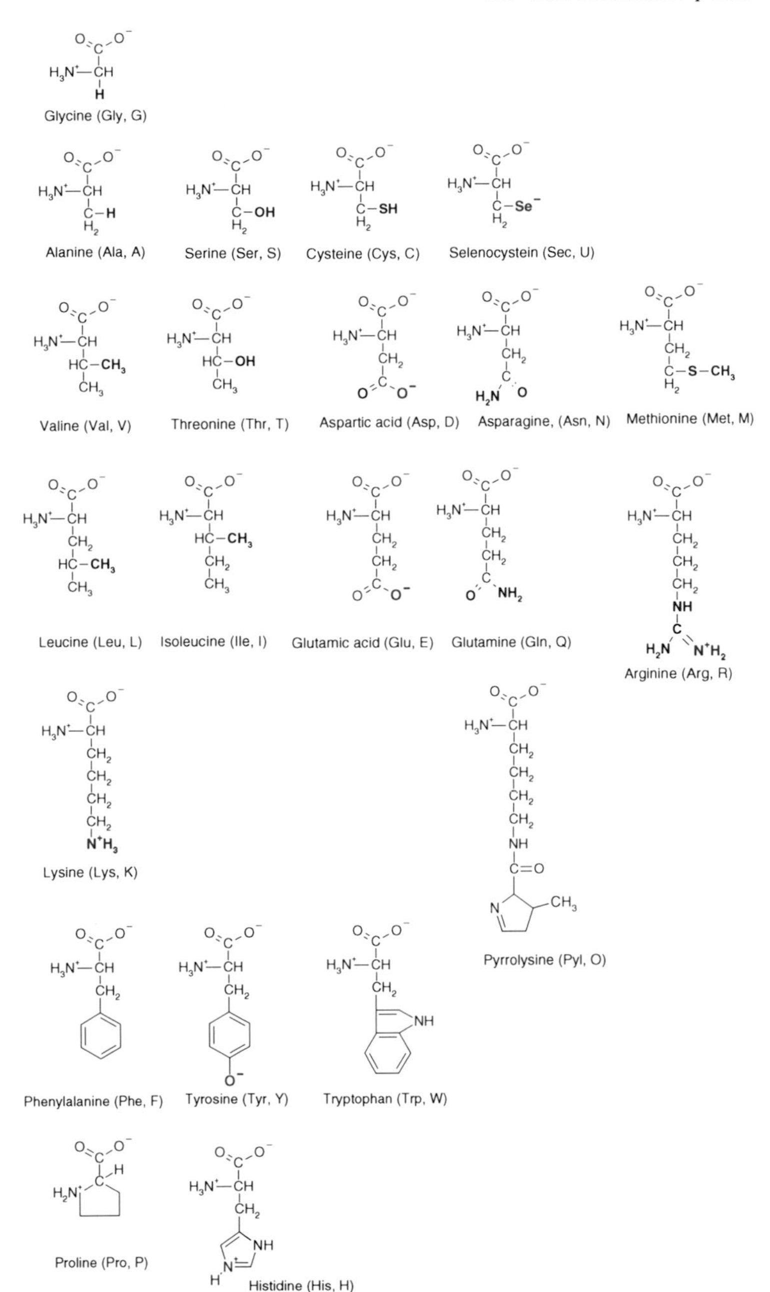

Figure 1.2. The 22 amino acids encoded by genes. Once incorporated into proteins, amino acids may be further modified. Pyrrolysine has been found only in bacteria, it is encoded by the "amber" stop codon UAG. Selenocysteine is encoded by the UGA "opal" stop-codon (see fig. A.1 on page 338). Acidic groups marked red, basic groups blue. Note that Thr and Ile have a chiral β- in addition to the α-carbon. Pyl has two chiral carbon atoms in the ring.

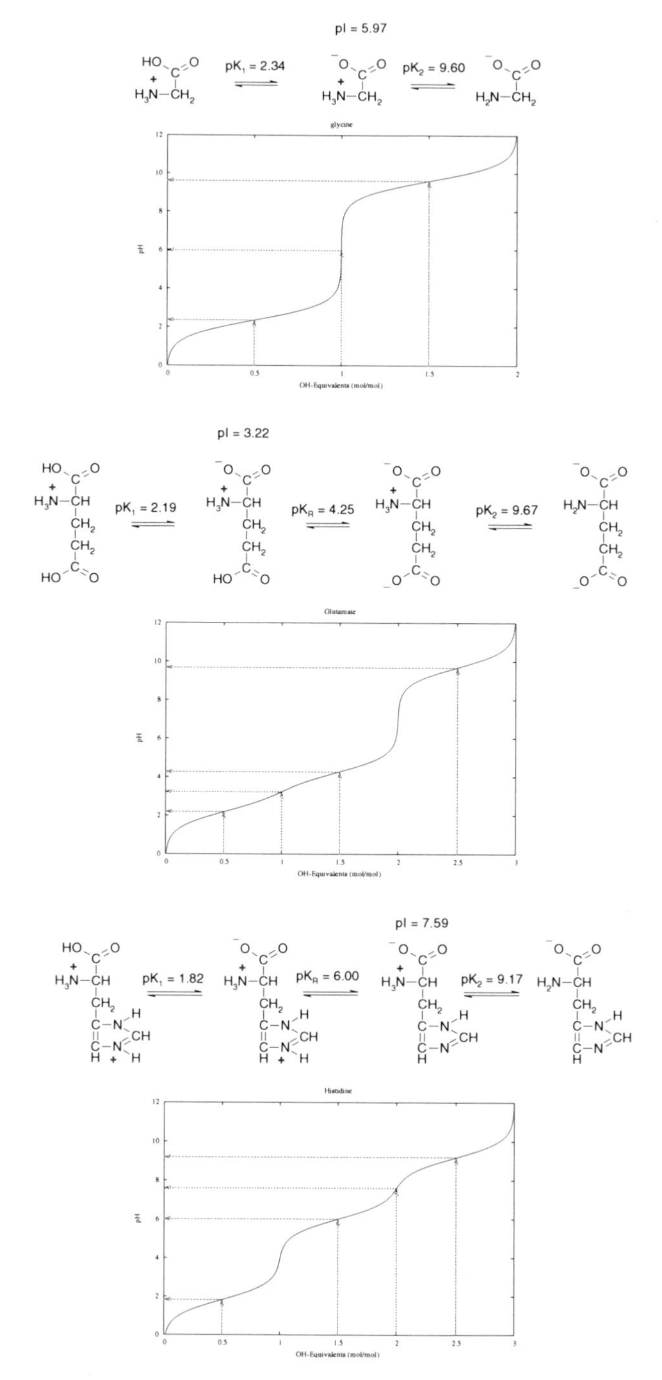

Figure 1.3. Titration curves of glycine, glutamic acid and histidine. At the isoelectric point amino acids carry an equal number of positive and negative charges, thus they have no net charge. The isoelectric point can be calculated as the average between the pK_a-values on each side of that uncharged form.

But how do we do it when there are 3 or more ionisable groups?

First we write down the various forms of the molecule, with the corresponding pK_a values between them. Then we count the positive and negative charges on each form and identify the electrically neutral one. The pI can be calculated as the average between the pK's on either side of it, just as it is done for molecules with only 2 ionisable groups.

Glutamic acid for example (see fig. 1.3, middle) at very low pH carries only one positive charge at the amino group (1st form). As the pH increases beyond 2.19, more and more protons are lost from the carboxy group on C'. This (2nd) form carries one positive and one negative charge, and is the neutral one. Beyond pH 4.25, a second proton is lost from the terminal carboxy group (3rd form, one positive, two negative charges) and beyond pH 9.67 a further proton is lost from the amino group (4th form, no positive and 2 negative charges). Thus the pI is calculated as the average of 2.19 and 4.25, which is 3.22.

The rational with histidine (see fig. 1.3, bottom) is similar. The electrically neutral form is the 3rd, and the pI is the average of 6.00 and 9.17, which is 7.59.

1.3 The one-letter code

In most cases amino acid names are abbreviated with the first three letters of their names. These abbreviations are easy to remember, however they use up unnecessary memory in computer data bases. The 20 common amino acids can be encoded by the 26 letters of the roman alphabet (leaving space for some rare amino acids), and then each amino acid in a protein sequence uses up only 1 rather than 3 bytes of storage space. Unfortunately, many amino acids start with the same letter (like Ala, Arg, Asp and Asn) thus we can not simply use the first letter to encode them.

The following list should help you to remember one-letter codes:

- Amino acids with a unique first letter: **C**ys, **H**is, **I**le, **M**et, **S**er, **V**al
- Where several amino acids start with the same letter, common amino acids are given preference: **A**la, **G**ly, **L**eu, **P**ro, **T**hr
- Letters other than the 1st letter are used for Asn (asparagi**N**), Arg (a**R**ginine), Tyr (t**Y**rosine)
- Similar sounding names: Asp (aspar**D**ic acid), Glu (glut**E**mate), Gln (**Q**tamine), Phe (**F**enylalanine)
- The remaining amino acids have letters that do not occur in their name: Lys (**K** close to L), Trp (**W** reminds of double ring), Sec (**U**), Pyl (**O**)

The letter **X** is used for "any" amino acid, **J** for Ile or Leu, the hyphen (-) for gaps.

Certain sequences of amino acids occur in several different proteins, were they serve a special function. Such **conserved motives** are usually named after their 1-letter amino acid abbreviations. Thus you may encounter KDEL-motives or DEATH-ATPases.

1.4 Biological function of amino acid variety

You may now ask why there are so many different amino acids. The answer is that these different molecules have different properties, that let them serve different functions in proteins (see also table 1.1).

One difference you have already learned about: There are amino acids whose side chains can bear positive or negative **charges**, while other side chains are always uncharged. Charged side chains have different pK_r, which can be influenced strongly by neighbouring amino acids, for example Cys (pK_r= 5–10), His (pK_r= 4–10) and the carboxylic acid group of Glu and Asp (pK_r= 4–7). This is important for proton transfer reactions in the catalytic centre of proteins. Ionisable groups also form the **ionic bonds (salt bridges)** which stabilise protein tertiary structure (see page 23).

Asp, Glu and His residues can chelate bivalent metal ions like Fe, Zn and Ca. This is important for enzymes with metal co-factors, in haemoglobin and in some regulatory proteins like calmodulin. Some amino acids are **hydrophilic** (= water friendly) because they carry polar groups (-COOH, $-NH_2$, -OH, -SH). Other amino acids are **hydrophobic** (= water fearing, fat friendly), with long aliphatic (Ile, Leu, Val) or aromatic (Phe, Trp) side chains. If these residues point into the solution, they force water molecules into a local structure of higher order (i.e. lower **entropy**), which is unfavourable. Burying these residues into the interior of the protein avoids this penalty, this is the molecular basis for **hydrophobic interactions**.

Some amino acids have **small** side chains (like glycine), others very big, **bulky** ones (like tryptophan). The small hydrogen residue of Gly not only fits into tight spaces (see section on collagen (page 202) for an important example), but because it has no β-carbon it can assume secondary structures (see page 16) that are forbidden for all other amino acids.

Proline has its nitrogen in a ring structure, which makes the molecule very stiff, limiting the flexibility of protein chains.

The SH-group of Cys, the unprotonated His and the OH-group of Ser and Thr are nucleophiles which are essential residues in the active centre of enzymes.

Table 1.1. Properties of the 21 amino acids encoded in a mammalian genome. Post-translational modification may change these properties considerably. The helix propensity measures the energy by which an amino acid destabilises a poly-alanine helix.

Amino acid	3-letter	1-letter	MW	pK_1 (-COOH)	pK_2 (NH_3^+)	pK_3 (Side group)	*p*I	Hydropathy	Abundance (%)
Alanine	Ala	A	89	2.34	9.69	-	6.01	+1.8	9.0
Arginine	Arg	R	174	2.17	9.04	12.48	10.76	-4.5	4.7
Asparagine	Asn	N	132	2.02	8.08	-	5.41	-3.5	4.4
Aspartic acid	Asp	D	133	1.88	9.60	3.65	2.77	-3.5	5.5
Cysteine	Cys	C	121	1.96	8.18	10.28	5.07	+2.5	2.8
Glutamic acid	Glu	E	147	2.19	9.67	4.25	3.22	-3.5	6.2
Glutamine	Gln	Q	146	2.17	9.13	-	5.65	-3.5	3.9
Glycine	Gly	G	75	2.34	9.60	-	5.97	-0.4	7.7
Histidine	His	H	155	1.82	9.17	6.00	7.59	-3.2	2.1
Isoleucine	Ile	I	131	2.36	9.68	-	6.02	+4.5	4.6
Leucine	Leu	L	131	2.36	9.60	-	5.98	+3.8	7.5
Lysine	Lys	K	146	2.18	8.95	10.53	9.74	-3.9	7.0
Methionine	Met	M	149	2.28	9.21	-	5.74	+1.9	1.7
Phenylalanine	Phe	F	165	1.83	9.13	-	5.48	+2.8	3.5
Proline	Pro	P	115	1.99	10.96	-	6.48	-1.6	4.6
Selenocysteine	Sec	U	168	2.16	9.40	5.20	3.68		rare
Serine	Ser	S	105	2.21	9.15	13.60	5.68	-0.8	7.1
Threonine	Thr	T	119	2.11	9.62	13.60	5.87	-0.7	6.0
Tryptophan	Trp	W	204	2.38	9.39	-	5.89	-0.9	1.1
Tyrosine	Tyr	Y	181	2.20	9.11	10.07	5.66	-1.3	3.5
Valine	Val	V	117	2.32	9.62	-	5.97	+4.2	6.9

Some amino acids confer properties to the protein which can be used in the laboratory: Met binds certain heavy metals which are used in X-ray structure determination and reacts with cyanogene bromide (Br−C≡N) to cleave the protein at specific sites. Cys and Lys are easily labeled with reactive probes. Aromatic amino acids, in particular Trp absorb UV-light at 280 nm, this can be used to measure protein concentration. In addition they show fluorescence, which can be used to measure conformational changes in proteins.

Exercises

1.1. Which of the following statements is/are *true* about amino acids?

1) All protein forming amino acids are chiral.
2) Threonine has two chiral centres.
3) Only L-amino acids occur in living organisms.
4) All amino acids have a pI-value.

1.2. Define the isoelectric point of a compound.

1.3. Glycine is chiral
because
the α-carbon contains only 3 different substituents.

1.4. Connect the following amino acids with their 1-letter code:

1) Valine	A) D
2) Tryptophane	B) V
3) Aspartic acid	C) W
4) Tyrosine	D) X
	E) Y

1.5. Connect the following properties with amino acids:

1) hydrophobic	A) Tryptophane
2) positively charged	B) Serine
3) small	C) Lysine
4) polar	D) Glycine
5) aromatic	E) Glutamine

1.6. Lysine has the $p\mathrm{K_a}$-values 2.18 (carboxy-group), 8.95 (α-amino group) and 10.53 (ϵ-amino group). The pI is ______.

1.7. The following compound is a dipeptide,
because
it contains two covalently linked amino acids

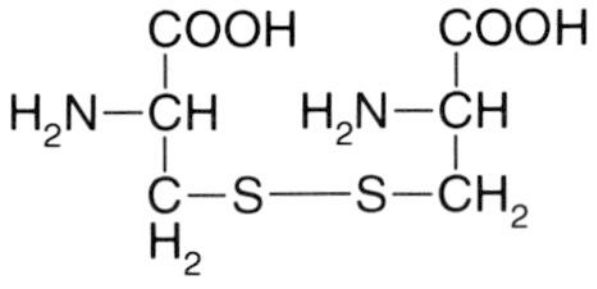

$$
\begin{array}{llll}
 & \mathrm{COOH} & & \mathrm{COOH} \\
 & | & & | \\
\mathrm{H_2N-} & \mathrm{CH} & \mathrm{H_2N-} & \mathrm{CH} \\
 & | & & | \\
 & \mathrm{C-S} & \mathrm{\!\!-\!\!-S-} & \mathrm{CH_2} \\
 & \mathrm{H_2} & &
\end{array}
$$

2

Protein structure

Nature is greatest in the smallest things.
(C. LINNAEUS)

2.1 Primary structure

The sequence of amino acids in a protein is called its **primary structure**. In biochemistry, this is always given starting with the N-terminal and ending with the C-terminal amino acid, because this is the order in which amino acids are added during protein synthesis in the cell (this process is discussed in detail in textbooks of molecular cell biology, e.g. [48]).

C_α, C', the nitrogen and the oxygen atom of the peptide bond form a single plane. The bond between C' and N is somewhat shorter than a normal C−N single bond, because of **mesomery** with the C=O double bond (see fig. 2.3).

No rotation is possible around double bonds, this is also true for the "partial" double bond between C' and the peptide nitrogen. Thus the R-groups of the amino acids can occur in *cis*- or in *trans*-configuration (see fig. 2.1). Because of the bulky R-groups, the *trans*-configuration is more stable for most amino acids (99.95 % probability). The exception is Pro, which occurs in *cis*-configuration much more frequently than other amino acids (6 % probability).

On the other hand, the N-C_α and C_α-C' bonds are normal single bonds, rotation around those is possible. The angles of rotation are named ϕ and Ψ respectively. Rotation in the peptide chain is limited by two factors. First, at certain angles ϕ and Ψ around one amino acid atoms of that amino acid would collide with atoms of the following amino acid. These angles are forbidden, by definition we assign the value 0° to the angle that would result in collision of C'_n with N_{n+1}. Rotation angles then can have values between −180°

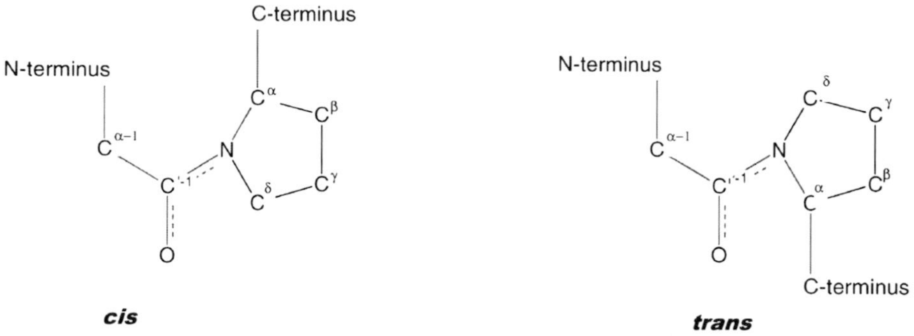

Figure 2.1. *cis-trans*-isomery around the peptide bond. Because the C'^{-1}-N-bond has the character of a partial double bond, rotation around this bond can not occur and *cis-trans* isomery results. For steric reasons the *trans*-configuration is much more probable than the *cis*-. Pro is unusual in that the cis-configuration has a probability of 5–6 %, which is about 100 times higher than with other amino acids.

and +180°. Additionally, size and charge of the R-groups can make certain positions more stable than others.

Thus in a plot of ϕ versus Ψ (RAMACHANDRAN-plot [37, 64], see fig. 2.6) there are regions which are sterically forbidden, there are fully allowed regions with no steric hindrance, and there are unfavourable regions which can be assumed by slight bending of bonds.

Proline is again a special case because the peptide nitrogen is part of a ring structure, this limits ϕ to values between −35° and −85°. As we will see in a moment, this has considerable consequences for protein secondary structure.

Because **glycine** has only a hydrogen as R-group, steric hindrance is much less of a problem than with other amino acids. Thus in a RAMACHANDRAN-plot Gly can be found in regions forbidden for other amino acids (see fig. 2.6e).

The great variety of proteins that can be observed today has arisen from a much smaller number of **ancestors** during evolution. This can be shown by comparing the primary structure of proteins, the more similar they are, the closer they are related.

Amino acid sequences of proteins and the nucleotide sequences of their genes are stored in data bases which are accessible from the world wide web, like the non-redundant protein sequence data base Owl (`http://www.bioinf.man.ac.uk/dbbrowser/OWL/`) and ExPASy (`http://au.expasy.org/`).

Multiple sequence alignment services are offered by BCM (`http://searchlauncher.bcm.tmc.edu/multi-align/multi-align.html`).

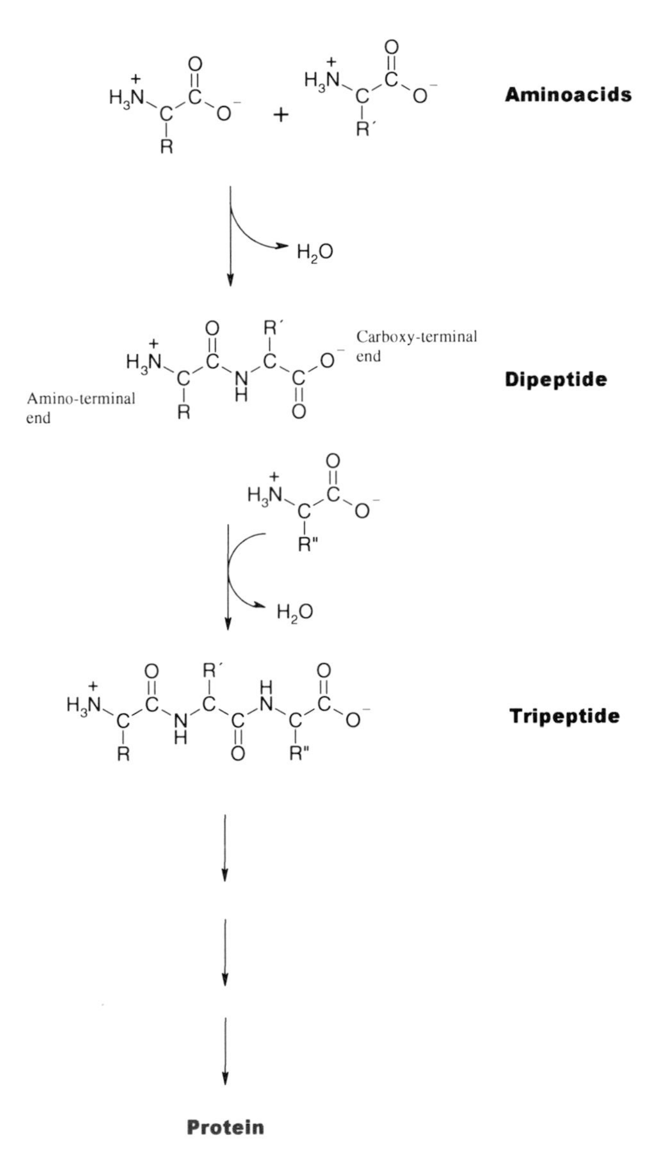

Figure 2.2. Polycondensation of amino acids to peptides and protein. Polycondensation is a reaction were organic molecules react with each other via their functional groups, producing small molecules (here: water) in addition to a macromolecule.

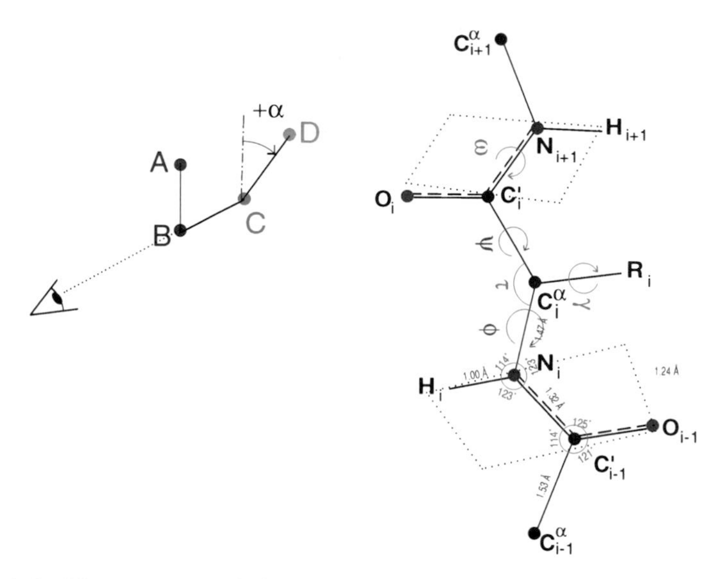

Figure 2.3. The geometry of the peptide bond. **Left**: Standard way to determine the dihedral angle of a bond (here between atoms B and C). Orient that bond into the paper plane, so that the neighbouring atoms (here A and D) point upwards. Then measure the angle formed, clockwise is positive, anticlockwise negative. **Right**: Because of mesomery, the dihedral angle of the bond between the carboxy-carbon (C') and the nitrogen (ω) is fixed to 180°, with N, H, C and O lying in a single plane. Slight deviations are possible, but rare. The bond angle of C^α (τ) is usually the tetrahedral angle 109.5°, but some flexing (± 5°) is occasionally found. Variable are the dihedral angles ϕ and Ψ, which determine the secondary structure of a protein (see next section).

> For taxonomic questions in general, `http://www.ncbi.nlm.nih.gov/Taxonomy/taxonomyhome.html` may be consulted.
>
> Several discussion groups in the `bionet` hierarchy on usenet deal with proteins, especially `bionet.molbio.proteins`.

2.2 Secondary Structure

Secondary structure describes the local conformation of the amino acids in the protein chain. It is stabilised by **hydrogen bonds** (see fig. 2.5) between the amino- and keto-groups of the peptide bonds, which carry a partial positive and negative charge, respectively (see fig. 2.3b). Although each hydrogen bond has only a relatively small bond energy, the sum of the bond energies over all hydrogen bonds in a protein is considerable.

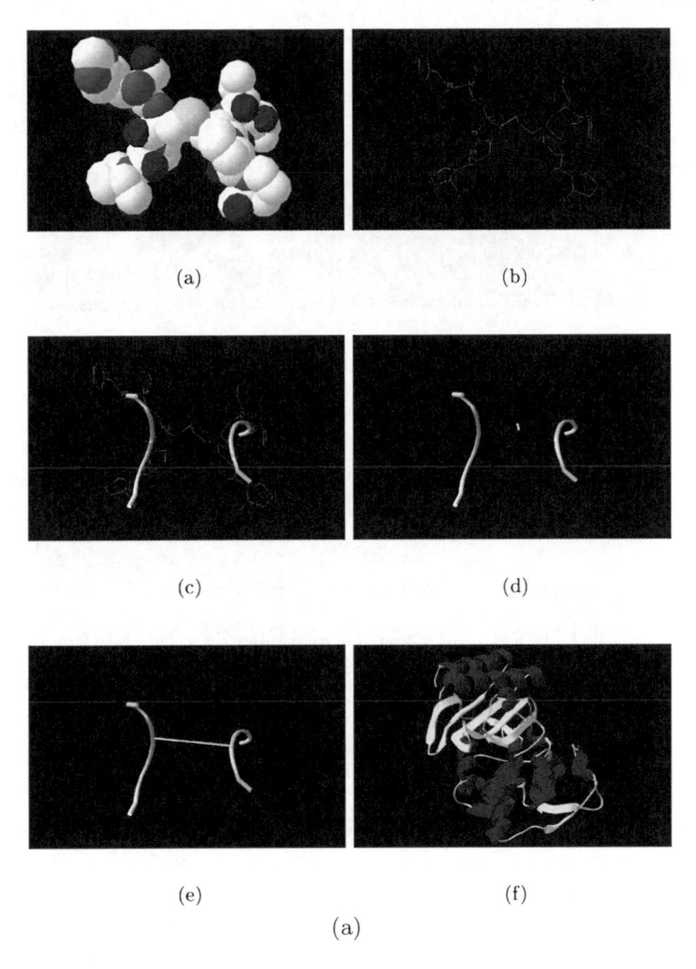

Figure 2.4. Interpretation of structure diagrams of proteins, using the pentapeptide HTCPP. **a**: Spacefilled diagrams show the true (VAN DER WAALS-) extension of a molecule, but even in short peptides clarity is lost. **b**: A wire diagram is clearer. The centre of the atoms are connected by thin lines, hydrogens are not drawn. Wire diagrams look a little like structural formulas in organic chemistry, however, atoms are shown in their true 3-dimensional arrangement. **c**: For larger proteins even wire diagrams would be too cluttered. Thus the atoms forming the protein backbone are connected by a thick line, which is used to represent the amino acid chain. **d**: If the protein contains disulphide bonds, showing only the backbone trace leaves the disulphide bond dangling in free space, thus **e**: this bond is often shown (incorrectly, but easier to interpret) connecting the backbone traces. **f**: A further abstraction is achieved by showing elements of secondary structure instead of the backbone trace: alpha helices as red helices and β-strands as yellow arrows, connected by the backbone trace of coils and turns (in grey). Other colouring schemes used in this book are N-terminal (red) to C-terminal (purple), different colours for different chains or Shapely colours for different amino acids. Shown here is β-lactamase (PDB-code 1m40, a protein that confers penicillin resistance to bacteria).

Figure 2.5. Hydrogen bonding in α-helix (left, cytochrome b_{562}, PDB-code 256B) and β-sheets (right, *E. coli* OmpA, PDB-code 1QJP). In an α-helix all hydrogen bonds between keto- and amino-groups in the protein backbone occur between neighbouring amino acids of the same helix. In β-sheets however all such hydrogen bonds occur between amino acids in different strands, alternating between the right and left neighbour.

There four particularly common structural motives [66]:

α-helix: The polypeptide chain is wound around an imaginary axis, with 3.6 amino acids per turn. Each turn is about 5.4 Å long, the pitch per amino acid is 1.5 Å. The angle between successive residues is about 100°, $\phi = -57°$, $\psi = -47°$ The R-groups point outward. α-helices are stabilised by hydrogen bonding between a carboxy-oxygen (partial negative charge) and the amino hydrogen (partial positive charge) 4 amino acids further along the chain. This "hydrogen bond loop" totals 13 atoms, hence α-helices are sometimes called 3.6_{13}-helices. In theory the turn of the helix could be clockwise or counterclockwise. However, in practice only **counterclockwise** rotation is observed. Such helices are called **right-handed**, because if you hold your right hand with the thump pointing from the N- to the C-terminus, the tips of the curled fingers move anticlockwise. Similarly, a clockwise helix would be called left-handed. Left-handed helices are improbable for L-amino acids, because the β-carbon and carbonyl-oxygen would collide.

β-strand: The polypeptide chain is stretched out ("extended"), no hydrogen bonding occurs between carboxy-oxygen and amino-hydrogens in neighbouring amino acids. However, if several strands lay next to each other, hydrogen bonds are possible between those chains instead. Such strands can be parallel (all carboxy-terminal ends are at the same side) or antiparallel. The R-groups point above and below the plane of the sheet, hydrogen bonds and ionic bonds between them may also occur. In schematic diagrams of protein structure each β-strand is drawn as a broad arrow. β-sheets are rarely flat, but twisted, the peptide bond planes of different strands form a right-handed coil. Large anti-parallel sheets may roll up into a **β-barrel** (see fig. 2.5).

turns: of 180° are important for example between the different strands of are antiparallel β-sheet. They can contain 4 amino acid residues (β-turn, frequent, with a hydrogen bond between $\mathrm{N{-}H}_i \cdots \mathrm{O{=}C}_{i-3}$) or 3 ($\gamma$-turn, rare). Turns often contain Gly (smallest amino acid) or Pro residues, the latter because of its specific value of ϕ, in addition the C^α and C^δ of Pro can undergo $\mathrm{CH}\cdots\pi$ interactions with neighboring aromatic amino acids which – although not as strong as regular hydrogen bonds – can stabilize the turn. Turns may also be found in the catalytic centre of an enzyme (with a $\mathrm{C\text{-}H}\cdots\pi$ bond between Pro and an aromatic substrate).

coils: any structure except those mentioned above. Note that amino acids in coils still have a defined position in the structure of a protein, thus the terms "random coil" or "unordered" sometimes found in the literature are incorrect. These areas have an important function too, because they add flexibility to the protein and allow for conformational changes, for example during enzymatic turnover.

In addition to the α-helix there are two other, much rarer helical conformations, the 3_{10}-helix with 3 residues per turn and a hydrogen bond between residues i and i+3 ($\phi = -50°$, $\psi = -25°$) and the π-helix with 5 residues per turn and a hydrogen bond between residues i and i+5.

There are some other structures which occur in only a few proteins, but have important functional roles. We will discuss those when we talk about some special proteins (e. g. collagen, see page 202). Each secondary structure can be characterised by the ϕ and Ψ angles in the protein backbone (see fig. 2.6).

Since β-strands do not contain any internal hydrogen bonds it is often assumed that they are stable only as part of a β-sheet. However, even single β-strands are stable due to entropic stabilisation. β-strands occupy the largest continuous permissible area in the RAMACHANDRAN-plot. In other words, in β-strands amino acids have considerable conformational freedom, equivalent to a large entropy. In addition, bulk solvent organisation is minimal in β-strands.

2.2.1 How do we determine secondary structure?

Three experimental methods are available for the determination of protein secondary structure, X-ray diffraction and electron microscopy of protein crystals and nuclear magnetic resonance (NMR) of proteins in solution.

X-ray crystallography

In a crystal the atoms and molecules have a very regular pattern. If an X-ray beam is passed through the crystal, the pattern acts as a grid and diffraction occurs. The diffraction pattern is recorded on a photographic film or (nowadays) a solid state detector. From this diffraction pattern it is possible in principle to calculate the position of the atoms in the protein molecule.

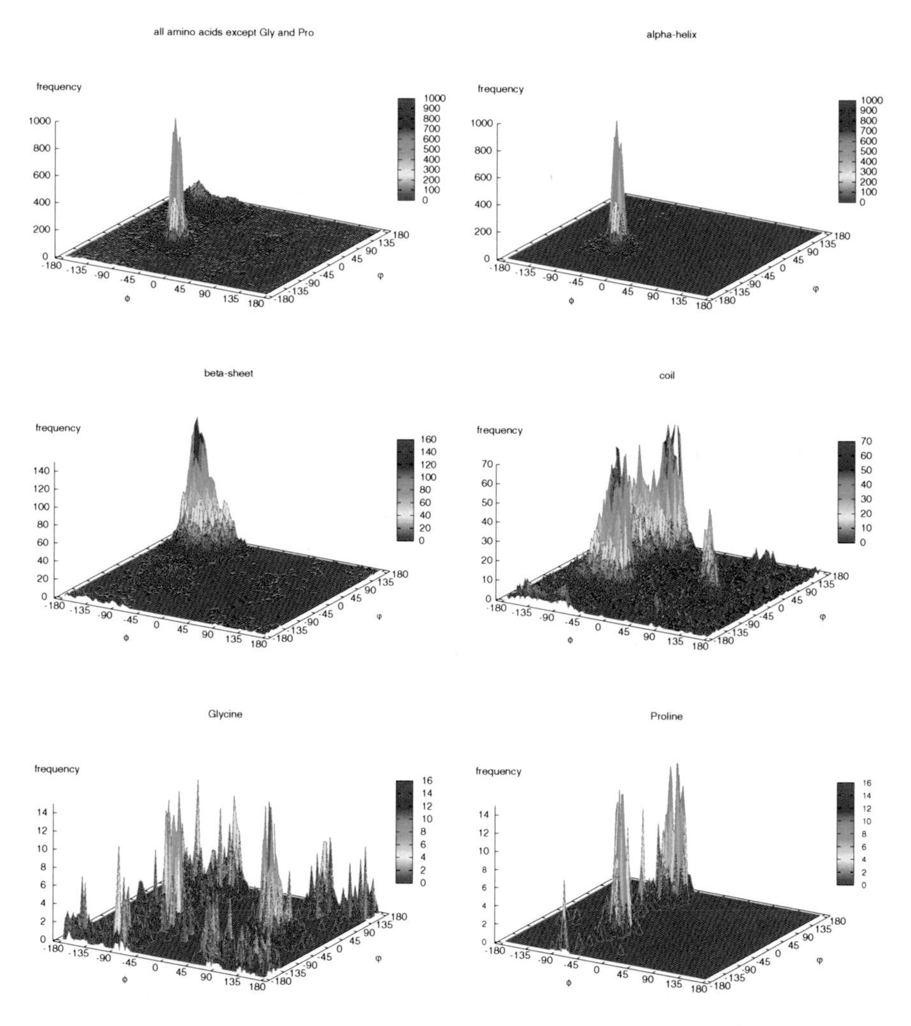

Figure 2.6. Actual dihedral angles found in 65 372 amino acids from 83 proteins discussed in this text. Amino acids of a particular secondary structure occupy a particular area of the RAMACHANDRAN-plot, Gly and Pro occur in conformations that other amino acids can not assume. Note the different scales of the plots.

Protein crystals need to be grown from a solution [50], but even for soluble proteins it is often difficult to get good crystals that give diffraction patterns with the required resolution to trace the peptide backbone (≈ 3 Å). Nevertheless, considerable progress has been made in recent years.

Electron Microscopy

For the proteins that occur in a biological membrane it is even more difficult to get crystals (the membrane has to be dissolved with detergents and the protein-detergent complex crystallised). However, such proteins can form 2-dimensional arrays ("2-D crystals") in the membrane, which can be investigated using the electron microscope. Because the molecules in such a 2-D crystal all have the same orientation, it is possible to calculate an "average" image of a protein molecule. Unfortunately, the resolution of such images is in most cases limited to 10–20 Å, which is not enough to trace the peptide backbone of the protein. Some larger structures like the proteasome can be investigated directly in the cell [52].

Nuclear magnetic resonance

Nuclei with an odd number of nucleons (^{1}H, ^{13}C, ^{31}P), or an odd number of both protons and neutrons (^{14}N) have an angular momentum called spin. This spin of a positively charged body results in the generation of a tiny magnetic field, which is measured by an NMR spectrometer [65]. Because part of the field is shielded by the electrons surrounding the atoms, NMR can detect the kind of bonds that an atom is involved in, and hence the secondary and tertiary structure of proteins.

With the instruments currently available in most NMR labs, the method is limited to proteins smaller than about 20 kDa. Better machines now coming onto the market will push that limit. Unfortunately, NMR instruments are very expensive and require a considerable sample size (several mg of pure protein).

However, it is possible to investigate the structure of proteins in solution, no crystallisation (which might result in artifacts) is required. One would therefore expect results of greater biological significance than those obtained by X-ray diffraction. A comparison of structures determined by the two methods shows fortunately that they are fairly similar, giving us faith in both.

Computer predictions

The biggest problem for the determination of protein structure is the vast amount of highly purified protein required (≈ 100 mg) for any of the above methods. Many proteins of particular biological interest occur in very small amounts in our body. In the last 20 years the sequences of many proteins have been determined by DNA sequencing (an almost complete sequence of the human genome was published in 2001 [16, 17, 78]). This gives us the chance to produce interesting proteins in larger quantities by genetic engineering, alleviating the sample size problem somewhat. However, experimental structure elucidation is still a difficult, tedious and expensive business.

This could be circumvented if it were possible to predict the secondary structure of a protein from its known primary structure. After all, a protein can fold unaided (at least under ideal circumstances). Thus all information necessary for folding must be contained in its primary structure (this is known as the ANFINSEN-hypothesis [4]).

This has in part to do with bond angles, for example Pro rarely occurs in β-sheets, because the nitrogen is part of a ring structure and ϕ can not assume the value required. On the other hand, bulky R-groups can prevent the tight packing of amino acids required for the formation of α-helices. Secondary structures are stabilised, if R-groups with opposite charge are brought close together, they are destabilised by R-groups with the same charge.

Attempts to use these principles for computer aided structure prediction have been made since the 1970s. The idea is to look at proteins with known secondary structure and to determine which amino acids (or amino acid combinations) occur most frequently in a particular secondary structure and to use this statistical information to predict the secondary structure of newly sequenced proteins. Some progress has been made in this field, and for soluble proteins the secondary structure at any amino acid can be determined with about 75 % accuracy. However, this means that a quarter of the amino acids in a protein will be assigned wrong secondary structures. Since the secondary structure of only a few membrane proteins has been solved (and the tables derived for soluble proteins may not apply), structure prediction for them is even more uncertain. Such predictions should therefore be viewed with due caution.

If the structure of a related protein (with similar sequence) has been solved, one can also try to "thread" the unknown protein onto this template. This method is somewhat more reliable than prediction by statistical methods and has become more and more important as the number of known protein structures available for comparison increased.

`http://www.predictprotein.org/` and `http://bioinf.cs.ucl.ac.uk/psipred/` are servers that offer protein structure prediction for submitted sequences, using different algorithms.

2.3 Tertiary structure

Tertiary structure describes the global conformation of a protein, in other words, the way in which the elements of its secondary structure are arranged in space. Tertiary structure is determined by interactions occurring between amino acids which can be long distances apart in the amino acid sequence, but which are brought close together in space by the way the protein folds. It is determined by ionic interactions between charged amino acid R-groups, by hydrophobic interactions (hydrophobic R-groups tend to be buried inside the protein, hydrophilic R-groups tend to occur at the surface, where they can interact with water) and by hydrogen and VAN DER WAALS bonds.

In transmembrane segments, hydrophilic amino acids are in contact with water (**snorkeling effect**) and hydrophobic amino acids are found in contact with the fatty acid tails of the lipids (**anti-snorkeling effect**). The lipid/water interface is formed by 3 amino acids with special properties: Trp, Tyr and Lys (the so-called **aromatic belt**). They have in common relatively long molecules which are hydrophobic, but have a hydrophilic (polarised or ionised) end and can make contact with lipids and water at the same time. Thus a transmembrane segment has a well-described position within the membrane and can not bob up and down.

Tertiary (and quaternary) structure can be stabilised by S-S bonds between Cys residues (R−SH + HS−R' → R−S−S−R' + 2 [H]). This is an oxidation (removal of hydrogen), which will not normally occur in the reducing environment of the cytosol. However, the environment inside the ER is oxidising. Thus **disulphide bridges** are found more frequently in cell surface and secreted proteins than in cytosolic ones.

Some proteins have several **domains**, that is individually folding regions connected by short segments. These individual domains can be isolated by gentle proteolysis, they may maintain not only their structure, but even their catalytic function. For example the chaperone Hsc70 has three domains: an ATPase-, a peptide-binding- and a regulatory domain. Gentle treatment with chymotrypsin will digest the links between those domains. The isolated ATP-ase domain will still hydrolyse ATP.

If one looks at many different proteins, one will find certain patterns in the way elements of secondary structures are arranged. These folding patterns are called **motives**. It is interesting to note that motives are much more stable during evolution than amino acid sequences. In other words, some proteins can be shown to be homologous by their folding patterns, even though they no longer have significant similarity in their amino acid sequence (for example the muscle protein actin, the enzyme hexokinase and the chaperone Hsc70).

According to their folding pattern, protein domains may be hierarchically classified into groups. Since such classification is somewhat subjective, different schemes have been suggested. One commonly used scheme is the **Structural Classification of Proteins (SCOP)** database `http://scop.mrc-lmb.cam.ac.uk/scop/`. Classification at present can not be done automatically, but requires expert knowledge.

The following taxa are used in SCOP:

Class Coarse classification according to the relative content of α-helix and β-strand.

Fold Major structural similarity, the proteins have identical secondary structure elements (at least in part) and the same topological connections. However, there may be considerable variation in peripheral regions of a domain. Similarities may arise from common origin or from convergent evolution.

Superfamily Domains have a common folding pattern and their functions are similar, but sequence identity may be low. The ATPase domains of actin, hexokinase and Hsc70 are an example for a superfamily. Common evolutionary origin is probable.

Family Proteins with high sequence homology ($> 30\,\%$ identity) and/or similar function. Proteins clearly have an evolutionary relationship. Identical proteins are subclassified by species.

For a basic understanding of protein folding only the class is relevant:

a) **all-α** Proteins which contain only α-helices, or where the content of β-strands is at least insignificant.

b) **all-β** Proteins which contain only β-strands, or where the content of α-helices is at least insignificant.

c) **α/β** Proteins which contain alternating or interspersed α-helices and β-strands. Mainly parallel β-sheets (beta-alpha-beta units).

d) **$\alpha + \beta$** Proteins which contain segregated α-helices and β-strands. Mainly antiparallel β-sheets.

e) **Multi-domain proteins** Folds consisting of two or more domains belonging to different classes

f) **Membrane and cell surface proteins** (excluding proteins from the immune system). Usually the transmembrane domains are α-helical, but β-barrels may also occur.

g) **Small proteins** Usually dominated by metal ligand, haeme, and/or disulphide bridges

h) **coiled coil proteins** α-helices wound around each other.

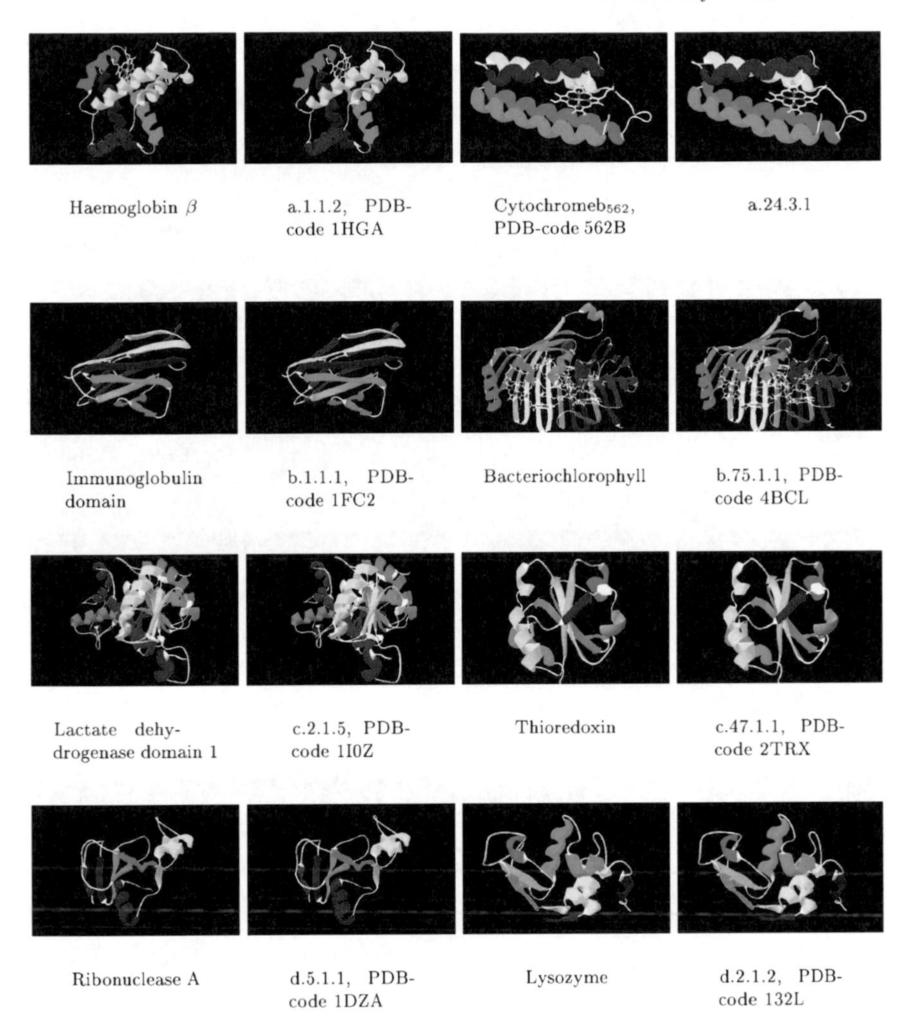

Figure 2.7. Examples for $\alpha, \beta, \alpha/\beta$ and $\alpha + \beta$ protein structures. There are two images for each protein (right and left eye view) to achieve a stereo effect. Some people can see the stereo image unaided (try looking "cross-eyed", at the right image with the left eye and vice versa), others need special goggles like those supplied with this book.

Additionally there are classes provided for low resolution structures, peptides, protein fragments and artificial proteins. However, since those are not natural entities we need not concern us here with them. Apart from SCOP there are also other approaches for protein classification, in particular CATH (`http://www.cathdb.info/`) and FSSP, (`http://www.sander.ebi.ac.uk/dali/fssp/`) but these yield largely similar results.

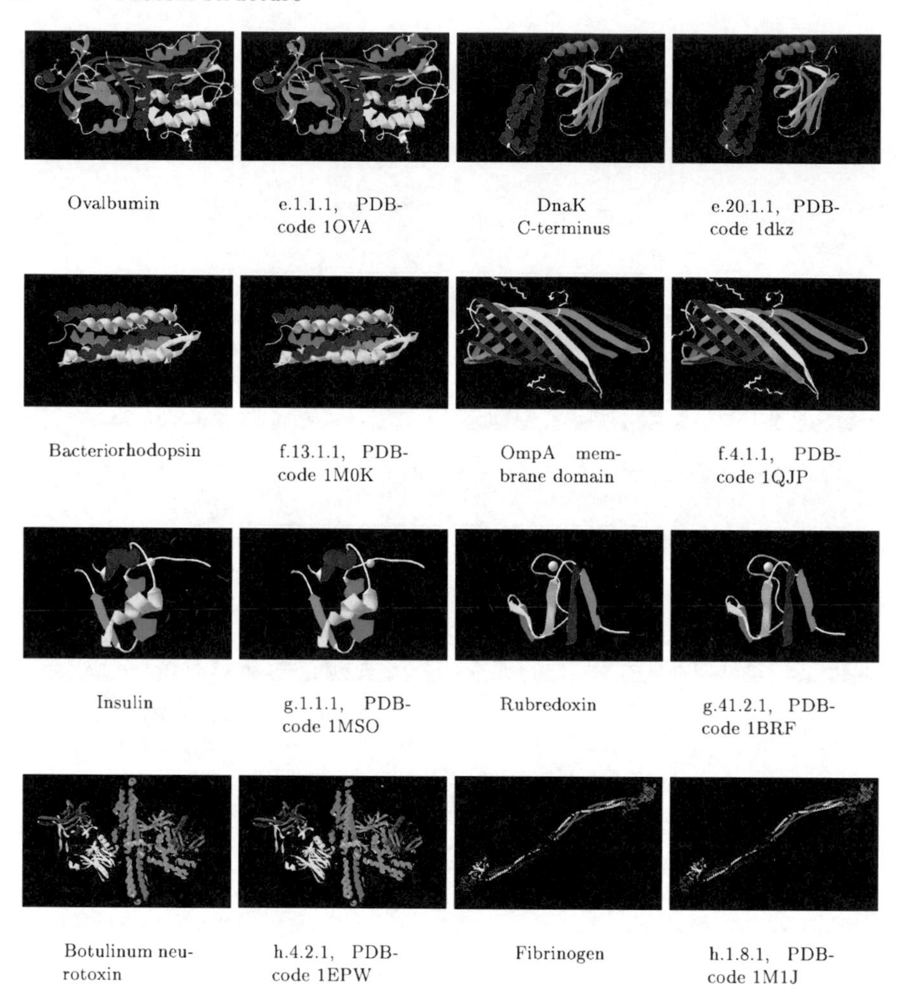

Figure 2.8. Stereo views of multidomain and membrane proteins, small proteins and coiled-coils.

Table 2.1. Number of entities in the SCOP-database (01/2005)

Class	folds	superfamilies	families
a) All alpha proteins	226	392	645
b) All beta proteins	149	300	594
c) Alpha and beta proteins (a/b)	134	221	661
d) Alpha and beta proteins (a+b)	286	424	753
e) Multi-domain proteins	48	48	64
f) Membrane and cell surface proteins	49	90	101
g) Small proteins	79	114	186
Total	971	1589	3004

Proteins can be described by a **set of concise classification strings (sccs)** according to their structure, for example b.2.1.1 (class b = all β, fold 2 = NAD(P)$^+$-binding ROSSMANN-fold domains, superfamily 1 = Alcohol dehydrogenase-like and family 1 = Alcohol dehydrogenase). Within families, proteins are sorted by species and isoform.

> Protein structures are stored in the Brookhaven Protein Data Bank (PDB) in a unified format that can be used by modelling software like protein explorer (`http://www.umass.edu/microbio/chime/explorer/index.htm`), DeepView (formerly known as Swiss-PDB, `http://www.expasy.ch/spdbv/mainpage.html`) or Rasmol (`http://www.umass.edu/microbio/rasmol/`). Coordinates may be obtained from PDBlite (`http://oca.ebi.ac.uk/oca-bin/pdblite`), OCA (`http://bip.weizmann.ac.il/oca-bin/ocamain`) or, if the EC-number is known, from `http://www.ebi.ac.uk/thornton-srv/databases/enzymes/`. The protein structures presented in this book were created with DeepView from data files obtained from OCA.

2.4 Quaternary structure

Quaternary structure describes how several polypeptide chains come together to form a single functional protein. Like tertiary structure it is determined by ionic and hydrophobic interactions between amino acid R-groups. Many proteins consist of several subunits. Depending on the number of subunits, we speak of monomers, dimers, trimers and so on. Depending on whether these subunits are identical or not, we put homo- or hetero- in front. Thus a heterodimer is a protein consisting of two different polypeptide chains. As we will see later, association of several subunits into a protein has important consequences for its function, which is often lost if the subunits are separated.

In some proteins several polypeptides come together to form a subunit, which repeats several times. Such subunits are called **protomers**. For example, haemoglobin is a diprotomer, each protomer consists of an α- and a β-chain (see fig. 7.2 on page 108).

2.4.1 Protein structure and the development of pharmaceuticals

Knowledge of the structure of enzymes and receptors is essential for the design of pharmaceuticals.

For example human immunodeficiency virus (HIV) protease is required for splitting an inactive precursor protein in the virus envelope into two active proteins. Without these proteins, the virus can not bind to its target cells, such a virus would be non-infectious.

HIV-protease is a homodimer, each subunit 99 amino acids long. They are held together by an antiparallel β-sheet formed by amino acids 1-4 and 96-99 of both subunits (see fig. 2.9).

This brings the catalytic aspartate residues (D^{25} in each subunit) together, thus forming the catalytic site of the enzyme. Several pharmaceuticals are

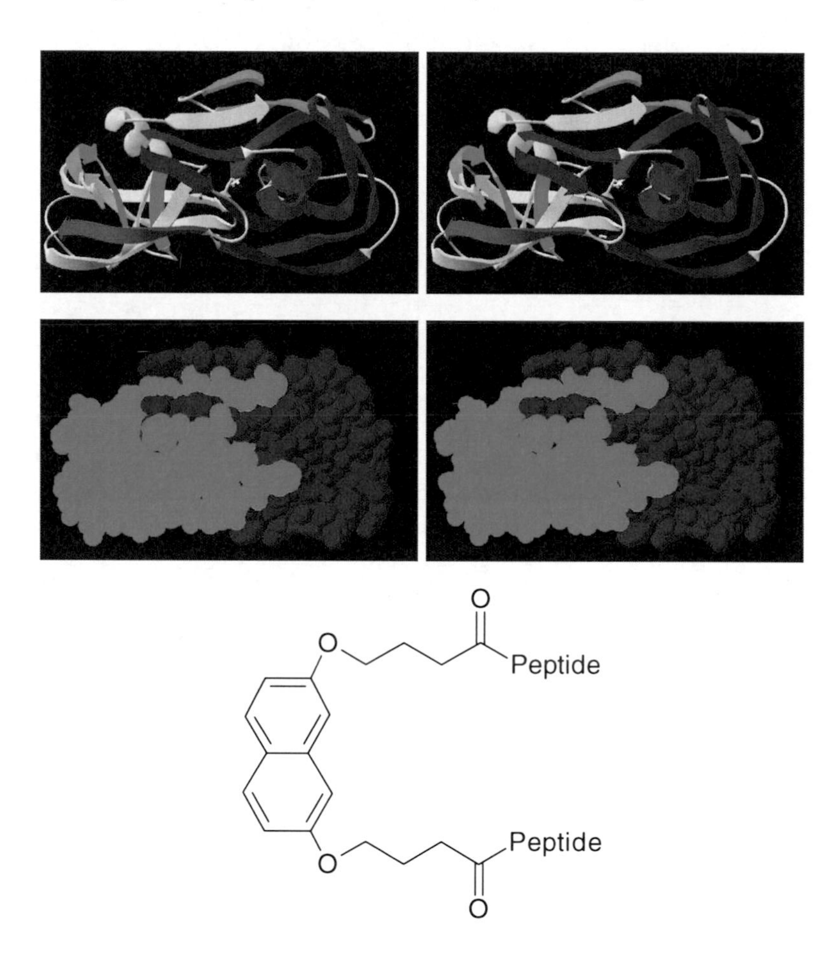

Figure 2.9. Top: Dimerisation of HIV-protease (PDB-code 1DAZ) occurs by the formation of an antiparallel β-sheet from the ends of both subunits. This brings the 2 catalytic aspartate residues (D^{25} and $D^{25\prime}$) close together. **Middle**: Space filling model of HIV-Protease. The β-sheet formed by the N- and C-terminal ends of both subunits is clearly visible.**Bottom**: Substances with two peptides linked by a stiff backbone can interdigitate into the dimerisation site of a monomer and prevent dimerisation.

on the market which bind to the catalytic centre of the enzyme, but these are beginning to loose their effectiveness due to the development of resistant virus strains. Also they are very hydrophobic compounds, which makes their pharmaceutical use difficult. A new class of protease inhibitors bind to the dimerisation site and prevent the formation of the active enzyme. Development of such substances requires an intimate understanding of the structure and function of an enzyme (see fig. 2.9 and [6] for further details).

2.4.2 Protein denaturation

Secondary, tertiary and quaternary structure of a protein are determined by relatively weak interactions, like hydrogen bonds or hydrophilic interactions. The energy of a hydrogen bond is only about 4 kJ/mol. Changes in environmental conditions can break such bonds, leading to the denaturation of proteins.

Heat An increase in temperature leads to increased molecular motion. This can result in breaking hydrogen bonds. If some hydrogen bonds break, the structure of the protein (for example an α-helix) is weakened, i.e. other hydrogen bonds become easier to break. Denaturation by increasing temperature therefore is a process that starts quite suddenly at a certain critical temperature and is completed at a temperature only marginally higher. Humans die if their core body temperature exceeds 42 °C, when key proteins loose their function.

Denaturants Many hydrogen bonds involve water. In such cases a group with a partial positive charge interacts with a water oxygen atom (which carries a partial negative charge), and the hydrogen atoms of the water (with a partial positive charge) interact with a negative group in the protein. If the water is removed by high concentrations of salt, urea or solvents like ethanol or acetone, proteins are denatured. Salts may also disrupt ionic bonds in a protein, by acting as counterions. Changes in *p*H can change the ionisation of amino acid R-groups, and disrupt ionic bonds.

2.4.3 Protein folding

Some proteins, like ribonuclease, can be denatured completely by heat or the addition of denaturants. If the solution is cooled, or if the denaturant is removed by dialysis, the protein spontaneously goes back to its native conformation, and enzymatic activity is restored. These experiments, carried out first by C. Anfinsen in the 1950s [4], prove that the secondary and tertiary structure of a protein is determined solely by its amino acid sequence. No external information is necessary.

This has an important consequence: A mutation of only a single amino acid may interfere with protein folding, resulting in loss of function. For example **cystic fibrosis** is a disease caused by lack of a transmembrane chloride channel (cystic fibrosis transmembrane conductance regulator, CFTR). See the chapter 17.2.1 on page 285 for a discussion of the pathomechanism of cystic fibrosis.

Another example would be collagen, where single amino acid mutations lead to osteogenesis imperfecta or EHLERS-DANLOS-syndrome (see page 203).

Folding starts already even while a protein is synthesised on a ribosome. It is a fairly fast process, synthesis and folding of a 100 amino acid protein in an *E. coli* cell is complete in less than 5 s (at 37 °C).

There are essentially three models to describe protein folding:

Framework model Protein adapt their secondary structure first, using the information contained in their sequence. Once the secondary structure is established, long-range interaction between amino acids stabilise the tertiary structure.

Hydrophobic collapse After synthesis proteins adapt a conformation were hydrophobic amino acids are buried inside and hydrophilic amino acids are on the surface of the protein. This structure is called "**molten globule**". Inside the molten globule long-range interactions between amino acids are established, allowing tertiary structure to form. Secondary structure is established last.

Nucleation/Condensation This model is a combination of the other two: proteins collapse and this reduction in the distance of amino acids leads to the formation of long range (tertiary structure) and short range (secondary structure) interactions at the same time. As a consequence, folding occurs in one big step without folding intermediates. For many proteins (especially those < 100 amino acids) this agrees well with experimental evidence.

Even if one considered only the 4 main structures in proteins (α-helix, β-strand, turn and coil) even a small protein of only 100 amino acids would have $4^{100} \approx 1.6 \times 10^{60}$ possible conformations. A random search for the conformation with minimum energy would take longer than the age of the universe (14×10^9 a $\approx 4 \times 10^{17}$ s) and the protein might get trapped in a local minimum. In reality, proteins assume their native conformation on a time scale of seconds. This apparent contradiction is known as LEVINTHALS paradox. However, the conformational freedom for protein folding is much smaller than it might appear at first sight. We have already discussed steric hindrance between the atoms of neighbouring amino acids, which lead to large non-permissible areas in the RAMACHANDRAN plot. But steric hindrance is also possible between amino acids further along the protein chain, unless the protein is in an extended conformation (upper left hand quadrant in the RAMACHANDRAN-plot). Thus it is also not possible to have a β-strand directly next to an α-helix without intervening coil. These restrictions no doubt explain the limited number of structural motives found in proteins.

Although in principle proteins are capable of folding on their own, the process is speed up by the presence of special proteins, molecular **chaperones** and **chaperonins**. This is particularly important in the cellular environment, where the high protein concentration gives plenty of opportunity for protein aggregation (unspecific interactions of hydrophobic residues). Both chaperones and chaperonins are covered in their own sections in chapter 14 on page 211ff.

2.5 Posttranslational modifications of proteins

The properties of proteins can be changed by posttranslational modification, in some cases this can be done (or undone) quickly in response to environmental stimuli, for example exposure to hormones. Such modifications can switch enzymes between active and inactive states. Posttranslational modification may be required also for the proper targeting of the protein to subcellular structures. The following reactions are of particular importance:

glycosylation: attachment of oligosaccharide trees to Ser/Thr-OH-groups or to the amino-group of Asn. Glycosylation occurs in the ER and GOLGI-apparatus and is covered in detail in chapter 16.3.2 on page 237.

disulphide formation: reaction between two sulphydryl- (SH−) groups to form an −S−S−bond. This reaction is important to stabilise the tertiary structure of proteins and occurs in the oxidising environment of the ER, but not in the reducing environment of the cytosol. Disulphide formation is covered in chapter 16.3.1 on page 235.

addition of hydrophobic tails that allow proteins to bind to a membrane (type 5 and type 6 membrane proteins, see chapter 16.2 on page 233.

phosphorylation either on Ser/Thr or on Tyr-OH-groups. This reaction is carried out by specific protein kinases, and can be reversed by protein phosphatases (see for example fig. 8.1 on page 122). It is used to regulate the activity of enzymes.

acetylation and deacetylation of ϵ-amino groups of Lys by acetylases (using acetyl-CoA as substrate) and deacetylases also seems to be important for regulation but we know much less about acetylation than about phosphorylation. In particular histones are subject to this regulation, deacetylation of histones results in chromatin silencing and telomere stabilisation. The tumour suppressor p53, TATA-box binding protein and tubulin are other

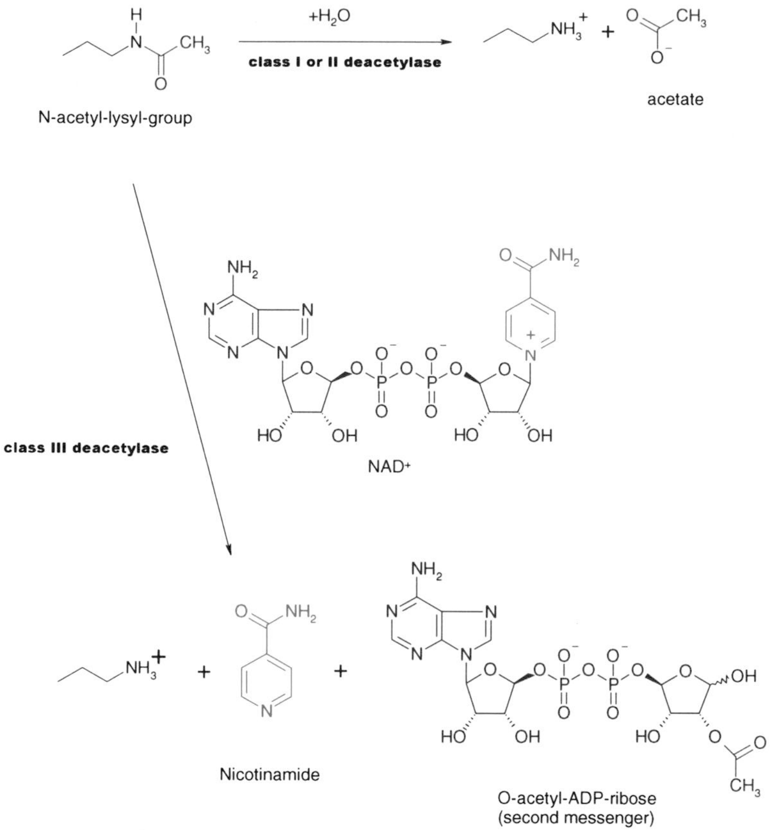

N-acetyl-lysyl-group
$+H_2O$
class I or II deacetylase
acetate
class III deacetylase
NAD+
Nicotinamide
O-acetyl-ADP-ribose
(second messenger)

Figure 2.10. Deacetylation of proteins. The acetyl-group can be split off either as acetate ion by class I and class II deacetylases, or it can be transferred to NAD^+, forming nicotinamide and the second messenger O-acetyl-ADP-ribose.

examples of proteins whose activity is regulated by acetylation. Deacetylases come in 3 classes, class I and II are Zn-dependent proteins which simply hydrolyse the acetyl-group, forming acetate. Class III deacetylases are NAD^+-dependent enzymes (see fig. 2.10), whose activity is regulated by the $NAD^+/NADH + H^+$-ratio in the cell and hence its nutritional state [22, 76]. The product of the yeast **silent information regulator 2 (SIR-2)** (homologues are *C. elegans* sir-2.1 and mammalian SIRT1) is an example for class III deacetylases. This might be the explanation for an old observation, that the life span of organisms is increased by caloric restriction. Polyphenols, like those contained in red whine seem to activate Sir2 in low and inhibit it in high concentrations [39], possibly explaining the protective effect attributed to the consumption of small amounts of red whine in some studies.

methylation using S-adenosylmethionine as methyl-donor and protein methyl transferases [12]. Methylation can occur on

carboxy-groups to form methyl esters. The reaction can be reversed by methyl esterases allowing regulatory processes similar to phosphorylation/dephosphorylation. Four types of methyl transferases are known:

Type I methyl transferases have been found only in prokaryotes and methylate Glu residues in membrane bound receptors. The effect of this modification on receptor function has not been elucidated.

Type II methyl transferases act on racemised or isomerised Asp and Asn residues, to label old and damaged proteins. Conversion to D-Asp and L-iso-Asn are dominant chemical damages in aging proteins. After methylation L-iso-Asn can spontaneously convert back to Asn, methylation of D-Asp seems to mark the protein for hydrolysis by the quality control system of the cell. Type II transferases occur in both eukaryotes and prokaryotes.

Type III methyl transferases act on isoprenyl-Cys at the C-terminal end of eukaryotic membrane-associated proteins. Many of these proteins are part of signal transduction cascades, the physiological consequences of methylation have not been determined yet.

Type IV methyl transferases act on Leu residues at the C-terminus of eukaryotic proteins. Again the physiological significance is unclear.

amino-groups to form methyl-amines. There are no known demethylases, so the modification is irreversible. Methyl-transfer can occur to

Met, Ala, Phe, Pro to form an α methylamine.

Lys to form ϵ-mono-, di- and trimethyl amines. The later results in a pH-independent positive charge.

Arg to form a N^G mono- or dimethylamine. The reaction has been observed in particular on RNA-binding proteins. Di-methylation can be symmetrical (type 2 methyl transferases) on both or asymmetrical (type 1 methyl transferases) on a single nitrogen. The initial single methyl-group can be added by both type 1 and -2 transferases.

His to form a τ-methyl amine.

Very little is known about the function of such methyl-groups.

sulphhydryl-groups to form methyl thioesters. this has been observed on structural proteins, functional significance is unknown.

No methylation of hydroxy-groups has ever been observed.

ADP-ribosylation using NAD^+ as donor. Inhibition of important cellular proteins by ADP-ribosylation is mechanism of some bacterial toxins, e.g. pertussis toxin (see fig. 16.11 on page 258).

glucosylation is the binding of single sugar molecules to proteins. This reaction can occur non-enzymatically in our blood on serum proteins (see section 7.3 on page 115 for a discussion of the medical relevance of this reaction). Similar to ATP-ribosylation it may also be caused by bacterial toxins (produced especially by the genus *Clostridium*) on Thr-residues of Rho-type small G-proteins, using UDP-glucose as substrate. Glucosylation of Rho-GTPases blocks their function and causes cell death. Similar to the ADP-ribosylating toxins the glucosylating ones belong into the class of A/B-toxins (see section 16.4.3 on page 257 for the mechanism by which they enter a cell).

splicing is the process of exscission of a protein sequence (**intein**), the sequences on either side (**N-terminal** and **C-terminal extein**) are then joined [58, 61]. The complex reactions required are catalysed by the intein, which frequently also contains homing endonucleases. It is hence a parasitic, infectious genetic element involved in horizontal gene transfer, possibly even across species borders. Inteins without endonuclease activity are known as **mini-inteins**. About 100 proteins are known to contain inteins, all from unicellular organisms but from all three kingdoms of life (archaea, eubacteria and eukaryota). These proteins are mostly enzymes involved in nucleic acid metabolism (DNA synthesis, repair and recombination, nucleotide biosynthesis). Inteins tend to disrupt highly conserved motifs, so the proteins containing them will be inactive until the intein is removed.

After expressing intein-containing proteins from thermophilic archaea — which have little enzymatic activity at 37 °C — in E. coli it was possible to investigate the exscission reaction simply by warming the purified protein to 95 °C. It turned out that all enzymatic reactions required can be performed by the intein itself, no other proteins, no cofactors and no metabolic energy are required. The reaction is intra-molecular, i.e. there is no swapping between the exteins from different protein molecules. The reaction mechanism of intein-mediated protein exscission is shown in fig. 2.11.

Since these inteins are fairly robust, their normal control-region can be replaced with binding sequences for various small molecules like tyroxine or oestrogens. Binding of the ligand results in a conformational change which activates the intein. In this way the conversion of an inactive pro-enzyme into the active enzyme can be experimentally controlled by adding the ligand.

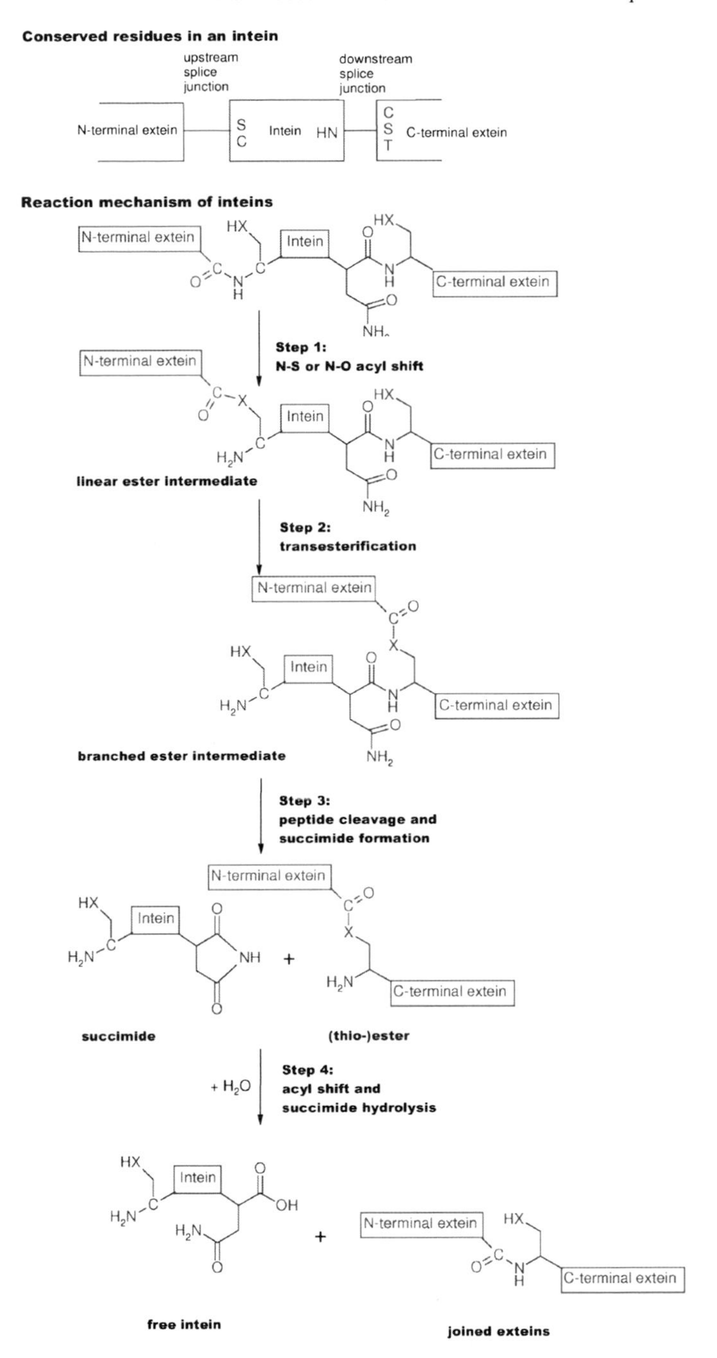

Figure 2.11. Reaction mechanism of inteins. **Top**: Inteins always have either Ser or Cys on the N-terminal and Asn at the C-terminal end. The C-terminal extein always begins with either Cys, Ser or Thr. **Bottom**: Reaction mechanism of inteins. The reaction does not require other proteins, co-factors or energy.

Inteins are now used in biotechnology as self-cleaving affinity tags that allow efficient purification of expressed proteins. Premature cleavage of the intein is prevented either by removing the essential Asn at the C-terminus of the intein (preventing step 3 of the intein-reaction, the branched ester formed in step 2 is then hydrolysed with thiol) or by cloning N- and C-terminal intein fragments to separately expressed proteins. If the proteins are mixed, intein structure and activity is restored ("splicing in *trans*").

Exercises

2.1. On which of the following amino acids can proteins not be post-translationally modified: a) Cysteine, b) Serine, c) Tyrosine, d) Alanine, d) Asparagine

2.2. Why can the isoelectric point of a protein not be predicted from its amino acid composition and the known isoelectric points of the amino acids?

2.3. Which of the following statements about protein structure is *false*?

a) The large number of different protein structures is explained by the free rotation around the peptide bond.

b) Disulphide bonds can stabilise the tertiary structure of proteins.

c) The α-helix is stabilised by intramolecular hydrogen bonds between amino- and carboxy-groups.

d) Hydrophobic interactions are important for the stabilisation of protein tertiary structure.

e) Formation of a quarternary structure requires the existence of at least two subunits.

2.4. Which of the following statements about protein structure is/are *correct*?

1) Hydrophobic amino acid side chains are buried inside globular proteins.

2) β-sheets are stabilised by hydrogen bonds between parallel or antiparallel strands.

3) Protein secondary and tertiary structure can be changed by high salt concentration.

4) Amino acids in coils have an undefined, random position.

2.5. For the peptide sequence ACNGSK which statement is *false*?

a) At physiological pH, this peptide has a positive net charge.

b) The peptide can enter into disulphide bond formation.

c) The peptide may carry N-linked sugars.

d) The peptide contains an aromatic amino acid.

e) The peptide is hydrophilic.

2.6. Which of the following amino acids occur frequently in the transmembrane segments of integral membrane proteins?

1) Leucine

2) Isoleucine

3) Valine

4) Glutamic acid

3

Purification and characterisation of proteins

Each cell contains several thousand different proteins. If we want to find out about their properties, we have to first purify them. How else could we be sure that a particular reaction is really caused by a particular protein?

The basic principles behind protein purification (and only those can be covered here) are easily understood, however, it is difficult if not impossible to predict which methods are most suitable for the purification of a given protein. Very rarely is it possible to purify a protein by a single method, usually the judicious combination of several purification steps is required. Protein purification is therefore much more art than science.

3.1 Homogenisation and fractionisation of cells and tissues

If you want to study, say, the enzymes present in liver, the first you have to do is to obtain fresh liver tissue. This is done at the abattoir, immediately after an animal has been slaughtered. The tissue is transported into the laboratory on ice, to minimise proteolytic damage.

Once in the lab, the tissue needs to be disrupted. This is a critical step: Cells should be broken open, but cell organelles should remain intact. Usually the tissue is minced first by hand, then cut into a fine pulp by rotating knifes (for example in a Warring blender like those used to make milk shakes in the kitchen) and finally homogenised by the application shearing forces in specialised equipment (POTTER-ELVEHJEM- or DOUNCE-homogeniser, French press). All these steps are performed on ice, buffer solutions are used to keep the *p*H at the required value, they usually also contain protease inhibitors, antioxidants and sucrose or mannitol to keep the osmotic pressure in the solutions at the same level as in the cell ($\approx$ 300–350 mosm). Ions like Na^+,

K^+, Ca^{2+} or Mg^{2+} are added as required by the enzyme to be isolated, some very sensitive enzymes also require the addition of their substrate to stabilise them.

Then the various cell organelles need to be separated from each other. This is done by fractionated centrifugation [26]. First connective tissue, undamaged cells and other debris are removed by a brief spin at low speed (10 min at 500 g[1]). In the next step nuclei and plasma membranes are spun down (10 min at 3000 g). Mitochondria, plastids and heavy microsomes require about 30 min at 20 000 g to be spun down, 1 h at 100 000 g is required for light microsomes. The remaining supernatant contains the cytosol.

These crude preparations are then subjected to further purification steps.

3.2 Precipitation methods

Protein precipitation is a very crude method for purification and rarely achieves enrichment by more than a factor of 2–3. However, it is quick and cheap. If applied to crude homogenisates, it may remove material that would interfere with later purification steps.

3.2.1 Salts

Proteins interact strongly with water, and only the hydrated form is soluble. If the available water concentration is reduced by the addition of salts or water miscible organic solvents (methanol, ethanol or aceton), proteins precipitate out of solution. It is essential to perform these reactions in the cold, proteins are rapidly denatured by precipitating agents at room temperature.

The precipitating agent is slowly added to the well stirred protein solution, until it reaches a concentration where the desired protein is just soluble. Precipitated material is removed by centrifugation, then more precipitant is added, until all desired material has precipitated. It is separated from the solution by centrifugation.

Proteins can be separated into two groups depending on their behaviour towards salt:

Globulins are more or less globular proteins, their electrical charge is evenly distributed across their surface area. Globulins are soluble in distilled water, and can be precipitated by high salt concentrations.

[1] The centrifugal acceleration is usually measured as multiple of the gravitational acceleration on earth ($\mathfrak{g} = 9.81\ m/s^2$).

Albumins have molecules with an elongated, rod-like shape with unsymmetrical distribution of electrically charged groups. As a result, albumins are insoluble in distilled water, because the molecules form head-to-tail aggregates held together by electrical forces. Low salt concentrations neutralise these charged groups and allow the albumins to go into solution, high salt concentrations precipitate them again.

From the salts ammonium sulfate is most often used, as it is cheap, non-toxic, highly water soluble and strongly ionised. Purified proteins can be crystallised by slow addition of ammonium sulfate, such crystals, suspended in the mother liquor, tend to be very stable if kept refrigerated. Many enzymes are sold in this form by the suppliers. Salts are removed from proteins by dialysis or gel filtration.

3.2.2 Solvents

Organic solvents (ethanol, methanol or acetone) need to be used even more carefully than salts to prevent irreversible denaturation of proteins, precipitation is usually done at subzero temperatures. Particular attention needs to be given to the fact that mixing of solvents with water generates heat. After precipitation the solvent is usually removed by lyophilisation (freeze-drying).

Polyethylene glycol (PEG) may also be used, it is less denaturing and produces no heat when mixed with water.

3.2.3 Heat

Proteins are irreversibly denatured by heat and precipitate out of solution. However, some proteins are more resistant than others, especially in the presence of their ligands. If crude protein extracts are heated to a carefully chosen temperature for a carefully chosen time, some of the extraneous proteins precipitate, whilst the relevant one stays intact. Precipitated material is removed by centrifugation.

3.3 Chromatography

Chromatography was invented by the Russian botanist MIKHAIL TSVETT for the separation of leaf pigments (on aluminium oxide columns). The method was extended to proteins by RICHARD WILLSTÄTTER. Proteins are bound to a solid support and then specifically eluted. Several types of interaction can be used (see also fig. 3.1):

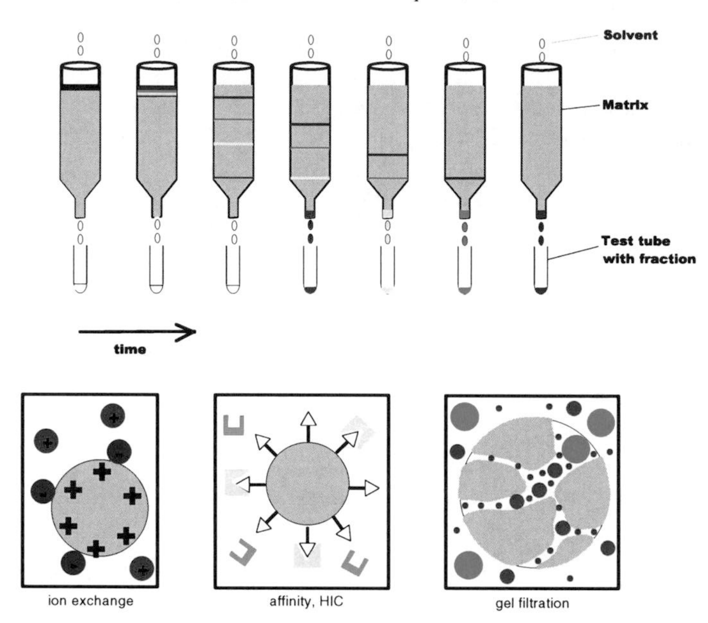

Figure 3.1. Principles of chromatography. The sample is moved by a solvent (mobile phase) past a matrix (stationary phase). Different sample molecules have different partition coefficients between mobile and stationary phase and are delayed differently. Separation can be by ionic interactions, specific interactions with a ligand and size exclusion. For details see text.

- The electrical charge of proteins depends on their amino acid composition and the *p*H of the medium. If the *p*H is lower than the isoelectric point of a protein, it will have a positive net charge, above the *p*I the net charge will be negative. If a protein is passed over a support with charged groups (for example sulfopropyl $-CH_2-CH_2-CH_2-SO_3^-$ or diethylaminoethan $-CH_2-CH_2-NH^+(CH_2-CH_3)_2$) it may or may not bind to it, depending on *p*H . Binding strength depends on the number of charges on the protein, weakly bound protein can be eluted with salt solutions of low concentration, for strongly bound proteins high salt concentrations are required (10 mM to 1 M). This method is called **ion exchange chromatography**, it is probably the most often used method for protein purification. It is also possible to use a *p*H gradient for elution (**chromatofoccusing**), this method has higher resolving power but is much more expensive and rarely used.
- **Gel filtration**, also called gel chromatography or size exclusion chromatography (SEC), is based on the different size and shape (STOKES-radius) of protein molecules. The matrix contains pores of different sizes. Small proteins can diffuse into all of them and are delayed, whilst the largest proteins do not fit into any pores and pass the gel without delay.

Since no binding of proteins to the gel occurs, this method is very gentle, however, it is applicable only to small volumes of concentrated samples.

- Proteins contain a variable amount of hydrophobic amino acids which leads to different binding strength to hydrophobic groups (for example butyl- or phenyl-groups) attached to a support matrix. For **hydrophobic interaction** chromatography proteins are passed over such a column dissolved in a salt solution (0.5 to several M), to maximise interactions with the support. They are eluted with a gradient of decreasing salt concentration, until even very hydrophobic proteins leave the column. In organic chemistry columns with octadecyl-groups (C_{18}, **"reverse phase chromatography"**) are used, but it would not be possible to elute proteins from such columns without denaturing them. C_{18}-columns are useful for small peptides however, which can be eluted with water/acetonitril mixtures.

- Proteins show specific interactions with some other molecules, enzymes with their substrates, receptors with their ligands, antibodies to antigens or glycoproteins with lectins. If such molecules are chemically bound to a support, those proteins that interact with them will be retained on the column, while all other proteins pass the column unhindered. The bound proteins can then be eluted with a ligand solution or by changing *p*H or ionic strength to reduce protein-ligand interactions. Because of the specificity of ligand-protein interactions, **affinity chromatography** can sometimes lead to one-step purification protocols. Biological ligands are often expensive and unstable, however, organic chemistry has produced a lot of compounds, which may to a protein look like its ligand, even though the structure may be quite different. Such compounds can be used with advantage for affinity chromatograhy, the use of Cibacron blue (see fig. 3.2) for the purification of NAD^+ or $NADP^+$ dependent enzymes is probably the most well known example.

- IMAC (immobilised metal affinity chromatography) uses columns where metal ions (mostly Ni, but also Co and some others) are held by chelating groups like nitrilotriacetate. Surface-exposed His-groups on proteins bind to these immobilised metal ions and can be eluted either by competition with increasing concentrations of imidazole or by stripping bound protein and metal ions from the column with strong complex forming agents like EDTA. This method is used chiefly with proteins genetically engineered to have a poly-His tail at either the N- or C-terminal end ("His-tag"), but can also be useful for natural proteins.

- The theoretical basis for **hydroxyapatite chromatography** is much less understood than for the above methods. Hydroxyapatite is a special crystalline form of calcium phosphate (also found in teeth and bones). Depending on chromatographic conditions the surface of these crystals bears an excess either of positively charged calcium or negatively charged phosphate ions. HA-chromatography is probably a combination of ion exchange- and

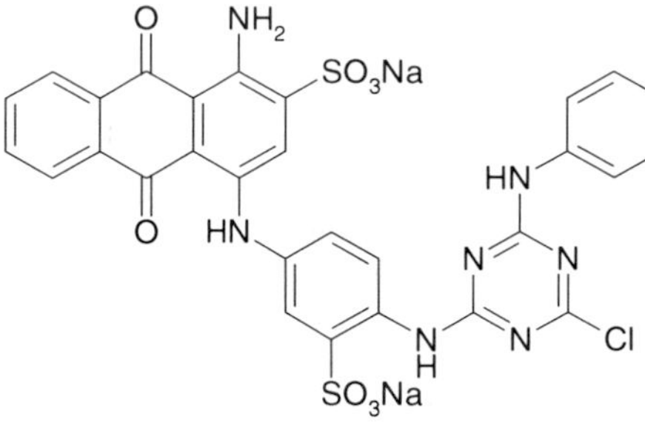

Cibacron Blue F3G-A (Procion blue)

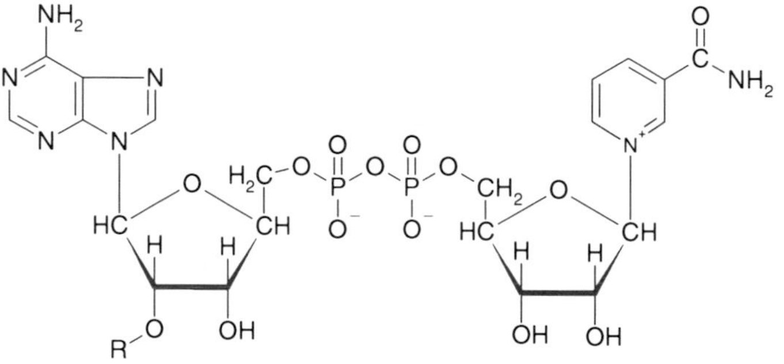

NAD(P)⁺

Figure 3.2. Cibacron Blue is used for the affinity purification of $NAD(P)^+$ dependent enzymes. Because unlike $NAD(P)^+$ the dye is very stable the column can be used for the isolation of such enzymes repeatedly.

affinity chromatography. Bound proteins are eluted with a potassium (or sodium-) phosphate solution of increasing concentration (10–500mM). Although results with HA chromatography are even more difficult to predict than with other methods, it is an essential tool for the isolation of some proteins.

Chromatographic methods are the workhorses in protein purification. Columns can be constructed for sample volumes from a few µl for analytical applications to several l for industrial scale preparative purification. The equipment for protein chromatography needs to be constructed from biocompatible materials (glass, certain plastics), steel may release heavy metal ions and is unsuitable (whilst it is in common use in organic chemistry). Sample vials, columns and fraction collectors should be kept at 4 °C during chromatographic runs. Fouling of the columns is prevented by filtering all samples and buffers through low-protein-binding membrane filters of 0.45 or 0.22 µm pore size.

3.4 Electrophoresis

Since proteins are charged molecules (if $p\text{H} \neq p\text{I}$), they move in an electrical field. Direction and magnitude of the electric force acting on the protein molecules depends on their amino acid composition and the pH of the buffer, the friction experienced depends on the pore size of the medium and the size and shape of the protein. Thus chromatographic separations can be designed to separate by **charge**, **isoelectric point** or by **Stokes-radius**.

Electrophoretic methods are normally used for analytical separations, however, equipment for small scale preparative work (up to a few mg) is also available.

Originally electrophoresis was carried out on paper- or nitrocellulose strips soaked in buffer. Nowadays this method is limited to special applications, instead gels of polyacrylamide are commonly used. The pore size of these gels can be adjusted by the polyacrylamide concentration. If very large pores are required, agarose or starch gels are used instead.

3.4.1 Native electrophoresis

Native electrophoresis [59] separates by charge (controlled by the pH of the buffer), size and shape of the protein molecules. Because of this, native gels are sometimes difficult to interpret.

A special case is the separation on paper- or nitrocellulose stripes, as the proteins move on the surface there are no pores to interact with. Thus separation is by charge only. Serum proteins are separated by this method in clinical laboratories.

3.4.2 Denaturing electrophoresis

Most common is the method described by Laemmli [46]. The protein is heated in a sample buffer which contains a mercapto-compound (β-mercaptoethanol or dithiotreitol (DTT)) to open $-$S$-$S$-$bridges in the protein, destroying its tertiary structure. The buffer also contains the anionic detergent sodium dodecylsulfate (SDS), which destroys the secondary structure of proteins and binds to them at a fairly constant ratio of about 1 molecule of SDS per 3 amino acids. For this reason the charge to weight ratio of all proteins is almost constant, and proteins experience constant acceleration in an electrical field, independent of their composition. However, the pores of the gel slow down big proteins more then small ones. Separation is therefore by size only. If the migration distance of a protein is compared to that of proteins of known size, its molecular weight can be estimated (see fig. 3.3).

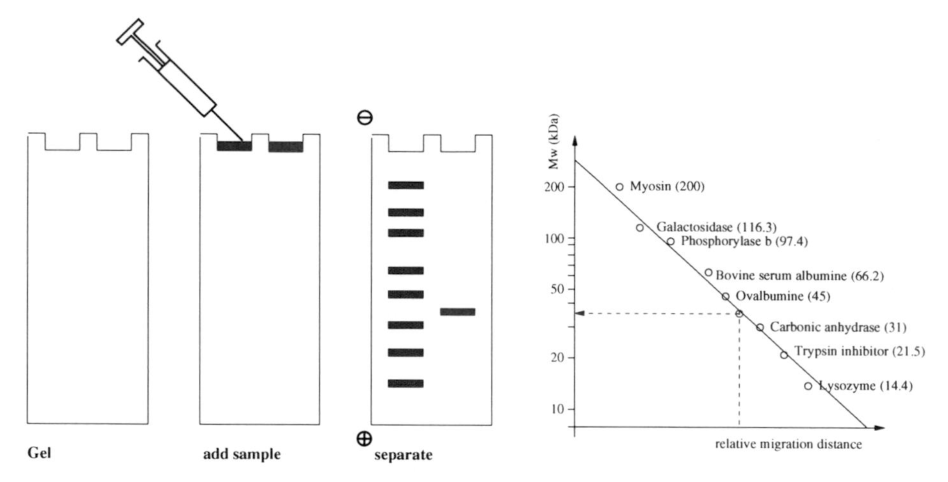

Figure 3.3. Gel-Electrophoresis. Proteins of known molecular weight are used to establish a standard curve that allows the MW of sample proteins to be determined.

A LAEMMLI-gel consists of two separate zones, which differ in their *p*H. In the *p*H 6.8 stacking gel the proteins are focussed into narrow bands, the *p*H 8 separating gels then separates the focussed bands (discontinuous electrophoresis, [59]).

This method is called SDS-polyacrylamide gel electrophoresis (SDS-PAGE). It is quick, relatively cheap and reproducible, and is the most often used type of electrophoresis in a biochemical research laboratory. In some cases the positively charged detergent CTAB may be used with advantage instead of the negatively charged SDS [8].

3.4.3 Isoelectric focusing and 2D-electrophoresis

If the gel contains several thousand different buffer molecules, with slightly different pK_a-values, these buffers can be sorted by an electrical field according to pK_a. This results in a continuous *p*H-gradient across the gel. If proteins are added to the gel, they move in the *p*H-gradient until they reach their *p*I, where they are uncharged and do no longer move. As a result proteins are sorted by *p*I. This method can separate proteins which differ only by a single charge, resulting in a difference of *p*I-value of 0.001.

Isoelectric focussing (IEF) can be performed with native proteins or with proteins whose tertiary and secondary structure has been destroyed with (non-ionic) caotropes like urea.

If after separation the IEF-gel is mounted across an SDS-PAGE-gel, the proteins which have already been separated by *p*I can be separated again by

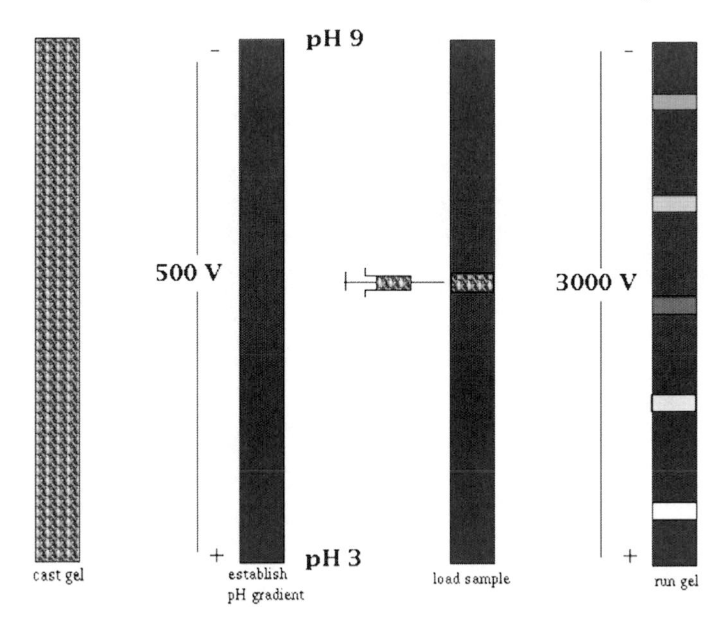

Figure 3.4. Isoelectric focussing. A pH-gradient is established across the gel by electrophoresis at relatively low voltages ($\approx 500\,\text{V}$). The sample is applied to this gradient and run at high voltages until all proteins have achieved their equilibrium position.

molecular weight. This "two-dimensional" electrophoresis can separate about thousand different proteins from a cell lysate.

Comparing the protein expression profiles from cells of different developmental stages, or of healthy and diseased cells, can turn up proteins involved in developmental regulation or in disease processes. With luck such a protein may be a useful drug target. Thus the field of **proteomics** (the proteom of a cell is the collection of all proteins it contains, like the genome is the collection of all its genes) has attracted considerable attention in the pharmaceutical industry.

3.5 Membrane proteins

Before membrane proteins can be separated from each other, they need to be taken out of the membrane. This process is called **solubilisation**. Proteins which are only attached to a membrane may be solubilised by elevated pH or salt concentration [27], or by caotropic agents like urea. However, proteins with transmembrane segments can be solubilised only by detergents. Detergents form soluble complexes with membrane proteins, any attempt to remove the

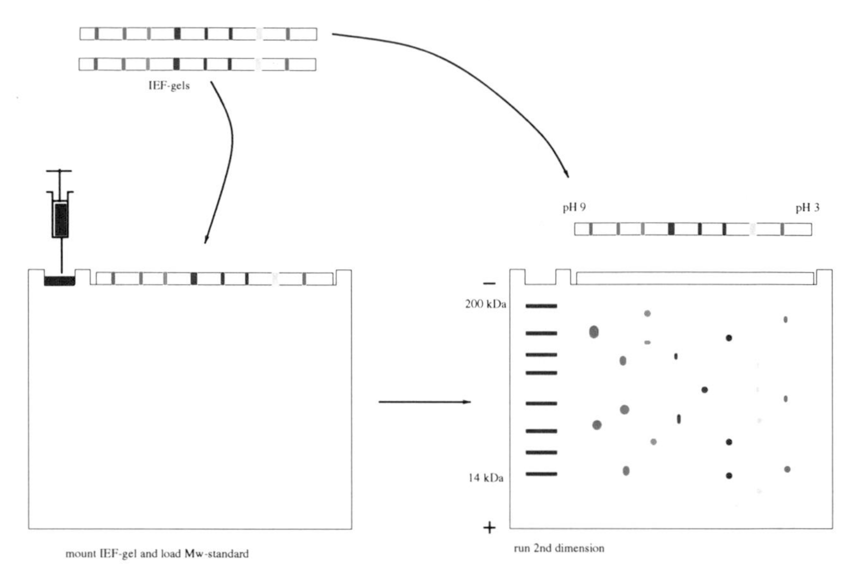

Figure 3.5. 2-dimensional electrophoresis. A protein mixture is first separated on an IEF-gel, which is then mounted across a LAEMMLI-gel. The second electrophoresis separates the bands of proteins with identical *p*I by molecular weight. About 1000 different proteins can be distinguished on a 2d-gel of mammalian cells.

detergent will usually lead to aggregation and precipitation of the proteins, which are not by themselves water soluble.

We distinguish nonionic, cationic and anionic detergents (see fig. 3.6). Of these the nonionic detergents tend to be the mildest, which is good for the preservation of protein function, but on the other hand their ability to dissociate protein complexes, a requirement for the purification of proteins, is also limited.

Purification of solubilised membrane proteins is in principle done in a similar way as with soluble proteins. Care must be taken however to have detergent present in all solutions, lest proteins aggregate.

The detergent/protein and detergent/lipid ratios are critical during solubilisation. Membrane proteins are surrounded by a ring of tightly bound lipids (**annular** as opposed to **bulk** lipids). If annular lipids are replaced by detergent, irreversible loss of protein function occurs, as annular lipids adapt their conformation to the irregular shape of the protein surface, many detergent molecules can't do this (see fig. 3.7). If the detergent/protein ratio is too small, several proteins will occupy a single detergent micelle and purification will not be possible. If the detergent/lipid ratio is too high, annular lipids will be stripped from the proteins and inactivation will occur. This target conflict can be solved by adding extraneous lipid during solubilisation.

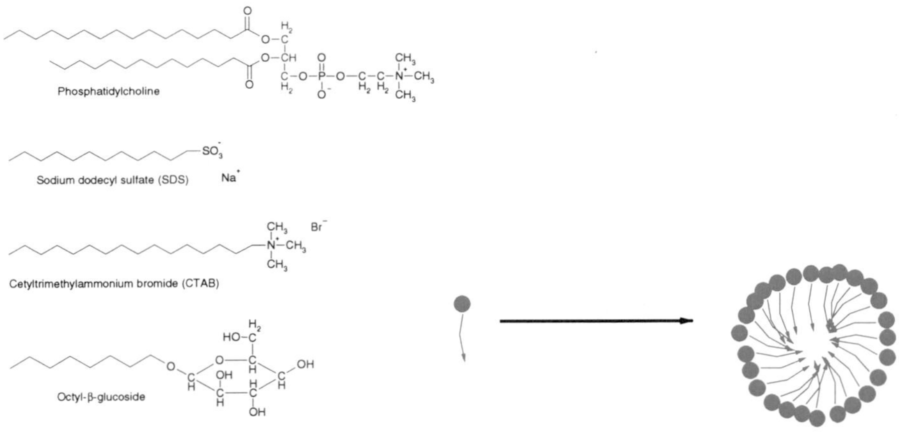

Phosphatidylcholine

Sodium dodecyl sulfate (SDS)

Cetyltrimethylammonium bromide (CTAB)

Octyl-β-glucoside

Figure 3.6. Left: Lipids and detergents. A typical lipid consist of the alcohol glycerol (red) esterified with two fatty acids (black). The remaining hydroxy-group is esterified with phosphoric acid, which carries a hydrophilic side chain, here choline (blue). All detergents have a lipophilic tail and a hydrophilic headgroup. In the case of SDS, the sulphonic acid group bears a negative charge at physiological pH(anionic), while CTAB is positively charged (cationic). OG is an example for a nonionic detergent, the hydrophilic headgroup is a glucose molecule. **Right**: Detergents form aggregates, so called micelles, above a certain concentration. In a micelle the lipophilic tails point to the centre, were they are shielded from water, the hydrophilic headgroups are on the surface and interact with water. The concentration at which micelles form (**critical micellar concentration (cmc)**) and the size of the micelles depend on the headgroup and the tail length of the detergent.

Once a pure membrane protein has been obtained, the protein needs to be reinserted into a membrane, so that its function may be studied. This is done by slowly removing the detergent in the presence of lipids [67]. Under suitable conditions, the lipids form closed membrane vesicles (**liposomes**), and the proteins insert into them. This process is called reconstitution and is probably the blackest of all arts in protein science.

Exercises

3.1. Which of the following statements is *false* about protein purification?

a) Gel filtration separates proteins by their STOKES-radius.

b) Hydroxyapatite is a useful matrix for chromatographic separation of some proteins.

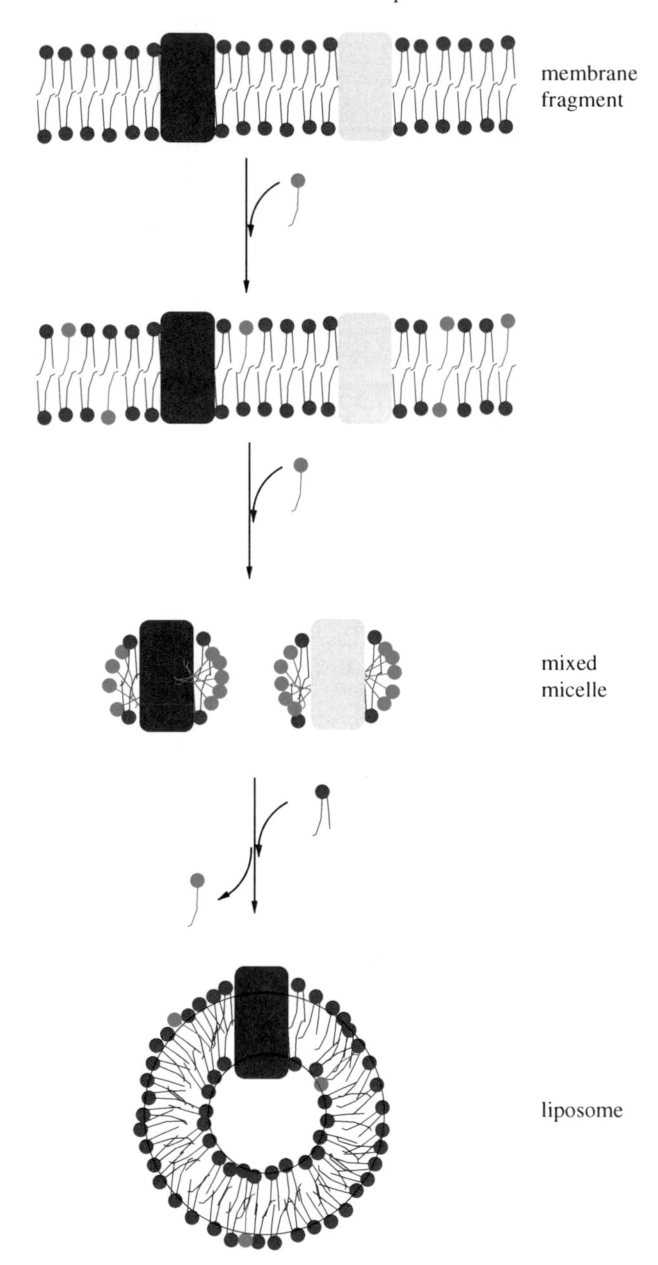

Figure 3.7. Solubilisation of membrane proteins by detergents (green). If detergent is slowly added to a membrane suspension, detergent molecules dissolve in the membrane plane, until saturation is reached. Then mixed detergent-lipid-protein micelles form. Proteins may stay active in such micelles unless the detergent replaces closely bound lipid molecules, which membrane proteins need to maintain their structure. Such mixed micelles contain only a single protein, and protein purification is possible. Once this has been achieved, proteins are reconstituted into liposomes by addition of lipid and removal of detergent.

c) Sodium dodecylsulfate polyacrylamide gel electrophoresis (SDS-PAGE) is a useful method to separate proteins by isoelectric point (pI).

d) Enzymes may be stabilised during purification by the presence of their substrate.

e) Detergent concentration for membrane protein purification must be chosen low enough so as not to remove annular lipids.

4

Protein sequencing

An important step in the characterisation of a new protein is the determination of its primary structure. If the protein is simply hydrolysed, the amino acids can be separated by chromatography and quantified. This gives the amino acid composition, but not the sequence. For this, the amino acids need to be removed one at a time. The classical way to do this is EDMAN-degradation.

The protein is bound to a matrix and is exposed to PITC and HCl in turn (see fig. 4.1), this is done automatically (in modern equipment the reactants are in the gas phase rather than in solution). Because the efficiency of the reaction is only about 80-95%, the signal diminishes each cycle. Under optimal conditions up to 50 amino acids can be sequenced, and only about 100 pmol of purified protein is required.

Larger proteins are split into peptides which are sequenced, these sequences are put together like a jigsaw puzzle. This can be a very time consuming process.

Today protein sequences are often determined by sequencing their genes, which is quicker and easier. Indeed, great effort has been spent in the last couple of years in sequencing the entire genome of bacteria, yeasts, plants, animals and humans. A fairly complete sequence of the human genome was published recently [16, 17, 78]. With these data available it is often sufficient to sequence the first 10 or 20 amino acids of a protein and then look for the gene sequence in a computer data base.

With all the data coming in from gene sequencing attention has now shifted to finding not only which proteins are expressed in a cell, and what their post-translational modifications are, but also how they interact with each other. First results were published in [29], they are also available from `http://yeast.cellzome.com`.

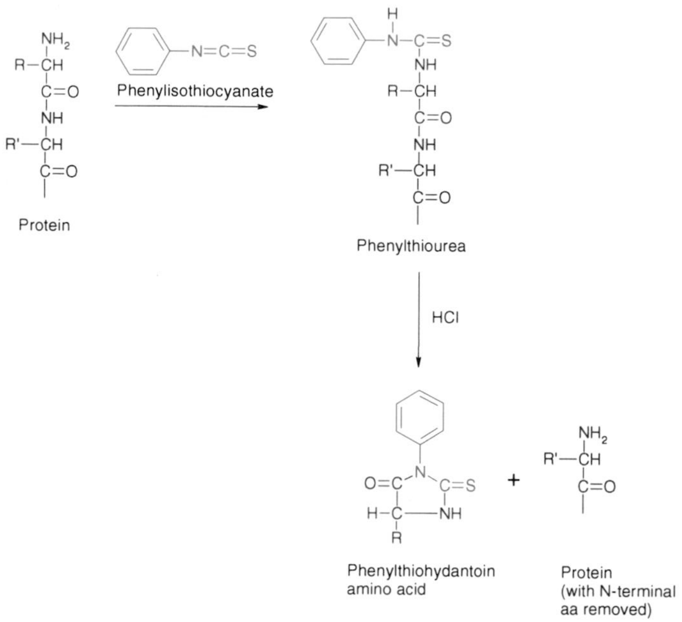
NH2
R–CH
C=O
NH
R'–CH
C=O
Protein
N=C=S
Phenylisothiocyanate
H
N–C=S
NH
R–CH
C=O
NH
R'–CH
C=O
Phenylthiourea
HCl
O=C
N
C=S
H–C
NH
R
+
NH2
R'–CH
C=O
Phenylthiohydantoin
amino acid
Protein
(with N-terminal
aa removed)

Figure 4.1. EDMAN-degradation of a protein. In each cycle one amino acid is cleaved of from the N-terminus and identified. The remaining protein can go directly to the next cycle.

4.0.1 Phylogenetic trees

Our current model of biological evolution says that all life on earth ultimately descends from a single primordial cell. Different organisms evolved from this cell by a series of mutations, which changed their shape, behaviour and ecology. One aim of biologists is to retrospectively understand the sequence of events, that is to draw evolutionary trees that show how the organisms that currently live, and those that we know as fossils, are related to each other. Classically, this has been done by looking at the anatomical structures of organisms and how they are related, for example, how mammalian inner ear bones developed from fish jaws.

You should understand by now that mutations in the sequence of the DNA of an organisms will change the sequence and expression patterns of its proteins. Thus an alternative method to construct evolutionary trees is to compare the sequence of the same protein in a variety of organisms.

Have a look at figure 4.2. You see an oversimplified example of such an evolutionary tree. Now assume, that you know only the sequences, but not the tree itself, which you wish to reconstruct.

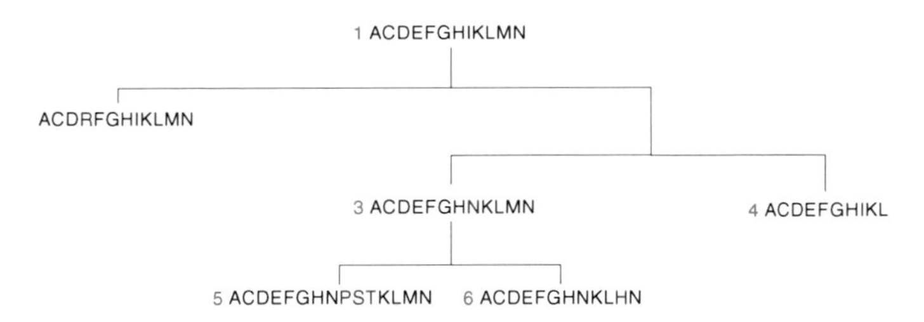

Figure 4.2. Phylogenetic tree of a hypothetical protein sequence. Substitutions, insertions and deletions create a variety of sequences from a single ancestor.

In order to do so, you have to compare the sequences and calculate a matrix of the number of changes between all sequences (note: a substitution or deletion counts as one change, even if several amino acids are involved).

	1	2	3	4	5	6
1	0	1	1	1	2	2
2		0	2	2	3	3
3			0	2	1	1
4				0	2	2
5					0	2
6						0

With this information you can try and reconstruct the evolutionary events. If you do so, you will notice a number of problems:

- There are several ways to draw the tree, in particular you can not tell which of the sequences is the original one, i.e. the root of the tree.
- In practice, sequences may be directly linked even though there are more than one mutation between them. That just tells you that either there were indeed two mutations that occurred at the same time, or, more likely, that there is a missing link between them.
- The same site may have mutated twice. Although this is an unlikely event, over 4×10^9 years of evolution, even rare events do happen.
- The rate of mutation may not be constant over time. Thus it is not necessarily possible to calculate the age of species from such trees.

A partial solution to these problems is to sequence more proteins, so that more information is available.

Note also that different proteins may have different mutation rates. Cytochrome c, a protein needed to burn food, for example will mutate only very slowly, because most mutations will result in a non-functional protein. On the other hand, cytochrome c is present in all aerobic organisms, from bacteria to man. Sequence alignment of this protein will tell you something about long term evolution (on the phylum or class level [21, 69, 71]), but organisms within an order or family may all share the same sequence.

Serum albumin is subject to much lower evolutionary pressure, as long as it is water soluble and has some hydrophobic pockets to bind hormones, it will do. Thus the serum albumins can be used to reconstruct short term evolution, for example how mammals evolved from each other. Long term evolution can not be read from such alignment, if only because many organisms do not have serum albumin.

An example for a phylogenetic tree obtained from protein sequences is found in Fig. 14.1 on page 201.

Part II

Enzymes

5

Enzymes are biocatalysts

Life is an ordered sequence of enzymatic reactions. (R. WILLSTÄTTER)

5.1 The nature of catalysis

A cube of household sugar (sucrose, saccharose) is indefinitely stable under common conditions, even though its oxidation would liberate a considerable amount of energy (see an introduction to general chemistry for a definition of the terms used):

$$C_{12}H_{22}O_{11} + 12O_2 \rightarrow 12CO_2 + 11H_2O \quad \Delta G^0 = -5700 \text{ kJ/mol} \tag{5.1}$$

Even with a match it is not possible to ignite the sugar. A little cigarette ash however makes igniting possible. The ash acts as a catalyst. Catalysts have the following properties:

- A catalyst does not change the equilibrium of a chemical reaction, but it increases the speed with which the equillibrium is attained.
- Catalysts act by lowering the activation energy of a chemical reaction. This allows a reaction to take place under milder conditions (lower temperature or pressure, *p*H closer to neutral) than would be possible without the catalyst.
- A catalyst is not used up in the reaction, it can be recovered quantitatively when the reaction has stopped. Even in those instances where catalysts form intermediate compounds with the reactants they are regenerated later (for example A + B + C → ABC → AB + C).

5.1.1 A brief history of enzymology

Most of the topics in this brief overview are covered in more detail in the following chapters. For further reading into enzymology [5, 18, 45] are suggested. If you have to deal with rate equations in special cases [70] is still useful.

Chemical catalysts

The properties of catalysts were described by the German-Russian chemist GOTTLIEB SIGISMUND CONSTANTIN KIRCHHOFF, who could show in 1812 that the hydrolysis of starch to glucose is accelerated by **acid**, and that the acid is not used up in the process[1]. In 1814 he showed that wheat extract has the same properties. The British chemist HUMPHRY DAVY discovered the catalytic properties of platinum in 1816, the chemist and physicist MICHAEL FARADAY speculated in 1825 that catalysts work by temporarily binding the reactants, thereby increasing their local concentration and orienting them properly. The famous Swedish chemist JÖNS JACOB BERZELIUS then coined the name **catalysis** in 1835 (from the Greek word for degradation).

Living organisms produce very powerful catalysts

The use of yeasts for the production of wine is a very old technique, first mentioned in the *Codex Hamurabi* around 2100 BC. But also the use of crude enzyme preparations rather than complete living organisms is old: In the 5th song of the *Iliad* by HOMER (around 600 BC) the use of ficin (fig tree extract) for curdling of milk in cheese production is mentioned. The French chemist REAMUR noted in the 18th century that the stomach juice of buzzards would digest meat, soften bone but leave plant material unchanged, the first demonstration of enzyme **specificity**. SPALLANZANI continued these experiments, he found in 1783 that the active ingredient of stomach juice lost its activity upon storage, showing that it was **unstable**.

The French chemists PAYEN and PERSOZ in 1833 extracted a substance from sprouting barley which was even more effective than acid in hydrolysing starch. This substance they could precipitate from the extract with ethanol, this was the first attempt to **purify** an enzyme. They called it **diastase** (from the

[1] CORNISH-BOWDEN cites on his web-site an earlier report by the English chemist ELIZABETH FULHAME, who published in 1794 a book "An assay on Combustion" in which she put forward the hypothesis that oxidations use water as a catalyst, the oxygen in water would react with the reducing agent, the resulting hydrogen with air oxygen so that water was re-formed. There are some references to this book in the 19th century literature, however today it appears to have been lost. No biographic data on ELIZABETH FULHAME seem to be available either.

Greek word for separation, as this substance separated soluble sugar from insoluble starch). The German scientist THEODOR SCHWANN isolated **pepsin** (from Greek "digestion") from stomach juice in 1834, which could hydrolyse meat much more efficiently than hydrochloric acid, the other known component of stomach juice.

Thus it became clear that catalysts in living organisms allow chemical reactions to proceed faster and under milder conditions than in the test tube. These substances were collectively called **ferments** (because the conversion of starch into sugar is required for fermentation).

At that time the prevailing theory in science was **vitalism**, the assumption that living organisms are distinguished from their surrounding by a "force of life" (lat. *vis vitalis*). Thus free ferments like diastase and pepsin could degrade compounds, but synthesis was possible only to organised catalysts inside the cell. WILHELM KÜHNE suggested in 1876 to limit the term **ferment** to those organised catalysts of living cells, and to use **enzyme** (from the Greek word for sour dough) for both the unorganised catalysts for example in digestive juices and the organised enzymes of living organisms.

The German scientists HANS and EDUARD BUCHNER tried in 1897 to preserve a cell-free juice pressed from yeast with sugar, so that they could market it for medical purposes. Much to their surprise they found that yeast juice could turn the sugar into alcohol just like whole yeast cells do, this observation effectively put an end to vitalism. It is now held that metabolism inside cells follows the same rules of chemistry and physics that govern reactions in a test tube (**reductionism**). This assumption has proven fruitful, however, it should always be remembered that these laws are applied in an environment of high order and organisation, which is unlikely to be created from the unordered surrounding. This is reflected in the rule "every cell comes from a cell" originally put forward by SPALLANZANI and PASTEUR. In this sense living organisms are indeed special.

Since the distinction between ferments and enzymes lost its significance these terms are now used synonymously.

Mathematical description of enzyme kinetics

In 1902 the French chemist VICTOR HENRI quantitatively described the relationship between substrate concentration and reaction velocity for single substrate reactions [35]. After PETER LAURITZ SÖRENSEN in 1909 pointed out the importance of hydrogen ion concentration for biological reactions and introduced the logarithmic pH-scale and the concept of buffering [72] (the theoretical basis was worked out later by HASSELBALCH [33]), the German chemist LEONOR MICHAELIS and his Canadian postdoc MAUDE LEONORA MENTEN were able to confirm HENRI's result with much higher precision (for a detailed discussion on the HENRI-MICHAELIS-MENTEN law see page 67ff), they also introduced the concept of initial velocity. G.E. BRIGGS & J.B.S. HALDANE generalised the HMM-law in 1925. W.W. CLELAND suggested a systematic approach to multi-substrate reactions and enzyme inhibition in 1963.

Fairly early it was realised that not all enzymes follow the hyperbolic HMM-law, in 1910 HILL [36] suggested a way to extend the HMM-equation to such cases. A mechanistic understanding of this "**co-operativity**" came with the papers of J. MONOD, J. WYMAN & J.P. CHANGEUX in 1965 [54] and D.E. KOSHLAND, G. NEMETHY & D. FILMER in 1966 [44].

The chemical nature of enzymes

GERARDUS JOHANNES MULDER coined the term "protein" in 1838. M. TRAUBE suggested in 1877 that enzymes were proteins, a notion that was not generally accepted until J.B. SUMNER crystallised urease in 1926 [73] and J.H. NORTHROP pepsin, trypsin and chymotrypsin in the 1930s (the first protein crystallisation (haemoglobin) was performed by ERNST FELIX IMMANUEL HOPPE-SEYLER in 1864). It was noted by BETRAND in 1897 that some enzymes required dialysable, non-proteinaceous cofactors, named **coenzymes**.

In 1955 F. SANGER reported the complete amino acid sequence of insulin and in 1957 J.C. KENDREW the crystal structure of myoglobin, proving that proteins were chemically well defined molecules. This was confirmed when in 1969 GUTTE & MERRIFIELD chemically synthesised an enzyme (ribonuclease, molecular weight 13.7 kDa) [30].

In 1980 THOMAS R. CECH discovered a RNA molecule in *Tetrahymena thermophila* which has enzymatic properties (**ribozyme**). Today we know that although the number of ribozymes is small, they are necessary for life. For example the actual peptide synthase in ribosomes is a RNA [20].

Indeed, RNA is probably a molecule with a lot of surprises in store. So it was recently discovered that the mRNA for D-glucosamine-6-phosphate synthase in certain bacteria contains a GlcN-6-P binding site, in the presence of this metabolite the mRNA digests itself, preventing unnecessary synthesis of the enzyme [82]. JACOB & MONOD have speculated that the earliest life on earth was based exclusively on RNA (**RNA-world**), and that DNA and proteins are later additions [41].

Enzyme mechanism

The German chemist EMIL FISCHER had suggested in 1894 that the specificity of enzymes was the result of a binding site in the enzyme, into which the substrate fitted like a key into a lock (**lock-and-key**-hypothesis). LINUS PAULING proposed in 1948 that enzymes act by stabilising the transition state of the substrate. D.E. KOSHLAND realised in 1959 that the substrate binding site was not preformed, but that during substrate binding the enzyme changed its conformation to fit the substrate molecule (**induced-fit**-hypothesis, [43]). Thus he could explain for example that in some enzymes with two substrates the second substrate is bound only after the first one.

The actual existence of the enzyme-substrate complex was demonstrated by B. CHANCE in 1943 when he measured the changes in the absorption spectrum of the haem group in peroxidase following the binding of hydrogen peroxide (reviewed in [10]).

5.2 Enzyme classification and EC code

Each cell contains several thousand different enzymes, each a specialist that performs one reaction on one substrate (or at least a small group of similar substrates). Originally each enzyme was given a name by its discoverer, and many such names are still in use today, like trypsin, pepsin or invertase. However, as the number of known enzymes increased, this system became untenable.

Today enzymes are named in a systematic way. The name of the substrate is followed by the name of the reaction performed on it and the ending -ase (following a suggestion of E. DUCLAUX in 1898). For example the enzyme that breaks down acetylcholine into acetic acid and choline in our synaptic gaps is called acetylcholine esterase. If several substrates are used, they are listed, separated by a colon, like NAD:Alcohol dehydrogenase. Water as substrate is however left out (thus it is acetylcholine esterase rather than acetylcholine:water esterase).

Note that according to International Union for Biochemistry and Molecular Biology (IUBMB) rules enzyme names should not be abbreviated. However, because of the frequently inconveniently long names of enzymes, this rule is most often honoured in the breech. Do make sure however that you know what an abbreviation means before you use it!

Enzyme Commission (EC) codes are also used for unambiguous designation of enzymes. The Enzyme Commission is a body of experts affiliated with the IUBMB. EC codes consist of four numbers, separated by points, like 1.2.3.4

The first number designates the class of reaction that is performed by the enzyme. There are 6 known classes of enzymes (see also fig. 5.1):

1. **Oxidoreductases** catalyse the transfer of electrons, hydrid ions or hydrogen atoms between molecules, for example NAD^+:alcohol dehydrogenase.

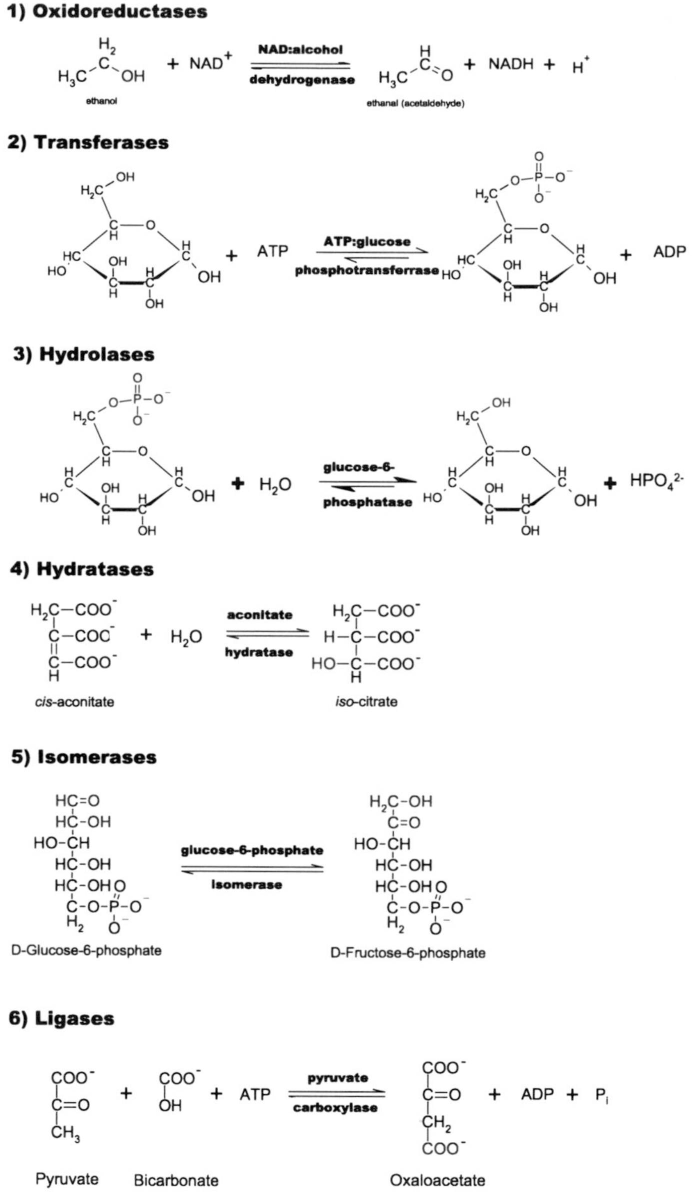

1) Oxidoreductases

ethanol + NAD^+ ⇌ (NAD:alcohol dehydrogenase) ethanal (acetaldehyde) + NADH + H^+

2) Transferases

glucose + ATP ⇌ (ATP:glucose phosphotransferase) glucose-6-phosphate + ADP

3) Hydrolases

glucose-6-phosphate + H_2O ⇌ (glucose-6-phosphatase) glucose + HPO_4^{2-}

4) Hydratases

cis-aconitate + H_2O ⇌ (aconitate hydratase) *iso*-citrate

5) Isomerases

D-Glucose-6-phosphate ⇌ (glucose-6-phosphate isomerase) D-Fructose-6-phosphate

6) Ligases

Pyruvate + Bicarbonate + ATP ⇌ (pyruvate carboxylase) Oxaloacetate + ADP + P_i

Figure 5.1. Examples for the reactions performed by the 6 known classes of enzymes. For details see text.

2. **Transferases** catalyse the transfer of functional groups between molecules, for example ATP:glucose phosphotransferase (glucokinase).

3. **Hydrolases** catalyse the transfer of functional groups to water, for example glucose-6-phosphatase.

4. **Lyases** form double bonds by removing functional groups from a molecule, or open them by adding functional groups. An example would be the aconitate dehydratase, which produces iso-citrate from aconitate in the KREBS-cycle.

5. **Isomerases** transfer functional groups within a molecule, for example the conversion of glucose-6-phosphate to fructose-6-phosphate by glucose-6-phosphate isomerase.

6. **Ligases** use the energy of ATP hydrolysis to form C-C, C-O, C-S or C-N bonds, e.g. the formation of oxaloacetate from pyruvate and bicarbonate by pyruvate carboxylase.

The second number in the EC code denotes the subclass, the third number further specify the reaction. Thus ATP:glucose phosphotransferase has the EC-code 2.7.1.1: it is a transferase (2), a phosphotransferase (7), transfer is to a hydroxy-group (1). The last number has no systematic meaning, it simply allows the unique identification of an enzyme.

The only problem with this system is that due to funding difficulties many enzymes have not yet been assigned EC numbers.

> On `http://www.bis.med.jhmi.edu/Dan/proteins/ec-enzyme.html`, or `http://www.chem.qmw.ac.uk/iubmb/` you can determine the EC-code for an enzyme or obtain information on an enzyme with given EC-code.
>
> Valuable information on kinetic constants, reaction conditions, substrates, inhibitors and physiological roles of enzymes can be found on BRENDA (`http://brenda.bc.uni-koeln.de/`, free only for academic users) or UniProt `http://www.uniprot.org`.

Exercises

5.1. Connect the following reactions with enzyme classes:

1) Ethanol + NAD^+ + $\rightleftharpoons$ ethanal + NADH + H^+	A) Ligase
2) Glucose + ATP $\rightarrow$ glucose-6-phosphate + ADP	B) Oxidoreductase
3) Glucose-6-phosphate $\rightleftharpoons$ fructose-6-phosphate	C) Isomerase
	D) Transferase
	E) Hydrolase

6

Enzyme kinetics

6.1 The Henri-Michaelis-Menten (HMM) equation

Enzymatic turnover of substrates requires the formation of an enzyme substrate complex and finally the dissociation of the enzyme-product complex:

$$E + S \rightleftharpoons ES \rightleftharpoons EP \rightleftharpoons E + P \tag{6.1}$$

The first to develop a mathematical model for this reaction was Victor Henri, [35] his work was later extended by Leonor Michaelis and his postdoc Maude Leonora Menten [53]. The model is therefore called the Henri-Michaelis-Menten equation (we will from now on abbreviate this to HMM-equation).

We start with the assumption that there is no product present, so that the back-reaction can not occur. Under these conditions we get:

$$E + S \rightleftharpoons ES \rightleftharpoons EP \rightarrow E + P \tag{6.2}$$

Because the conversion of ES to EP does not involve binding or release steps, its speed is independent of the concentration of reactants. Thus we can further simplify:

$$E + S \underset{k_{-1}}{\overset{k_{+1}}{\rightleftharpoons}} ES \xrightarrow{k_{+2}} E + P \tag{6.3}$$

If the conversion of ES to E + P is much slower than the binding of S to E (**rapid equilibrium assumption**), the reaction velocity (rate of product formation or substrate consumption) will be proportional to the concentration of ES:

$$v = \frac{d[P]}{dt} = k_{+2} * [ES] \tag{6.4}$$

By the law of mass action, the concentration of ES depends on the concentration of E and S:

$$K_a = \frac{k_{+1}}{k_{-1}} = \frac{[ES]}{[E] * [S]} \tag{6.5}$$

$$[ES] = K_a * [S] * [E] \tag{6.6}$$

Strictly speaking, we would have to use the concentration of free substrate, [S], for these calculations. This concentration we do not know however, all we know is the total substrate concentration $[S]_t = [S] + [ES]$. However, if $[S]_t \gg [E]_t$, formation of the ES complex does not appreciably change the concentration of free substrate. In experiments, we can therefore use $[S]_t$ as an approximation for $[S]$. This assumption can not be used for the enzyme however.

K_a is called **association constant** for the binding reaction, and is a measure for the affinity of an enzyme for its substrate. Its unit is M^{-1}.

This allows us to calculate Θ, the fraction of enzyme molecules that has substrate bound to it:

$$\Theta = \frac{[ES]}{([ES] + [E])} = \frac{K_a * [E] * [S]}{(K_a * [E] * [S] + [E])} = \frac{K_a * [S]}{(K_a * [S] + 1)} = \frac{[S]}{([S] + \frac{1}{K_a})} \tag{6.7}$$

$\frac{1}{K_a}$ is called the **dissociation-constant** (K_d) of an enzyme. K_d is a concentration, its unit is M. If $[S] = K_d$, then $\Theta = K_d/(K_d + K_d) = 1/2$. K_d is the smaller, the higher the affinity of an enzyme is for its substrate.

According to our model the conversion of ES $\rightarrow$ E + P is the rate limiting step in the reaction. Thus even if $[S]$ is so high that all enzyme molecules occur as ES ($\Theta = 1$, which occurs at $[S] = \infty$), reaction velocity will be limited, namely to $k_{+2} * [E]_t$. This limit is called **maximal velocity** V_{max}.

Reaction velocity is measured in mol/s, this unit is called katal (kat). Still in common use is the enzyme unit (1 U = 1 µmol/min = 16.7 nkat).

It follows that when $[S] < \infty$, the reaction velocity must be smaller than V_{max}, and that

$$v = V_{max} * \Theta = \frac{V_{max} * [S]}{K_d + [S]} \tag{6.8}$$

This is the equation of a hyperbola (see fig. 6.1).

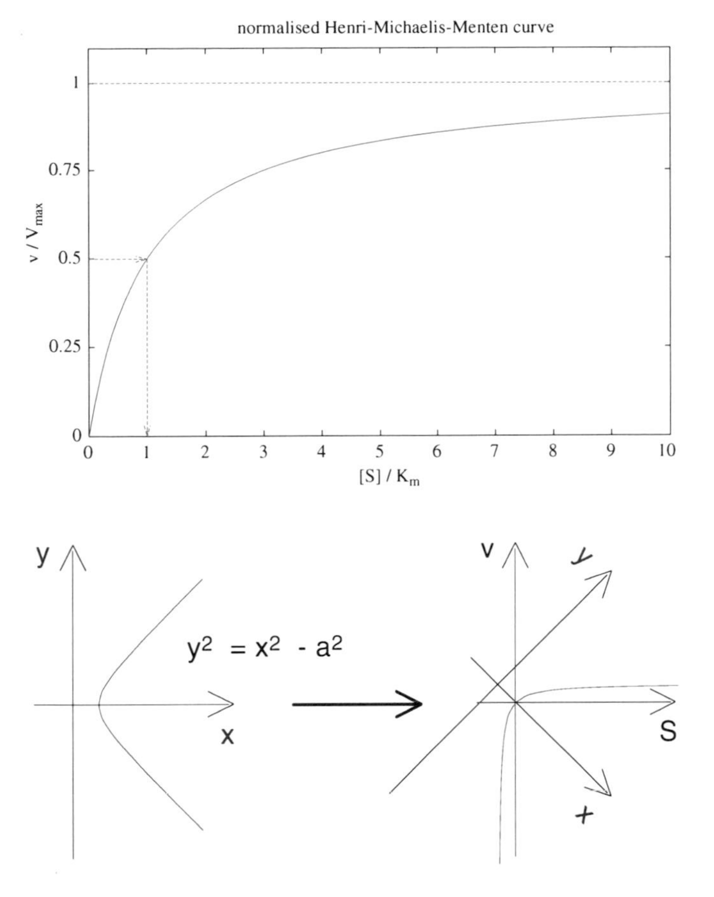

Figure 6.1. **Top**: Velocity of an enzyme reaction as function of the substrate concentration $[S]$. The curve is a hyperbola. K_m can be obtained as the substrate concentration at which $v = 1/2\ V_{max}$. **Bottom**: The graph of the HMM-equation is obtained from a rectangular hyperbola by a 45° rotation of the coordinate system, followed by a translation to bring the vertex of the hyperbola to the origin.

6.2 A more general form of the HMM-equation

For our derivation of the HMM-equation we have made some simplifying assumptions:

- We have assumed that the reaction rate is determined by a single rate limiting step, and got $V_{max} = k_{+2} * [E]_t$. This condition is not necessary true, but even if it is false, the HMM-equation is still applicable. However, we can no longer interpret k as a single first order rate constant. Instead, we speak of k_{cat} the **turnover number** of an enzyme (number of substrate molecules turned over by each enzyme molecule per second). k_{cat} may be a complicated function of several partially rate limiting reaction constants.

- We have assumed that $k_{+1} \gg k_{+2}$. This too need not be true, the HMM-equation still holds even when $k_{+1} \ll k_{+2}$, but the interpretation becomes different. In the general case, we have to replace K_d with the MICHAELIS-constant $K_m = \frac{k_{+2}+k_{-1}}{k_{+1}}$ which is not inversely proportional to the affinity of the enzyme for its substrate. K_m is the substrate concentration where $v = V_{max}/2$, K_d is the substrate concentration where $\Theta = 1/2$.

Thus in the general case, the HMM-equation should be written as:

$$v = \frac{k_{cat} * [E]_t * [S]}{\frac{k_{+2}+k_{-1}}{k_{+1}} + [S]} \tag{6.9}$$

$$= \frac{V_{max} * [S]}{K_m + [S]} \tag{6.10}$$

The turnover number k_{cat} can assume very different values for different enzymes. Some enzymes, like Hsc70 or RecA, turn over less than 1 substrate molecule per second, while a molecule of catalase (which destroys H_2O_2 in our cells) can turn over 40×10^6 molecules per second.

The **efficiency constant** of an enzyme is defined as the ratio of k_{cat}/K_m, its unit is that of a second order reaction constant ($M^{-1}s^{-1}$).

Enzymes, like all catalysts, do not change the equilibrium of a reaction, they merely increase the speed at which it is obtained. Under equilibrium conditions the forward and reverse reactions proceed with equal speed, hence

$$\left(\frac{k_{cat}}{K_m}\right)_f [E][S] = \left(\frac{k_{cat}}{K_m}\right)_r [E][P] \tag{6.11}$$

$$K_{eq} = \frac{(k_{cat}/K_m)_f}{(k_{cat}/K_m)_r} \tag{6.12}$$

which is known as the HALDANE-**relationship**.

Note that at very low substrate concentration ($[S] \ll K_m$) the HMM-hyperbola can be approximated by a straight line, with the equation

$$v = \frac{k_{cat}}{K_m} * E_t * [S] \tag{6.13}$$

Catalytic perfection and multienzyme complexes

The rate limiting step in an enzymes reaction can be either the association of enzyme with its substrate, or the conversion to and release of product. The association velocity is described by the association rate constant k_{+1}, which is in the order of $1 \times 10^9\,\mathrm{M}^{-1}\mathrm{s}^{-1}$. This is the rate at which substrate can diffuse and bind to the enzyme in a dilute, aqueous medium at physiological temperatures. If the efficiency constant is of the same order of magnitude (note that it has the same unit!) the enzyme is called **catalytically perfect**, because the total reaction velocity is limited only by diffusion of the substrate to the enzyme (**diffusion controlled reaction**), the enzyme can turn over the substrate as fast as it is delivered.

The limitation of the efficiency constant by diffusion is overcome by **multi-enzyme complexes**, where the product of one enzyme is fed directly to the next enzyme as substrate, without time consuming diffusion steps in between. This not only increases the overall speed of a biochemical pathway, it also may prevent the destruction of an unstable intermediate by some side reaction.

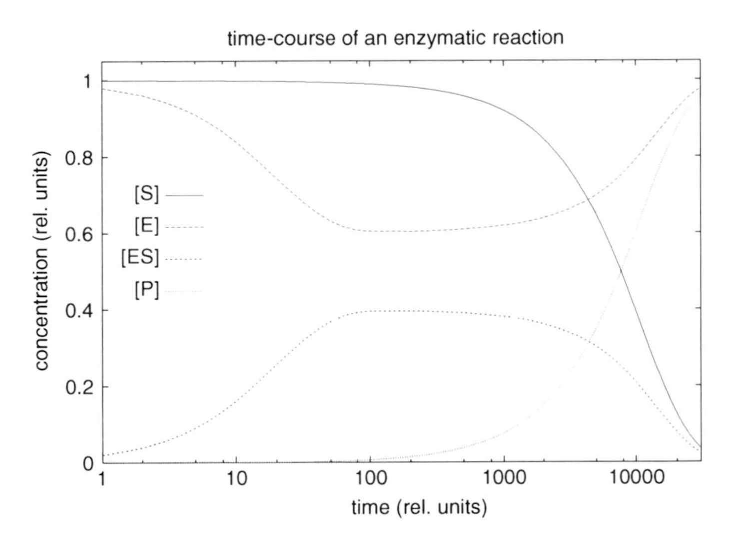

Figure 6.2. In a closed system, an enzyme catalysed reaction proceeds in 3 phases: Immediately after mixing the reactants, $[ES]$ builds up, and product starts to form. This phase is investigated by methods of rapid (non-steady state) kinetics, which are beyond the scope of an undergraduate course. After a brief time (usually a few ms) the rate of breakdown of ES becomes equal to the rate of formation, and $[ES]$ approaches its equilibrium value. In this phase, the rate of product formation is almost linear, and follows HMM-kinetics. As $[S]$ is depleted, $[ES]$ and v drop, $[P]$ and $[S]$ asymptotically approach their equilibrium values.

Advanced material: Derivation of the HMM-equation in its general form

In the previous section we have derived the HMM-equation for the reaction

$$\mathrm{E + S} \underset{k_{-1}}{\overset{k_{+1}}{\rightleftharpoons}} \mathrm{ES} \xrightarrow{k_{+2}} \mathrm{E + P}$$

from equilibrium considerations by assuming that the establishment of the binding equilibrium is much faster than the conversion of the ES complex to E and P. This was how HENRI derived the HMM-equation [35], but for some enzymes the assumption of $k_{+1} \gg k_{+2}$ does not hold. BRIGGS & HALDANE [7] have derived the HMM equation for the general case in 1925:

The enzymatic reaction can be described by a system of differential equations:

$$\begin{cases} \dfrac{d[S]}{dt} = k_{-1}[ES] - k_{+1}[S][E] \\ \dfrac{d[E]}{dt} = (k_{-1} + k_{+2})[ES] - k_{+1}[S][E] \\ \dfrac{d[ES]}{dt} = k_{+1}[E][S] - (k_{-1} + k_{+2})[ES] \\ v = \dfrac{d[P]}{dt} = k_{+2}[ES] \end{cases} \tag{6.14}$$

Such differential equations describe the rate of change of a parameter over time. The integrated form would describe the parameter over time. In many processes it is easy to write the differential, but very difficult to get the integrated form.

A considerable simplification of this system of differential equations may be achieved if we limit ourself to the steady state phase of the reaction, where ES is formed as rapidly as it is broken down, i.e. $\frac{d[ES]}{dt} = \frac{d[E]}{dt} = 0$ (see fig. 6.2). This is known as the **steady-state-assumption**. Note that such an equilibrium phase exists independent of ratio of k_{+1}/k_{+2}. Then above equations simplify to

$$k_{+1}[S][E] = (k_{-1} + k_{+2})[ES] \tag{6.15}$$

Because $[E]_t = [E] + [ES] = \text{const}$ (the total enzyme concentration is the sum of free enzyme and ES-complex concentrations and is constant over time) we can also write eqn. 6.15 as:

$$k_{+1}[S]([E]_t - [ES]) = (k_{-1} + k_{+2})[ES] \tag{6.16}$$

$$k_{+1}[S][E]_t - k_{+1}[S][ES] = (k_{-1} + k_{+2})[ES] \tag{6.17}$$

$$k_{+1}[S][E]_t = (k_{+1} + k_{-1} + k_{+2})[ES] \tag{6.18}$$

$$[ES] = \frac{k_{+1}[E]_t[S]}{k_{+1}[S] + k_{-1} + k_{+2}} \tag{6.19}$$

$$= \frac{[E]_t[S]}{\frac{k_{-1}+k_{+2}}{k_{+1}} + [S]} \tag{6.20}$$

$$= \frac{[E]_t[S]}{K_m + [S]} \tag{6.21}$$

If we replace $[ES]$ in the system of differential equations with eqn 6.20, and consider that the maximal velocity is achieved when all enzyme is converted to the enzyme-substrate complex ($[ES] = [E]_t$), we get

$$v = \frac{k_{+2}[E]_t[S]}{K_m + [S]} = \frac{V_{max}[S]}{K_m + [S]} \tag{6.22}$$

Thus the rapid equilibrium approach of HENRI and the steady-state approach of BRIGGS & HALDANE yield the same result, except that K_d is replaced by K_m.

Linearisation of the HMM-equation

Because the hyperbola of the HMM equation approaches V_{max} only very slowly, it is quite difficult to determine V_{max} accurately from such a graph. Nowadays non-linear curve fitting on a computer can be used to estimate K_m and V_{max} from v versus $[S]$ data, but the more traditional way is a linear transform of the data. The most common transform is that attributed to LINEWEAVER & BURK. If $1/v$ is plotted vs $1/[S]$ instead of v vs $[S]$, the data points should form a straight line, intersecting the y-axis at $1/V_{max}$ and the x-axis at -$1/K_d$ (see fig. 6.3).

The HMM-equation can also be expressed in LINEWEAVER-BURK-coordinates:

$$\frac{1}{v} = \frac{1}{V_{max}} + \frac{K_m}{V_{max}} * \frac{1}{[S]} \tag{6.23}$$

Note that any transformation of data results in a change of their error distribution (see fig. 6.4). Thus calculating K_m and V_{max} by linear regression

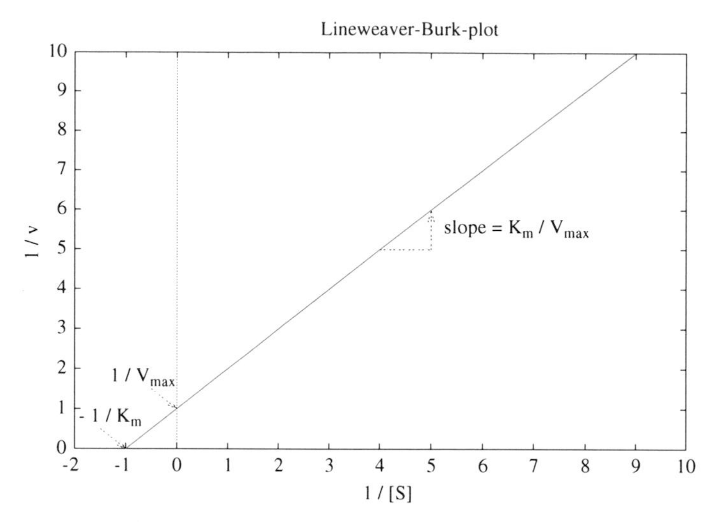

Figure 6.3. LINEWEAVER-BURK-transformed version of fig. 6.1. Data points form a straight line in this transform, V_{max} and K_m can be obtained easily.

from a LINEWEAVER-BURK-plot is less precise than non-linear curve fitting to the original data. Since computers are now widely available, transforms should no longer be used for this purpose, but can still be quite useful for the presentation of results.

Not all enzymes follow HMM-kinetics (see the discussion of co-operativity on page 108ff), and in such cases the data usually do not form a straight line in a LINEWEAVER-BURK-plot. Thus linear transforms are an easy way to judge whether or not the correct model has been chosen.

Apart from the LINEWEAVER-BURK-transformation there are two other transformations which achieve the same purpose. These are:

$$\frac{[S]}{v} = \frac{[S]}{V_{max}} + \frac{K_m}{V_{max}} \tag{6.24}$$

$$v = V_{max} - K_m * \frac{v}{[S]} \tag{6.25}$$

These linearisations, known as HANES- and as EADIE-HOFSTEE-plots respectively, lead not only to a change in error distribution (although somewhat less than the LINEWEAVER-BURK-transform), but also do not separate dependent and independent variable onto different axes. For this reason they

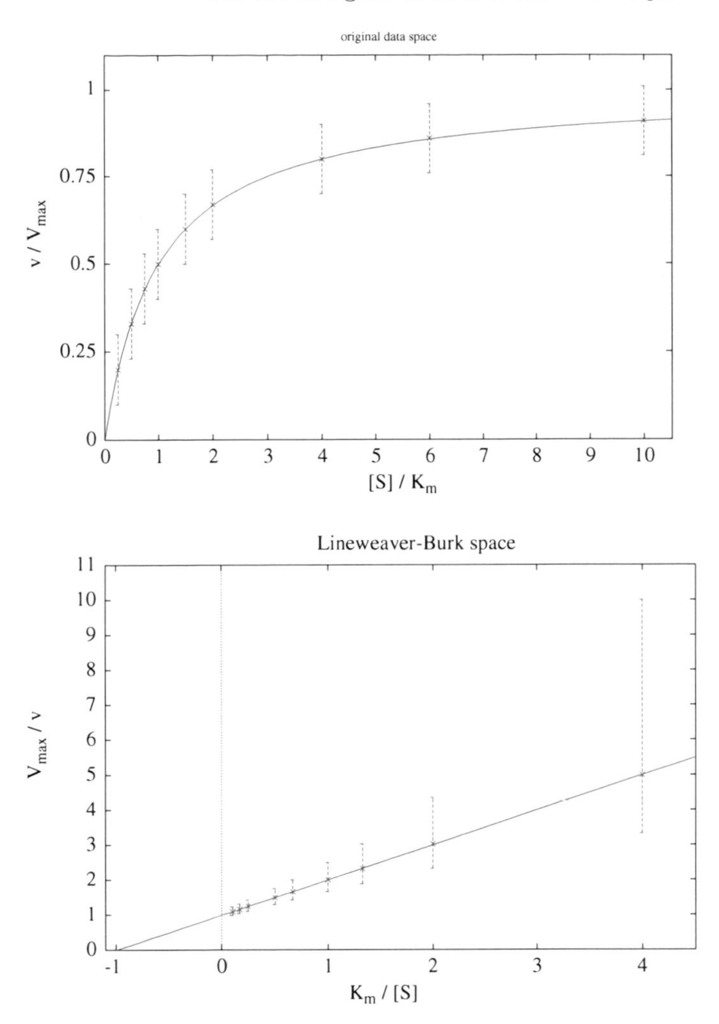

Figure 6.4. Linearisation of data leads to a change in error distribution. **Top:** Original data. Each data point has a standard deviation, assumed here to be constant for all data points. **Bottom:** LINEWEAVER-BURK-transform of the same data. The standard deviation has become unsymmetrical, additionally data points with small $[S]$ have disproportionately larger errors than those with large $[S]$. This will adversely affect the fitting of a regression line.

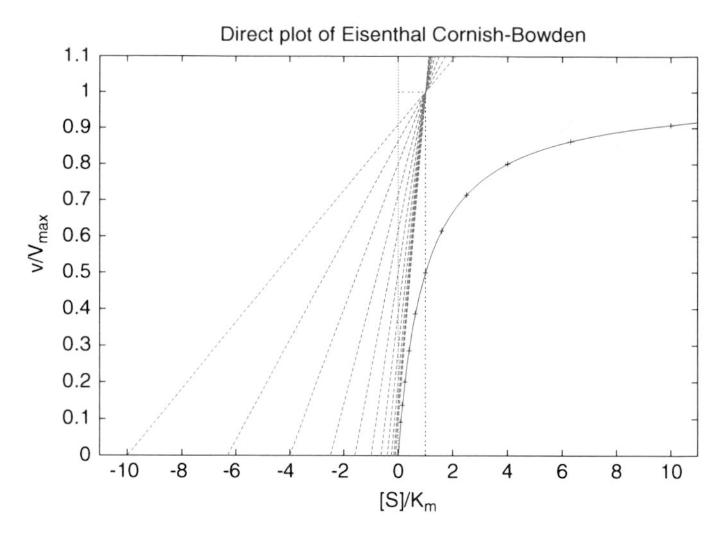

Figure 6.5. The direct plot (introduced by EISENTHAL & CORNISH-BOWDEN) is an alternative way to plot enzyme kinetics data.

are rarely used, medical students may safely ignore them[1]. Because of the close relationship between ligand binding and enzyme activity all three linearisation methods may also be used in binding experiments for plots of $[ES]$ vs $[S]$, however, in this case they have different names. The plot of $[ES]$ vs $[ES]/[S]$ is most commonly used and known as SCATCHARD-plot (related to the HANES-plot). The y-intercept gives the maximum binding capacity, the slope is K_d.

The "direct plot" of EISENTHAL & CORNISH-BOWDEN

If the negative substrate concentrations on the x-axis is connected by straight lines to the corresponding measured velocities on the y-axis, all these lines should meet in a single point, with the coordinates K_m/V_{max} (see fig. 6.5). Since the data are not transformed, the associated statistical problems do not occur. With K_m and V_{max} known, the HMM-curve through the data can be plotted [23].

With actual experimental data however the lines do not intersect in a single point, but in a number of closely-spaced points. The median of their

[1] Historically, all three linearisation methods were originally suggested by WOOLF, however, due to injuries received in a road accident he was unable to publish his results [31]. He had shown the manuscript to his friend HALDANE, who referred to them in his seminal book on enzyme kinetics [32]. However, at that time those methods had little impact, they were later independently developed by the authors whose name they bear today.

coordinates can be used to determine K_m and V_{max}, the spread (minimal and maximal values) gives a good indication of the error margins.

6.2.1 Experimental pitfalls

Initial velocity

We have deduced the HMM equation under the condition that no product is present and that therefore the speed of the back reaction is zero. Of course this will be the case only the moment we start the reaction. If we let the reaction proceed, the substrate concentration will decrease, and the product concentration increase[2] Both effects will lower the velocity of the forward reaction, and product will not linearly increase with time. Rather, a hyperbolic curve will be seen if $[P]$ is measured as a function of time (see fig. 6.6). This curve can be approximated by a straight line at the beginning. The slope of this line is called the **initial velocity** of the reaction, kinetic results are valid only if measurements were taken under these conditions.

Alternatively, if the reaction is irreversible and product inhibition does not occur (conditions that can be forced to be true by removing the product in a second reaction), one can use the **average substrate concentration** $[\overline{S}] = ([S]_0 + [S]_t)/2$ over the measurement time t as an approximation for the "true" substrate concentration $[S]$ [47]. This is especially useful if the decrease in $[S]$, rather than the increase in $[P]$ is followed, because of the statistical problems involved in measuring small differences.

If enzyme reactions are to be followed over longer time periods, the substrate can sometimes be kept constant by the use of a **regenerating system**. For example, if the reaction velocity of an ATPase is to be measured, [ATP] can be kept constant by adding phospho*enol*pyruvate (PEP) and pyruvate kinase (PK) to the system (ADP + PEP $\xrightarrow{\text{PK}}$ ATP + pyruvate). Thus any ADP produced would be converted back to ATP. Of course in this case the speed of P_i production would have to be measured, as overall no ADP is produced nor ATP destroyed. A second option would be to measure the produced pyruvate.

[2] HENRIS version of the HMM equation tries to take care of the increasing product concentration by treating the product as a competitive inhibitor of the reaction:

$$v = \frac{V_{max}([S]_0 - [P])}{1 + \frac{[S]_0 - [P]}{K_s} + \frac{[P]}{K_p}} \tag{6.26}$$

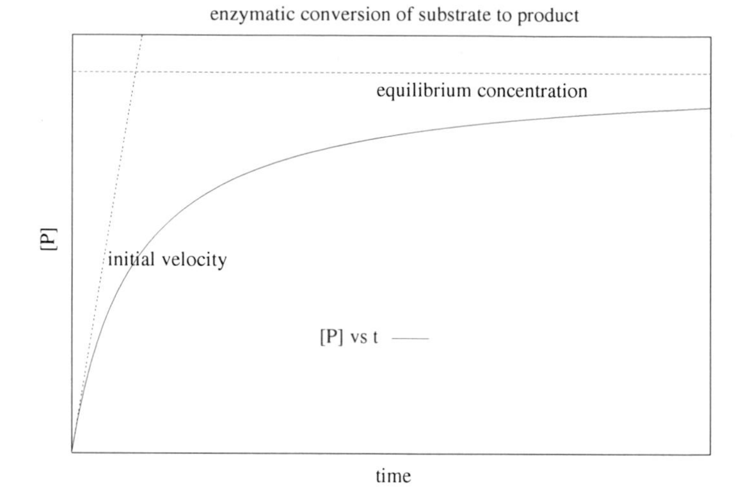

Figure 6.6. Increase of $[P]$ over time in an enzymatic reaction. The reaction velocity ($\frac{d[P]}{dt}$) is highest at the $t = 0$, it approaches zero as the ratio of $[P]/[S]$ approaches the equilibrium. The reaction velocity at $t = 0$ is called "initial velocity".

Measurement error

As mentioned not all enzymes follow HMM-kinetics. It is thus important to experimentally distinguish those that do from those that don't. This is made more difficult by the fact that all experimental data are subject to measurement error. For this reason enough data points must be collected over a large enough concentration range to decide whether or not the kinetics of an enzyme follows the HMM-equation and, if so, reliably calculate V_{max} and K_m. As a rule of thumb at least 10–12 data points should be collected, and they should evenly cover the range of 0.1–5 times K_m, resulting in reaction velocities between 9 and 83 % of V_{max}. It is also important to calculate error estimates for V_{max} and K_m (make this a general rule: results without error estimates are almost useless).

The integral of the HMM-equation

The HMM-equation, which relates the substrate concentration with the resulting reaction velocity ($d[P]/dt$ or $-d[S]/dt$) can be integrated, yielding the Henri-equation:

$$v = -\frac{d[S]}{dt} = \frac{V_{max} * [S]}{K_m + [S]} \quad (6.27)$$

$$V_{max} dt = -\frac{K_m + [S]}{[S]} d[S] \quad (6.28)$$

$$V_{max} \int_{t_0}^{t} dt = -\int_{[S]_0}^{[S]} \frac{K_m + [S]}{[S]} d[S] \quad (6.29)$$

$$= -K_m \int_{[S]_0}^{[S]} \frac{d[S]}{[S]} - \int_{[S]_0}^{[S]} d[S] \quad (6.30)$$

$$V_{max} t = -K_m \ln(\frac{[S]}{[S]_0}) - ([S] - [S]_0) \quad (6.31)$$

$$= 2.303 K_m \log_{10}(\frac{[S]}{[S]_0}) + ([S] - [S]_0) \quad (6.32)$$

$$= 2.303 K_m \log_{10}(\frac{[S]}{[S]_0}) + [P] \quad (6.33)$$

$$-\frac{1}{K_m}\frac{[P]}{t} + \frac{K_m}{V_{max}} = \frac{2.303}{t} \log_{10}(\frac{[S]_0}{[S]}) \quad (6.34)$$

$$= y \quad (6.35)$$

Thus a plot of y as a function of $[P]/t$ yields a straight line with y-intercept V_{max}/K_m, slope $-1/K_m$ and x-intercept V_{max}. Several data points should be obtained, starting at a concentration significantly higher than K_m and letting the reaction proceed until $[S] \ll K_m$. In actual practice the HENRI-equation is rarely used, as results tend to be unprecise.

Stability of enzyme

Enzymes may be unstable, especially in dilute solutions. Thus if an enzyme stock solution is prepared in the morning, it may give different results from those obtained with the same solution after several hours of storage in the evening.

To minimise such problems, enzyme stock solutions are prepared fresh each day, they are stored in an ice bath during the experiment, and it is good practice to establish enzyme stability in control experiments. If very dilute solutions are required, inert carrier proteins may be added. Gelatin is very effective, it also does not contain any aromatic amino acids, which could absorb UV-light in photometric assays.

Control experiments required

Enzyme kinetic experiments should **always** include controls for the turnover of substrate without enzyme (all reagents except enzyme solution, which is replaced by buffer) and for turnover of other compounds that might be present (all reagents except substrate). Otherwise serious mis-interpretation of experimental results is possible.

For example if the breakdown of *para*-nitrophenylphosphate (pNPP) by alkaline phosphatase is studied as a function of *p*H, it might appear that the phosphatase becomes more active at acidic *p*H. This increased breakdown of pNPP however is also seen without the enzyme, it is caused by H^+-ions acting as catalysts.

6.3 Inhibition of Enzymes

The usual view of enzymes (and receptors) is that they bind very specifically their substrates, but nothing else. If this were the case it would be impossible to influence disease with pharmaceuticals, because most of them work by binding to enzymes and receptors and blocking their action. Binding can be **reversible** or **irreversible**; reversible binding results in **inhibition**, irreversible binding in **inactivation** of the enzyme or receptor (the latter will be dealt with in the next section).

Reversible binding of an inhibitor to an enzyme can have four mechanisms:

- Binding of inhibitor prevents binding of the substrate, and therefore its turnover. This is called **competitive inhibition.**
- Binding of the inhibitor occurs only to the enzyme-substrate complex, not to the free enzyme. Binding of inhibitor does prevent substrate turnover. This (very rare) mechanism of inhibition is called **uncompetitive**.
- Binding of inhibitor still allows binding of substrate (and *vice versa*), but prevents its turnover. This is called **non-competitive** inhibition.
- **Partially non-competitive** inhibition is similar to non-competitive inhibition, in that substrate and inhibitor can bind to the enzyme at the same time. However, in the case of partially noncompetitive inhibition the EIS complex is still catalytically active, but the turnover will be slower than that by the ES complex.

Note that some textbooks use the words non-competitive and partially non-competitive with a different (outdated) meaning, which again somewhat differs from textbook to textbook. The nomenclature used here is the most systematic and was originally suggested by W.W. CLELAND [13, 14, 15].

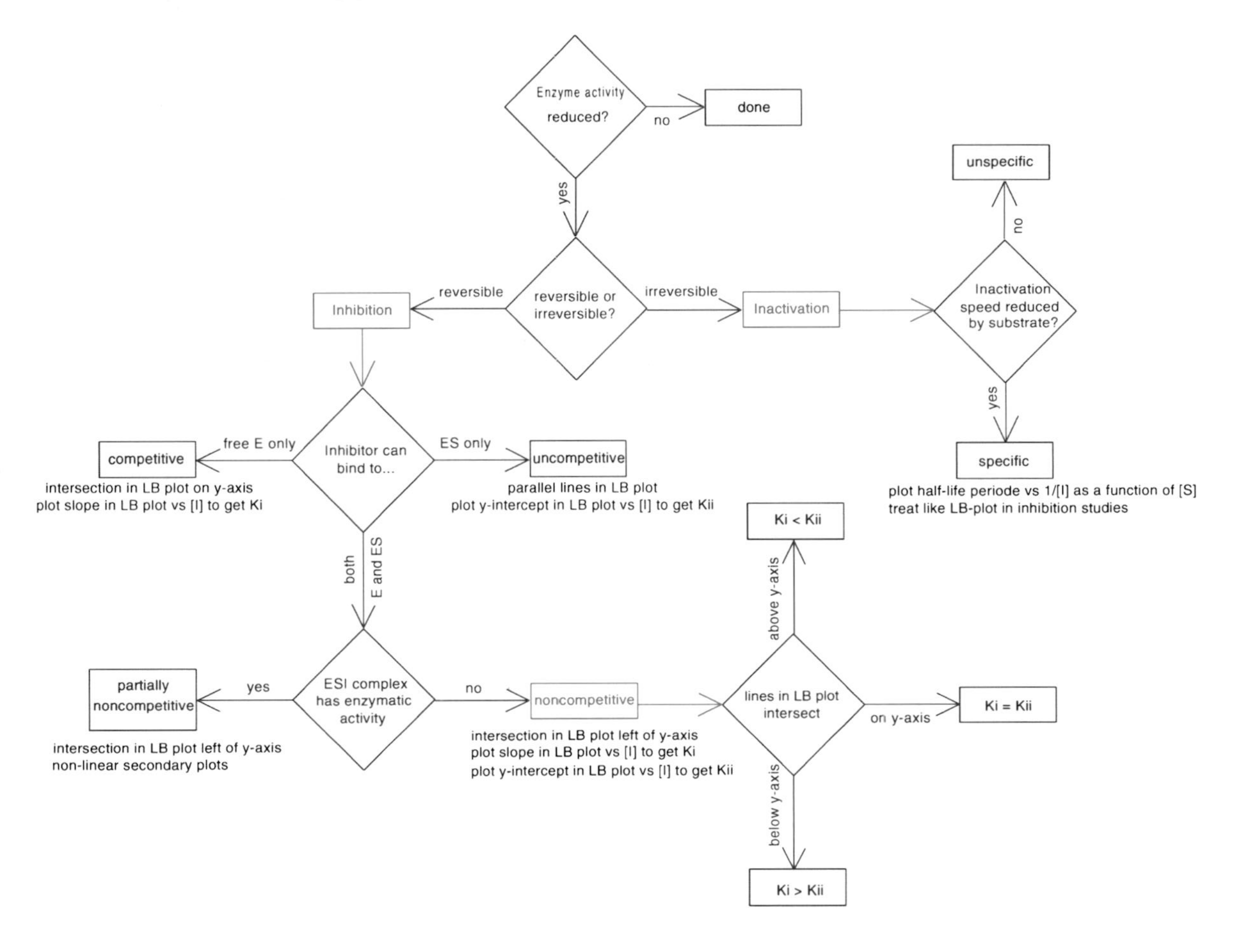

Figure 6.7. Decision tree for enzyme kinetics: Summary of the various forms of inhibition and inactivation.

We will now look at these four mechanisms in more detail.

6.3.1 Competitive inhibition

Competitive inhibition occurs when substrate and inhibitor compete for binding to the enzyme, i.e. only either the substrate or the inhibitor may be bound to the enzyme at any given time:

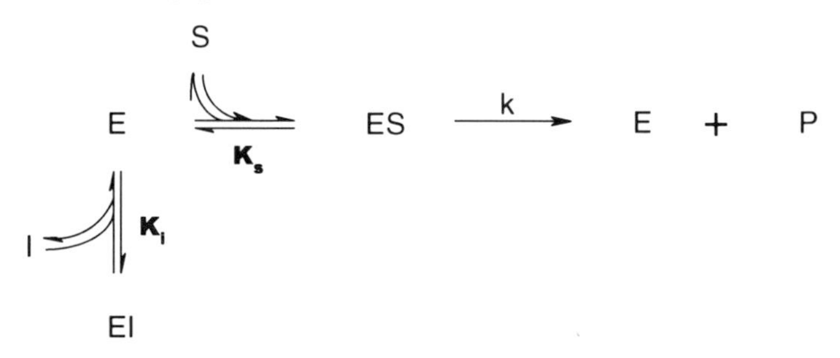

From a competitive inhibition some people conclude that the inhibitor binds at the substrate binding site and thereby prevents the binding of substrate. This conclusion however is not necessarily correct, as shown by figure 6.8.

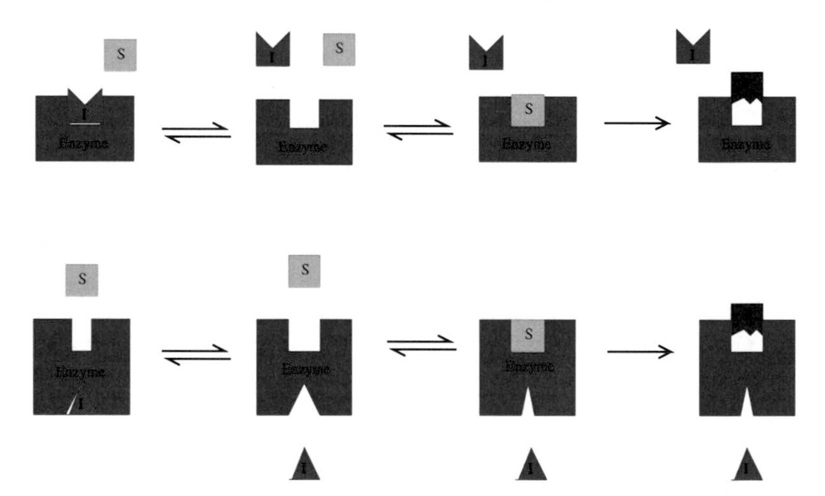

Figure 6.8. Two possible molecular mechanisms of competitive inhibition. **Top**: Substrate and inhibitor share the same binding site on the enzyme. **Bottom**: The enzyme has different binding sites for substrate and inhibitor, but binding of substrate changes the enzyme conformation so that inhibitor can no longer bind and *vice versa*. In both cases only either the substrate or the inhibitor can be bound to the enzyme at any given time, this results in competitive inhibition.

In the presence of a competitive inhibitor the HMM-equation (and its LINEWEAVER-BURK-transform) needs to be modified as follows:

$$v = \frac{V_{max} * [S]}{(1 + \frac{[I]}{K_i}) * K_m + [S]} \tag{6.36}$$

$$1/v = \left(1 + \frac{[I]}{K_i}\right) * \frac{K_m}{V_{max}} * \frac{1}{[S]} + \frac{1}{V_{max}} \tag{6.37}$$

It is obvious that in the LINEWEAVER-BURK-plot only the x-intercept ($-1/K_m$), but not the y-intercept ($1/V_{max}$) is changed by the presence of inhibitor. This results in the graphical patterns in fig. 6.9.

A famous example for the clinical use of competitive inhibition is the treatment of methanol-poisoning. Methanol is converted in our body to methanal (formaldehyde) by the enzyme alcohol dehydrogenase. Methanal then attacks the cells of the optic nerve, leading to blindness. The only known treatment for this condition is to infuse the patient with ethanol, which competes with the methanol for the alcohol dehydrogenase and prevents the formation of methanal. Treatment has to start early, before a lot of methanal has been formed, and the blood level of ethanol needs to be carefully controlled for several days, until the methanol has been excreted by the kidneys.

6.3.2 Uncompetitive inhibition

Uncompetitive inhibition is rare and almost limited to multisubstrate enzymes. It is caused by an inhibitor binding only to the ES-complex, not to free enzyme:

S
E ⇌ ES —k→ E + P
K_s
K_{ii} ⇅ I
ESI

The HMM-equation and its LINEWEAVER-BURK-transform for this case are:

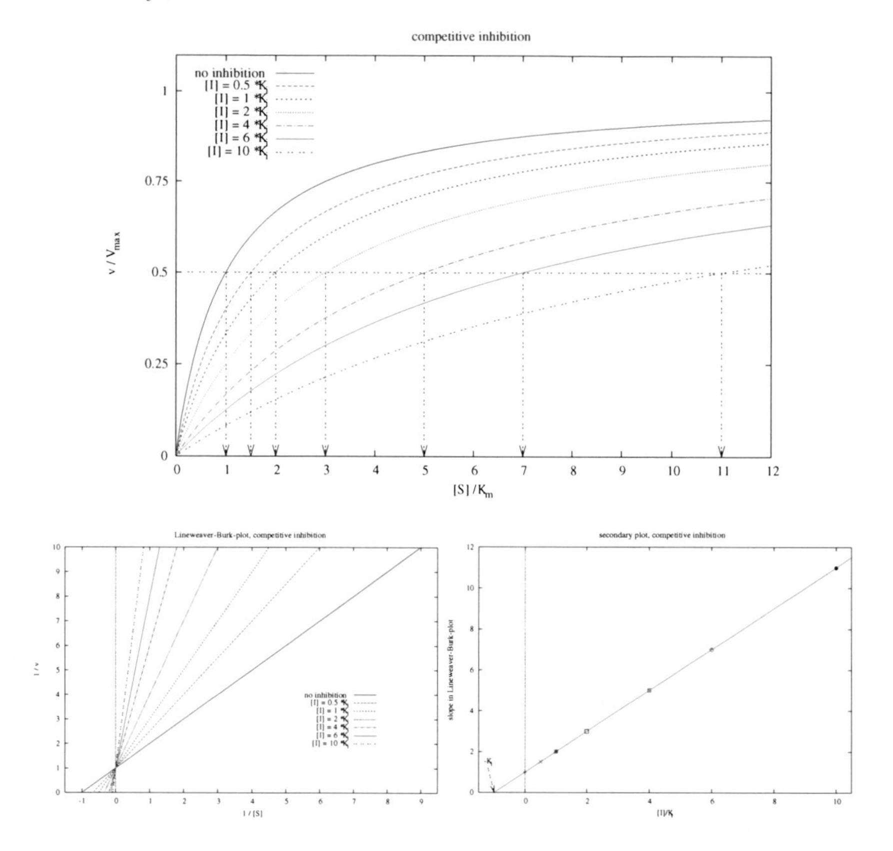

Figure 6.9. Top: v vs [S] curves for different concentrations of a competitive inhibitor. V_{max} remains constant, but the apparent K_m increases with [I]. As a result, higher [S] is required to reach the same velocity in the presence of inhibitor. **Bottom**: In the LINEWEAVER-BURK-plot competitive inhibition results in a family of lines whose slope increases with $[I]$, but whose y-intercept ($1/V_{max}$!) remains constant. Plotting the slope or x-intercept vs inhibitor concentration results in straight lines.

$$v = \frac{V_{max} * [S]}{K_m + (1 + \frac{[I]}{K_{ii}}) * [S]} \tag{6.38}$$

$$1/v = \frac{K_m}{V_{max}} * \frac{1}{[S]} + \left(1 + \frac{[I]}{K_{ii}}\right) * \frac{1}{V_{max}} \tag{6.39}$$

Both V_{max} and the apparent K_m are changed in uncompetitive inhibition, in the LINEWEAVER-BURK-plot we observe a characteristic pattern of parallel lines (fig. 6.10).

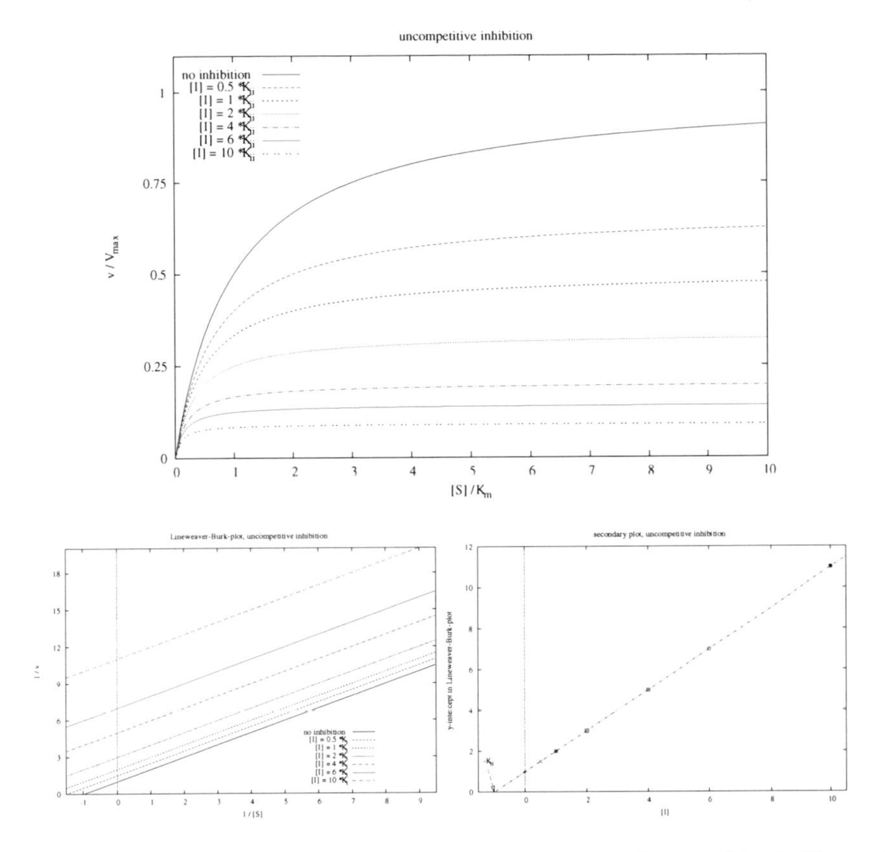

Figure 6.10. Top: Uncompetitive inhibition results in a change of both K_m and V_{max}. **Bottom**: In the LINEWEAVER-BURK-plot uncompetitive inhibition is characterised by a family of straight, parallel lines. Only the y-intercept, but not the slope change. Plotting their y-intercept vs $[I]$ results in a line with x-intercept -K_{ii}.

The occurrence of uncompetitive inhibition can offer significant insights into enzyme mechanism: Na^+/K^+-ATPase (see page 266) is located in the plasma membrane of most animal cells; it uses the energy from the hydrolysis of ATP to export 3 Na^+ from, and to import 2 K^+ into the cell. Na^+/K^+-ATPase belongs into the class of E_1E_2-ATPases, it occurs in a Na-transporting E_1-conformation with high, and a K-transporting E_2-conformation with low affinity for ATP. Mant-ATP is a fluorescent derivative of ATP, which binds exclusively to the E_2-conformation, and only if low ATP-concentrations are present to saturate the E_1-binding site. Thus the inhibition pattern is uncompetitive with respect to ATP, and E_1- and E_2-binding sites must coexist. The simplest explanation is that active Na^+/K^+-ATPase consists of 2 protomers, which co-operate during turnover.

6.3.3 Non-competitive inhibition

Non-competitive inhibition occurs when inhibitor and substrate can bind to the enzyme at the same time, but the presence of substrate is not required for inhibitor binding:

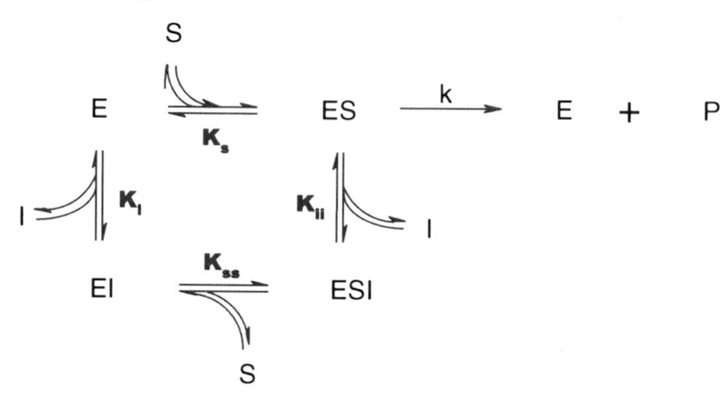

In this respect non-competitive inhibition looks like a mixture of competitive and uncompetitive inhibition, and this is also apparent from the equations describing this case:

$$v = \frac{V_{max} * [S]}{(1 + \frac{[I]}{K_i}) * K_m + (1 + \frac{[I]}{K_{ii}}) * [S]} \tag{6.40}$$

$$1/v = \left(1 + \frac{[I]}{K_i}\right) * \frac{K_m}{V_{max}} * \frac{1}{[S]} + \left(1 + \frac{[I]}{K_{ii}}\right) * \frac{1}{V_{max}} \tag{6.41}$$

Both apparent K_m and V_{max} can be affected, but the precise pattern depends on the ratio of K_i and K_{ii} (see fig. 6.11).

K_{ss} is calculated from the **law of micro-reversibility** (which can be derived from the definition of K_s, K_i, K_{ii} and K_{ss}):

$$\frac{K_s}{K_{ss}} = \frac{K_i}{K_{ii}} \tag{6.42}$$

6.3.4 Partially non-competitive inhibition

In the previous cases the EIS-complex was catalytically inactive. Partially competitive inhibition occurs when this complex also has activity, which may be smaller, equal to or even higher than that of the ES-complex:

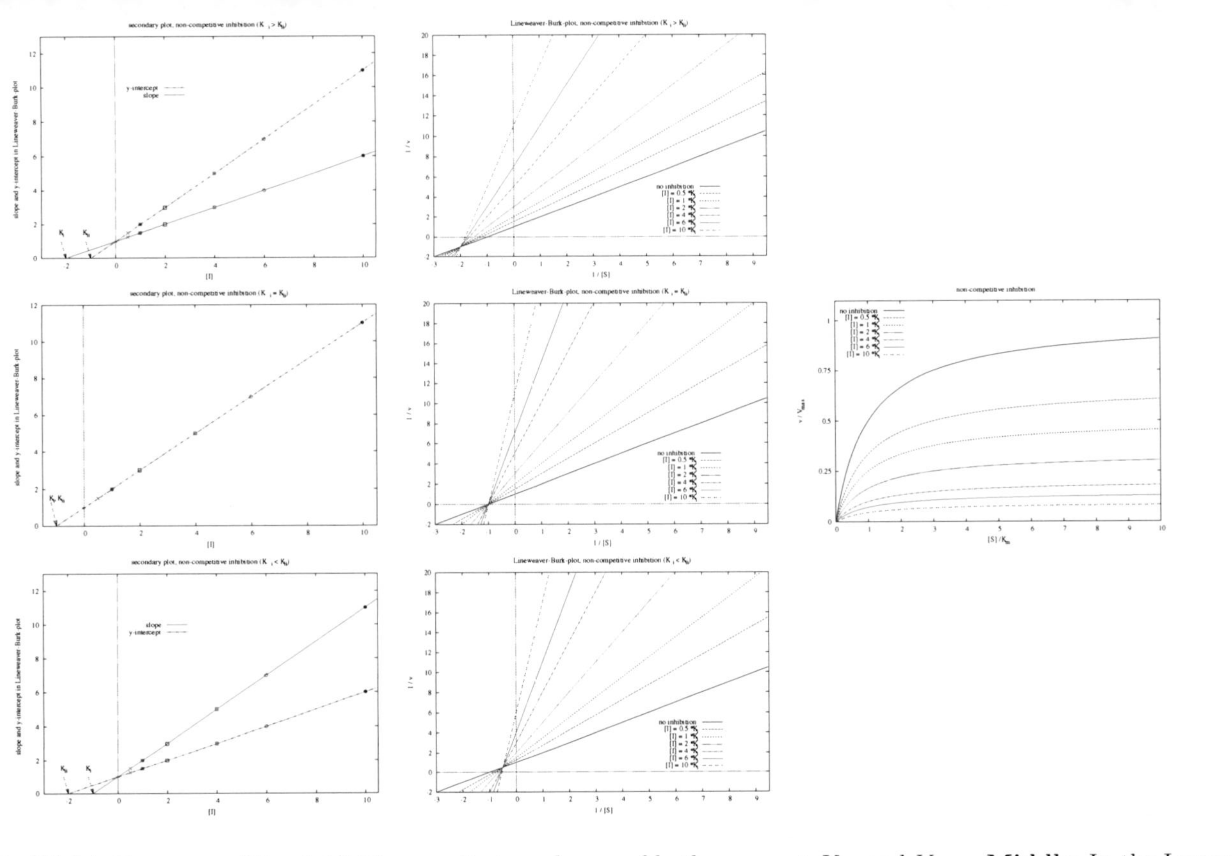

Figure 6.11. Right: Non-competitive inhibition results in a change of both apparent K_m and V_{max}. **Middle**: In the LINEWEAVER-BURK-plot both slope and y-intercept change with $[I]$. The lines intersect in one common point to the left of the y-axis, however, this intersection may be below, on or above the x-axis, depending on whether K_i is smaller, equal or bigger than K_{ii}. **Left**: Secondary plots of y-intercept and slope of the lines in the LINEWEAVER-BURK-plots vs $[I]$. From the x-intercepts of these lines K_i and K_{ii} can be obtained.

Table 6.1. If the x- or y-intercepts or the slopes of the lines in a LINEWEAVER-BURK are plotted as function of the inhibitor concentration (secondary plot), the resulting straight lines can be used to estimate K_i and K_{ii}. With the advent of personal computers, this method should no longer be used, but the diagrams are still useful for data presentation. Partially non-competitive inhibition does not result in straight lines in secondary plots, so this inhibition type is not included in the table.

	slope vs. [I]			y-intercept vs. [I]			x-intercept vs. [I]		
inhibition type	x-intercept	y-intercept	slope	x-intercept	y-intercept	slope	x-intercept	y-intercept	slope
competitive	$-K_i$	$\frac{K_m}{V_{max}}$	$\frac{K_m}{K_i V_{max}}$	—	$\frac{1}{V_{max}}$	0	$-K_i$	K_m	$\frac{K_m}{K_i}$
uncompetitive	—	$\frac{K_m}{V_{max}}$	0	$-K_{ii}$	$\frac{1}{V_{max}}$	$\frac{1}{K_{ii} V_{max}}$	—	—	—
non-competitive	$-K_i$	$\frac{K_m}{V_{max}}$	$\frac{K_m}{K_i V_{max}}$	$-K_{ii}$	$\frac{1}{V_{max}}$	$\frac{1}{K_{ii} V_{max}}$	—	—	—

S
E ⇌ ES —k→ E + P
K_s
I K_I K_{II} I
K_{ss}
EI ⇌ ESI —k'→ EI + P
S

Note that competitive, uncompetitive and non-competitive inhibition are only special cases of partially non-competitive inhibition.

A detailed treatment of partially non-competitive inhibition is however beyond the scope of this course. Just for the sake of completeness, here is the rate equation:

$$v = \frac{(V_{max1} + \frac{V_{max2}*[I]}{K_{ii}}) * [S]}{K_m * (1 + \frac{[I]}{K_i}) + (1 + \frac{[I]}{K_{ii}}) * [A]} \tag{6.43}$$

6.4 Inactivation of enzymes

Inactivation occurs when a substance binds so tightly to an enzyme, that it can no longer dissociate. Important clinical examples are Aspirin (acetyl salicylate) – which blocks the synthesis of prostaglandins, important mediators of inflammation and pain – and the antibiotic penicillin (see fig. 6.13 on page 91).

Another example for enzyme inactivating compounds are organo-phosphates, which covalently bind to Ser-residues in proteins. Ser occurs in the catalytic centre of many hydrolases, for example proteases and esterases. Phospho-esters are used as pesticides (Parathion) and chemical weapons (VX, Tabun), because they inactivate acetylcholine esterase, an enzyme that breaks down acetylcholine into acetic acid and choline in the synaptic gap of our motor-neurons. The result is a constant contraction of all muscles in the body, resulting in painful death from suffocation and heart failure. Treatment is by competitive inhibition of the acetylcholine receptor in the postsynaptic membrane by its antagonist atropine.

A special case are the so called **suicide inactivators**. These substances are substrate analogues which are enzymatically converted to the inactivating

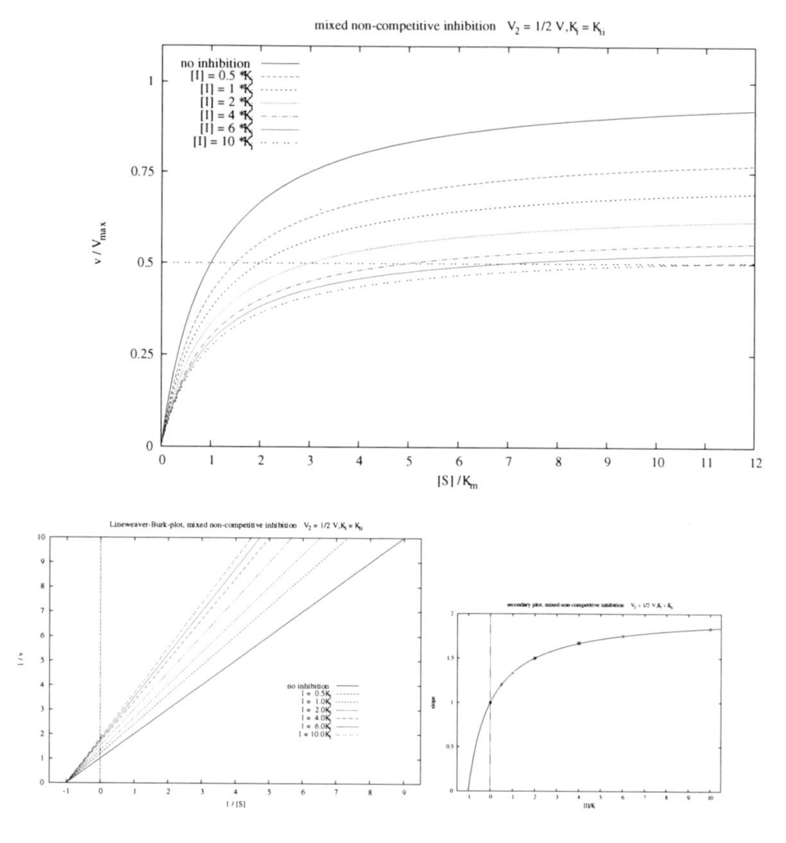

Figure 6.12. Partial non-competitive inhibition. Unlike non-competitive inhibition the maximal activity in the presence of an infinite concentration of inhibitor is higher than 0. Although the LINEWEAVER-BURK-plot looks similar to that of non-competitive inhibition, secondary plots are non-linear.

compound. This makes such inactivators very specific for a particular enzyme, and ideal drugs. Formally, the reaction of suicidal inhibitors resembles the reaction with substrates, except that the enzyme is not regenerated at the end:

$$E + I \underset{k_{-1}}{\overset{k_{+1}}{\rightleftharpoons}} E \cdot I \xrightarrow{k_{+2}} E - I \tag{6.44}$$

If the association to the loose complex is faster than the conversion to the final stable complex, then the reaction rate will be determined by the second step, which is a first order reaction, described by the following equations:

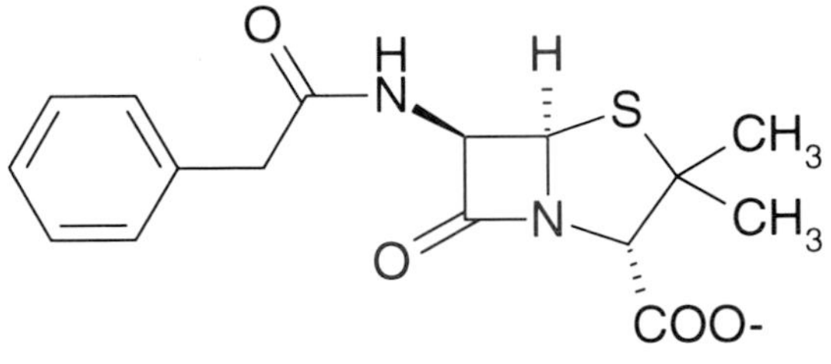

Figure 6.13. Structure of penicillin, an important antibiotic. The β-lactam ring (marked red) opens in the substrate binding site of D-alanine transpeptidase to form an ester with an essential serine residue of the enzyme. Hence formation of the bacterial cell wall is prevented and bacteria lyse during cell division. Recall that 4-membered ring systems are reactive because the bond angle is forced to 90° rather than the tetrahedral angle of 109° (BAYER-tension).

$$\frac{d[E]}{dt} = k_{+2} * [E] \tag{6.45}$$

$$[E]_t = [E]_0 * e^{-k_{+2}*t} \tag{6.46}$$

$$\ln([E]_t) = \ln([E]_0) - k_{+2} * t \tag{6.47}$$

Thus plotting the natural logarithm of the remaining enzyme activity ($[E]_t$) against time (t) will result in a straight line, the slope of which is the reaction rate constant k_{+2}. If k_{+2} is plotted as a function of the inactivator concentration, a hyperbola is obtained. Then k_{+2max} is the inactivation rate at infinite inactivator concentration, when $[E \cdot I] = [E]_0$. The concentration of inactivator that results in half-maximal reaction rate is called K_i.

Like with enzyme turnover, the LINEWEAVER-BURK-transformation can be used, by plotting $1/k_{+2}$ (the **relaxation time** τ, time required for the enzyme activity to drop to $1/e \approx 36.8\,\%$ of the original) or, alternatively, the **half life period** $\mathbf{t}_{1/2} = \ln(2)/k_{+2}$ (time required for the remaining activity to drop to 50 %) against $1/[I]$.

The presence of substrate then may result in a protection of the enzyme against inactivation, which can be either competitive (substrate and inactivator exclude each other from binding to the enzyme), non-competitive (substrate and inactivator can bind to the enzyme at the same time, but EIS can not turn into the stable complex) or uncompetitive (substrate can bind only to EI, but EIS can not form the final complex).

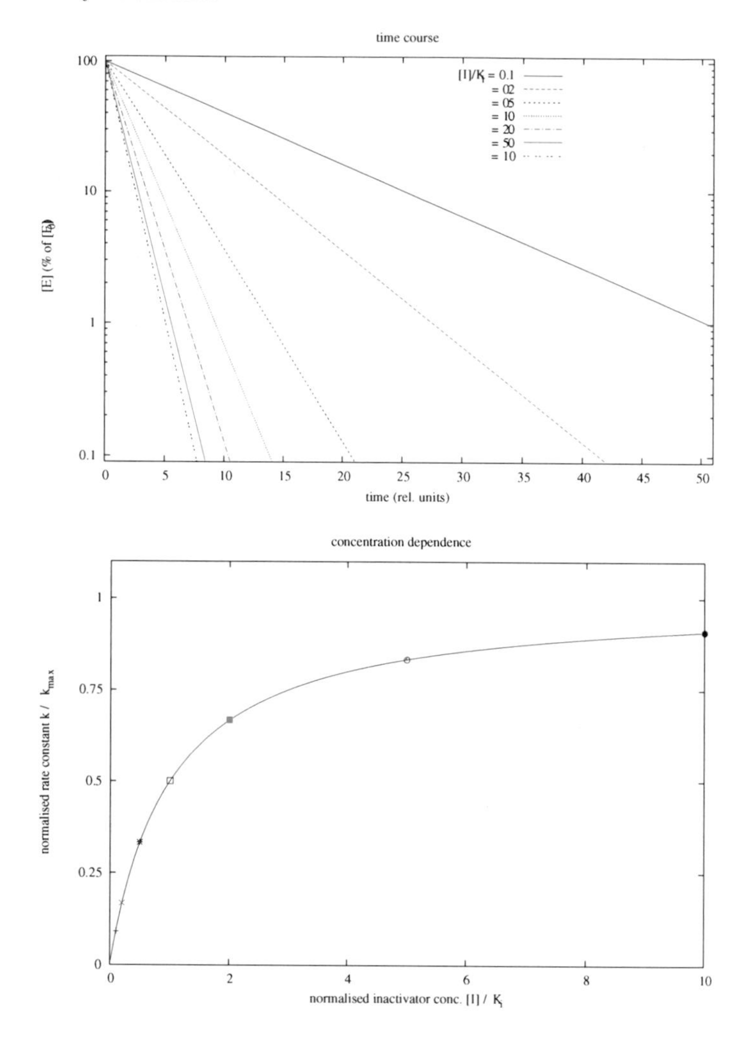

Figure 6.14. Top: Time course of the inactivation of an enzyme at various concentrations of inactivator. Note that the y-axis has a logarithmic scale. **Bottom:** If the slopes are plotted against the inactivator concentration, a hyperbola is obtained.

6.5 Enzymes with several substrates

We have looked at the kinetics of enzymes with a single substrate and a single product. However, many enzymes in our body have several substrates and/or products. Although a detailed kinetic treatment of these enzymes is beyond the scope of this course, we should at least look briefly at the mechanism of such enzymes. Again we use the nomenclature and formalism suggested by W.W. CLELAND [13].

6.5.1 Nomenclature

To graphically represent the reaction mechanism of enzymes, CLELAND suggested a special type of diagram, where the reaction is represented by a long horizontal arrow. Vertical arrows denote entering substrates or leaving products. A conversion inside the enzyme, without entering or leaving substances is denoted by a pair of brackets and called an **inner complex**. Thus our single substrate, single product enzyme reaction:

$$E + S \rightleftharpoons ES \rightleftharpoons EP \rightleftharpoons E + P \tag{6.48}$$

would look like this in a CLELAND-diagram:

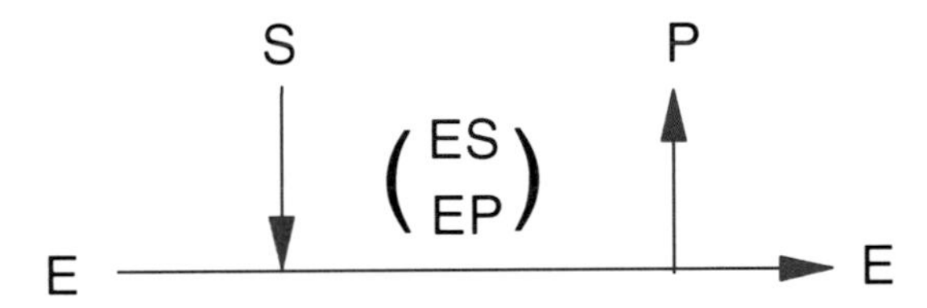

For enzymes with two substrates it may be irrelevant whether S_1 binds before or after S_2. Such a mechanism is called **random-bi**, and the CLELAND-diagram would look like this:

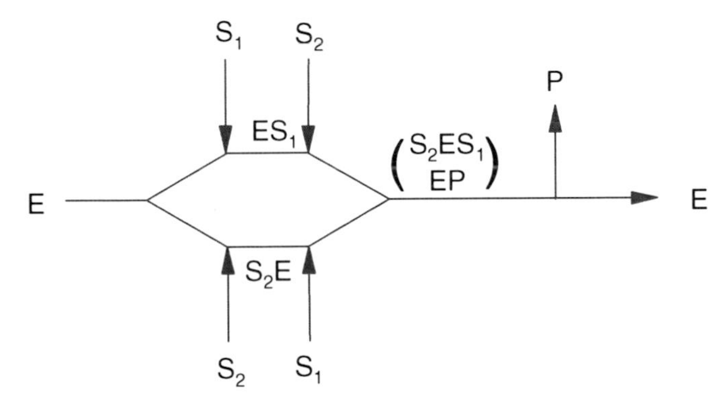

If the enzyme has three substrates rather than two, it would be called **random-ter**.

On the other hand, both substrates may have to bind in a predetermined order, S_2 can not bind to the enzyme unless S_1 has already bound. Such a mechanism would be called **ordered bi**:

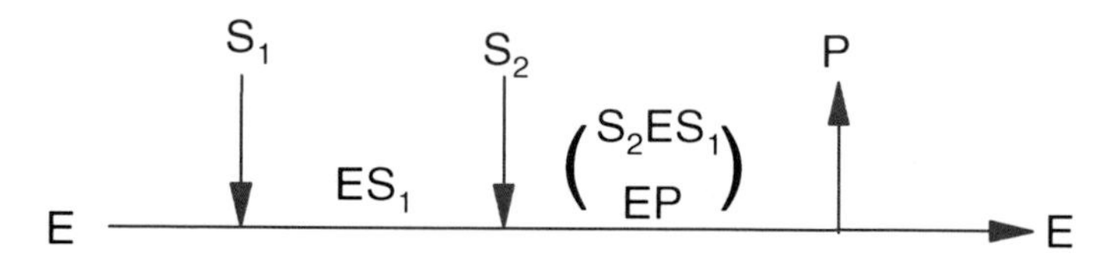

A third possibility, realised with many transferases, is that the first substrate binds to the enzyme, transfers a functional group to the enzyme and is released as the first product. Only then can the second substrate bind and accept this functional group, to be released as second product. This mechanism is called **ping-pong**:

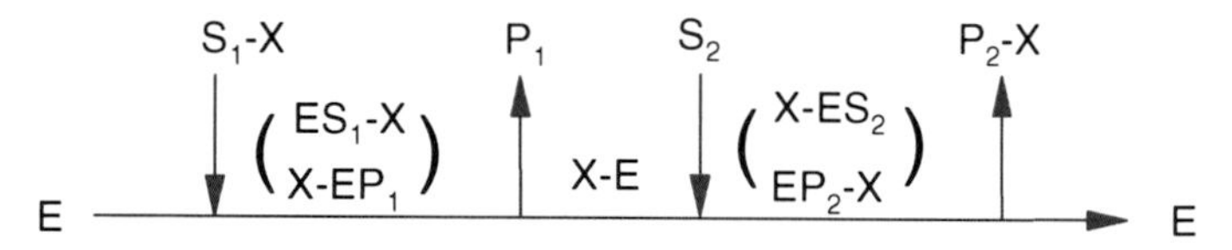

Of course, enzymes may have more than one product, and those may be released in random or sequential order. CLELAND-diagrams are constructed the same way as for substrates, and the nomenclature is identical too. For example, a random-bi-bi mechanism would refer to an enzyme with two substrates and two products, which can be bound and released in any order.

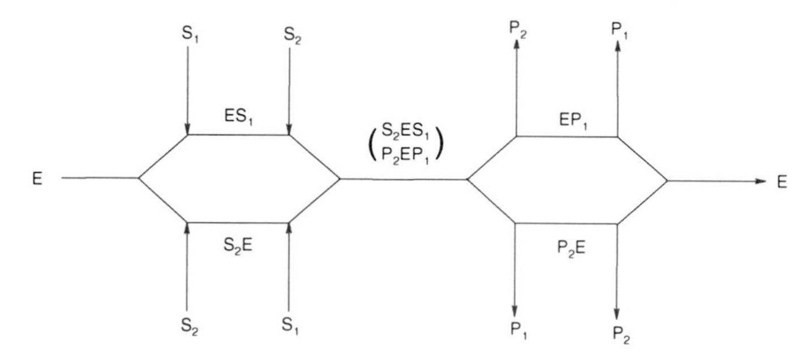

6.5.2 How do we determine the mechanism of multi-substrate enzymes?

The experimental procedure is surprisingly similar to the approach taken to characterise inhibitors. The reaction velocity is measured as the function of the concentration of one substrate, while the concentration of the second is held constant. This experiment is repeated with different concentrations of the second substrate, thus a number of HMM-curves is obtained with different parameters depending on $[S_2]$. In the LINEWEAVER-BURK-transformation the different reaction mechanisms result in characteristic patterns, secondary plots are then used to determine the K_m-values for the substrates.

Enzymes with ping-pong mechanism give LINEWEAVER-BURK-plots with parallel lines, quite similar to uncompetitive inhibition. If the lines intersect, a ternary complex is formed (random-bi or ordered-bi).

Similar approaches can be taken with multi-substrate enzymes, but the work involved increases geometrically with the number of substrates.

6.6 Enzyme mechanism

In describing genetic mechanism, there is a choice between being inaccurate and incomprehensible. (FRANÇOIS JACOB)

We have extensively discussed *what* enzymes do, now we will take a look at *how* they do it.

An enzyme will orient the reactants properly. For this reason if the reaction of a non-chiral substrate leads to a chiral product only one of the possible isomers will be produced (for example aconitase in the citric acid cycle).

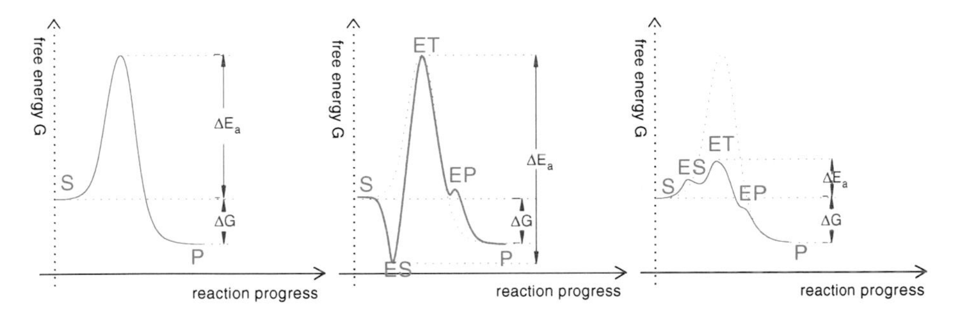

Figure 6.15. Left: A chemical reaction without a catalyst. The product has a lower free energy (G) than the substrate, so the reaction should proceed. However, because the substrate (S) first needs to be activated to a transition state (T) with high energy, reaction rate is slow. Catalysts accelerate the reaction rate by lowering the activation energy (ΔE_a) . Because they do not influence ΔG, they also have no influence on the equilibrium position. **Middle**: If the substrate binding site in the enzyme were complementary to the substrate (**lock-and-key model**), enzymes would stabilise the substrate and increase the activation energy for the reaction. The reaction rate would be lowered. **Right**: If the binding site is complementary to the transition state (**induced fit model**) a small amount of activation energy is required to form the initial ES complex, but the binding energy (ΔE_b) between enzyme and transition state mostly compensates for the activation energy required for the formation of the transition state. The change in free energy resulting from the conversion of the transition state into product then drives the formation of the EP complex, breaking the E-T bonds in the process. A further small amount of activation energy is then required to release the product from the enzyme.

Substrates are held in the optimal position for much longer than they would be during random collisions. Effectively, the concentration of substrates in the catalytic centre of an enzyme is considerably higher than that in solution, and second-order reactions between molecules are converted to first-order reactions inside a molecule.

Acidic and/or basic side chains of the enzyme may participate in **acid/base catalysis**, their pK_a-values are adjusted by the environment to the needs of the reaction. For example in the xylanase from *Bacillus circulans* Glu-172 acts as a proton donor and has a pK_a of 6.7, while Glu-78 with $pK_a = 4.6$ is ionised and stabilises the positively charged intermediate. Calculation of pK_a-values in proteins (rather than in water) is possible, if the 3D-structure has been solved [57].

Redox-reactions may be catalysed by metal centres in proteins, their redox-potential again can be adjusted to the needs of the reaction by the protein environment. For example free Fe^{3+}/Fe^{2+} has a standard potential of $+771\,mV$, for protein bound iron the potential can reach from $+365\,mV$ (cytochrome f) down to $-432\,mV$ (ferredoxin).

The binding site may be hydrophobic or hydrophilic, depending on the needs of the reaction.

In order to form the strong bond between the transition state of the substrate and the enzyme, bonds in the substrate may be strained and become susceptible to break (**rack mechanism**).

The induced-fit hypothesis of enzyme mechanism [43] has an important practical application: If an antibody is raised against a stable transition state analogue, it may have enzymatic properties. Such **catalytic antibodies** can then be engineered with molecular biology methods for improved catalytic properties. That way it is possible to generate enzymes for reactions that do not occur in nature, a process of considerable commercial interest.

Although some catalytic antibodies have indeed been obtained, results with this approach have been largely disappointing. While enzymes can accelerate a reaction by a factor of 1×10^8 to 1×10^{11}, the best catalytic antibodies reach about 1×10^6. Thus other possible mechanism of enzyme catalytic action are currently investigated, based in particular on quantum mechanics. It is postulated that electrons or protons can tunnel through the energy hill, rather than crossing it. This tunnelling may be aided by vibrational movement in the protein (**vibration assisted tunnelling**). This approach however is still in its infancy.

6.6.1 Advanced material: molecular mechanism of serine-proteases and -esterases

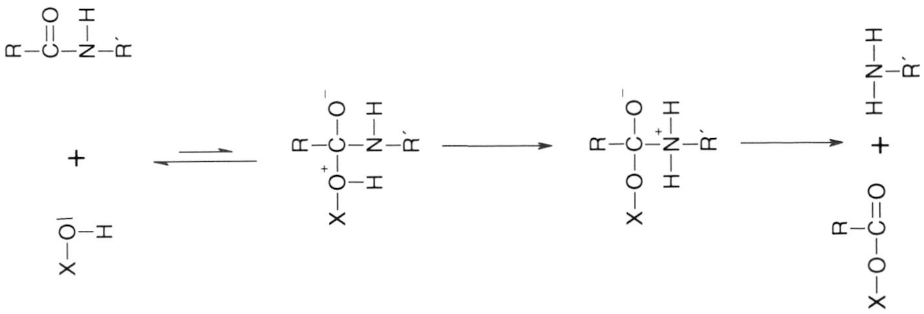

Proteins and esters are relatively stable molecules, because their hydrolysis involves the formation of an unstable intermediate, that easily reverts back to the original compound. Acids and bases can increase the speed of hydrolysis by catalysing the conversion between the first and second intermediate, this is the reason that such compounds are caustic to our skin.

Many proteases (like chymotrypsin) and esterases (like acetylcholine esterase) contain serine in their active centre. This serine forms a so called "catalytic triad" with two other residues, histidine and aspartic acid. The His acts as a general base and removes the proton from the Ser-substrate complex, the resulting positive charge on the His is stabilised by the negative charge on the aspartate. The His then transfers the proton to the nitrogen on the leaving group. Thus no positive charge develops on the Ser-oxygen, and its nucleophilicity is increased. The negative charge on the carbonyl-oxygen is stabilised by hydrogen bonding to additional Ser- and Gly-nitrogens (called the "oxanion hole").

6.7 Enzyme precursors and their activation

Some cells produce and secrete enzymes which would damage the cell if those enzymes were already active inside the cell. A typical example are the digestive proteases like pepsin and trypsin. Obviously these must be produced in an inactive form, secreted and then activated.

Such enzymes are secreted as pro-enzymes consisting of two domains, the enzymatic domain and a regulatory one, which also may carry a marker that is recognised by the protein export machine in the cell, and which may act as a chaperone during protein folding. This domain is cleaved of during or after secretion.

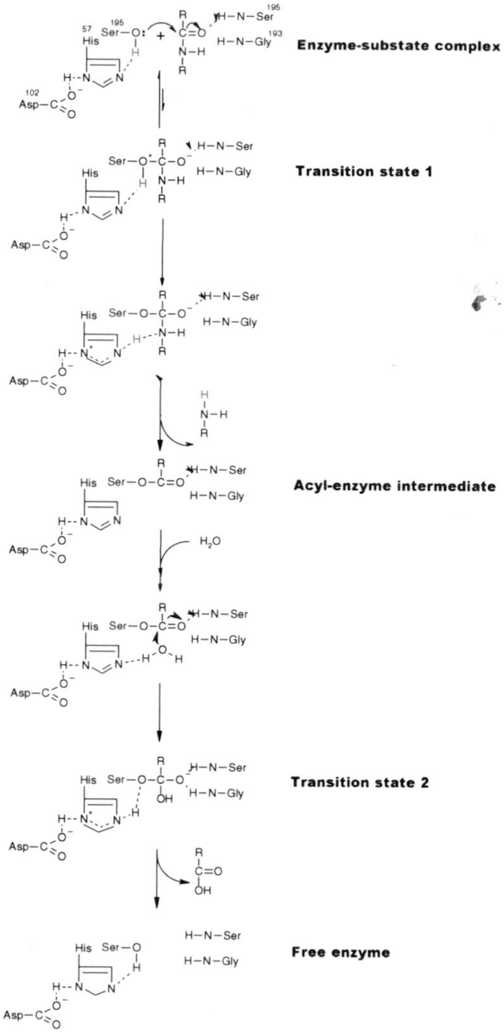
Enzyme-substate complex
Transition state 1
Acyl-enzyme intermediate
Transition state 2
Free enzyme

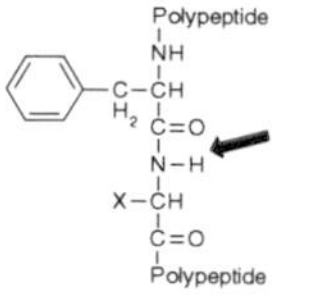
Polypeptide
Polypeptide
Substrate for chymotrypsin

Figure 6.16. Reaction mechanism of Ser-proteases. For details see text.

In the case of the digestive proteases the cleavage process is autocatalytic, i.e. pepsin molecules in the stomach juice (or trypsin molecules in pancreas juice) cleave of the regulatory sequences of newly secreted pro-enzymes.

In chymotrypsin the structural changes during activation have been elucidated by X-ray crystallography. Cleavage of the N-terminal 15 amino acid pro-sequence allows the newly freed amino group of Ile-16 to form a salt bridge with Asp-194 (see fig. 6.17), the resulting conformational changes activate the enzyme. Activation can be reversed by increasing the pH and thus abolishing salt bridge formation, the enzyme then returns to its inactive conformation. Somewhat unusually the pro-peptide and the enzyme stay together after activation, because of a disulphide bridge between Cys-1 and -122.

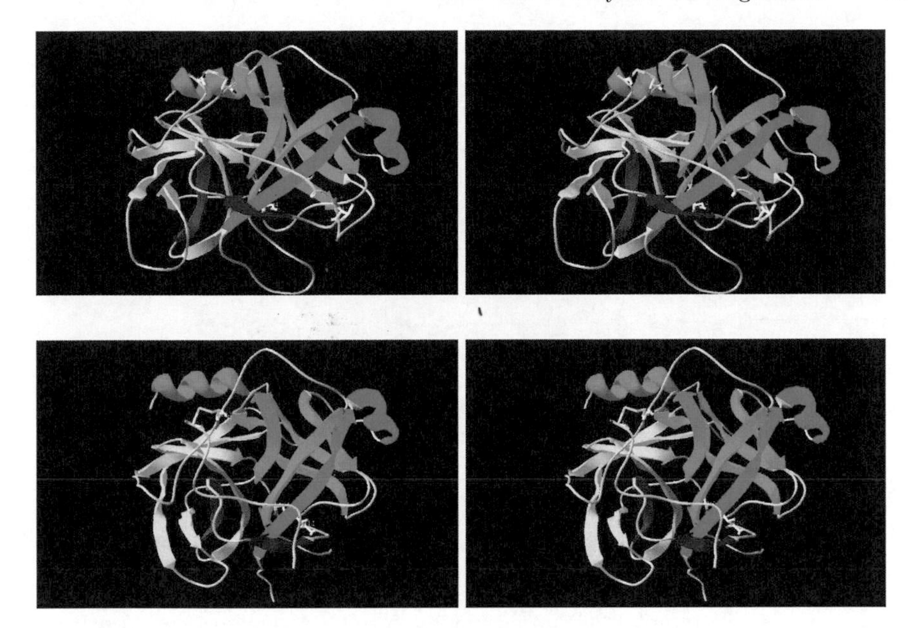

Figure 6.17. Activation of α-chymotrypsinogen (top, PDB-code 1CGI) to α-chymotrypsin (bottom, PDB-code 1OXG) by proteolytic cleavage at Ile-16. The free amino-group of Ile-16 can form a salt bridge to Asp-194, leading to substantial conformational changes. The pro-peptide stays bound to chymotrypsin because of a disulphide bond between Cys-1 and -122.

6.8 Use of enzymes for diagnostics

Many enzymes occur only in certain specialised cells in our body. If those cells are damaged, enzyme molecules are released into the circulation. Thus determination of enzyme activities is an important diagnostic tool in the clinical laboratory. For example creatine kinase (CK) is prevalent in muscle cells. Destruction of heart muscle cells in an acute myocardial infarct or of skeletal muscle cells in DUCHENNE muscular dystrophy leads to the appearance of CK in the blood. Interestingly, different organs contain different isoforms of CK, so that it is possible to tell from which organ an increased blood CK activity originated.

6.8.1 The coupled spectrophotometric assay of WARBURG

The concentration of substrates and enzymes in clinical samples is often determined using the coupled spectrophotometric assay which was introduced by O. WARBURG [80].

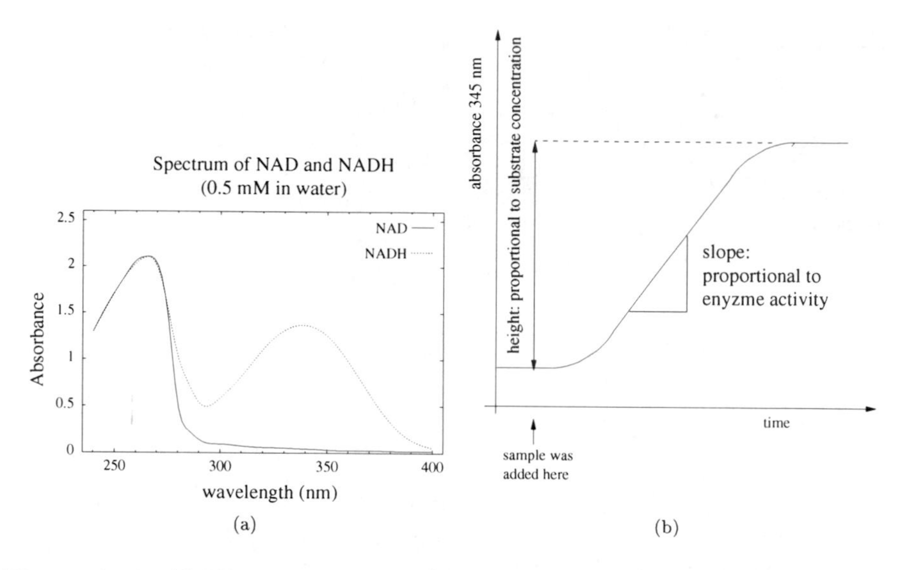

Figure 6.18. A) UV-spectra of NAD^+ and $NADH + H^+$ (in water). Both have an absorbance maximum at 260 nm, attributable to the adenine-ring. $NAD(P)H + H^+$ has an additional absorbance maximum at 340 nm, which can be used to distinguish it from $NAD(P)^+$. B) Change of UV-absorbance over time during a coupled spectrophotometric test. Substrate- and enzyme concentrations can be determined.

This test is based on the observation that $NADH + H^+$ and $NADPH + H^+$ absorb UV-light at 345 nm, while NAD^+ and $NADP^+$ do not (see fig. 6.18). These coenzymes are used by many dehydrogenases. Thus the activity of dehydrogenases can be determined by the rate at which the absorbtion of UV-light of a sample changes. On the other hand the concentration of substrates of such dehydrogenases can be determined by the total change of absorbtion.

There are many enzymes that do not themselves use NAD^+ or $NADP^+$, but which can be coupled to an enzyme which does. The creatine kinase discussed above is an example:

$$\text{Creatine} + \text{ATP} \xrightleftharpoons{\text{Creatine kinase}} \text{Creatinephosphate} + \text{ADP} \tag{6.49}$$

$$\text{ADP} + \text{Phosphoenolpyruvate} \xrightleftharpoons{\text{Pyruvate kinase}} \text{Pyruvate} + \text{ATP} \tag{6.50}$$

$$\text{Pyruvate} + \text{NADH} + \text{H}^+ \xrightleftharpoons{\text{Lactatedehydrogenase}} \text{Lactate} + \text{NAD}^+ \tag{6.51}$$

Clinical laboratories do this type of assay fully automated and with reagent kits bought premixed from suitable manufacturers.

Exercises

6.1. Which of the following statements about HMM-kinetics is *correct*?

a) The velocity of an enzymatic reaction is usually limited by the formation of the enzyme-substrate complex.

b) The maximal velocity of an enzymatic reaction depends on the enzyme concentration.

c) The MICHAELIS-constant K_M is defined as the concentration at which half of the enzyme molecules have substrate bound.

d) K_M is a characteristic constant for each enzyme/substrate pair.

e) V_{max} is achieved at a substrate concentration of 2 times K_M.

6.2. Which of the following statements is *false*? If an enzymatic reaction proceeds under condition of substrate saturation,

a) the reaction rate is directly proportional to the enzyme concentration.

b) virtually all the enzyme molecules have substrate bound.

c) is the reaction rate proportional to the substrate concentration.

d) the reaction rate depends on the speed at which the ES-complex is turned over into E + P.

e) the reaction rate will be determined by the law of mass action.

6.3. At a substrate concentration of 3 times K_M the reaction velocity will be

a) $0.25 * V_{max}$.

b) $0.33 * V_{max}$.

c) $0.50 * V_{max}$.

d) $0.75 * V_{max}$.

e) V_{max}.

6.4. In the figure the enzymatic reaction rate (v) of a substrate X is plotted as a function of $[X]$. Which change(s) is/are required to get from curve 1 to curve 2?

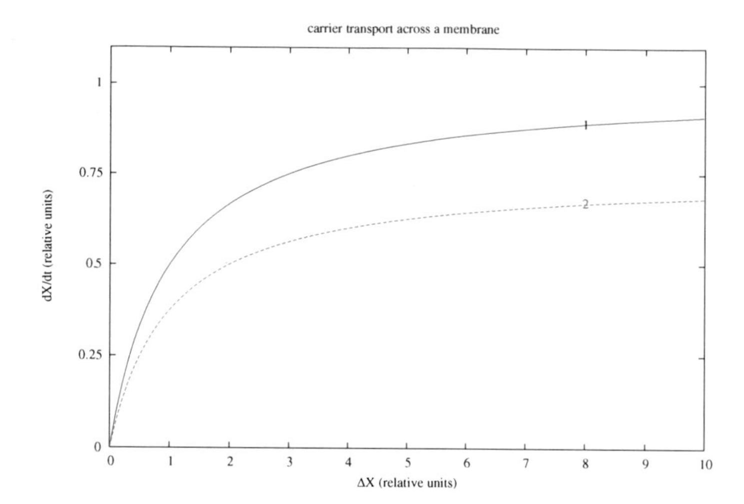

a) Increase the total concentration of X.

b) Decrease the number of enzyme molecules.

c) Increase the affinity of the enzyme for X.

d) All of the above.

e) None of the above.

6.5. 10 µg of a pure enzyme with a molecular mass of 50 kDa turn over 10 µmol/min of substrate. The turnover number is ________

6.6. Two enzymes compete for the same substrate. Most of the substrate will be turned over by the enzyme with:

a) the higher molecular weight.

b) the higher activity and lower K_M-value.

c) the lower activity and higher K_M-value.

d) a pI-value nearer to the pH of the solution.

e) the higher specificity for the substrate.

6.7. Isoenzymes

a) have several identical subunits.

b) react with the same substrate, but form different products.

c) have identical pIs.

d) are obtained from the same gene by different posttranslational modifications.

e) Are proteins with different primary structure that perform the same reaction on the same substrate.

6.8. You are supposed to measure the kinetics of an enzyme reaction. Critically discuss possible experimental pitfalls and the methods to control for them.

6.9. Which of the following statements about enzyme inhibition is *false*?

a) Enzyme inhibition is always reversible.

b) Acetylsalicilate (Aspirin®) works by competitively inhibiting enzymes that produce prostaglandins (pain mediators).

c) In a LINEWAVER-BURK-plot of non-competitive inhibition all lines intersect in a common point to the left of the y-axis.

d) Partially competitive inhibition occurs if the ESI-complex still has enzymatic activity.

e) In uncompetitive inhibition the inhibitor binds only to the ES-complex, not to the free enzyme.

6.10. Which of the following statements is/are *correct* about competitive inhibition?

1) In equilibrium the degree of inhibition depends on how long the inhibitor was in contact with the enzyme.

2) The inhibitor reacts with the substrate under formation of a substrate-inhibitor complex.

3) Competitive inhibition requires that the inhibitor binds in the substrate binding site.

4) V_{max} may be reached if [S] is high enough.

6.11. The mechanism of the following reaction is called: ________________

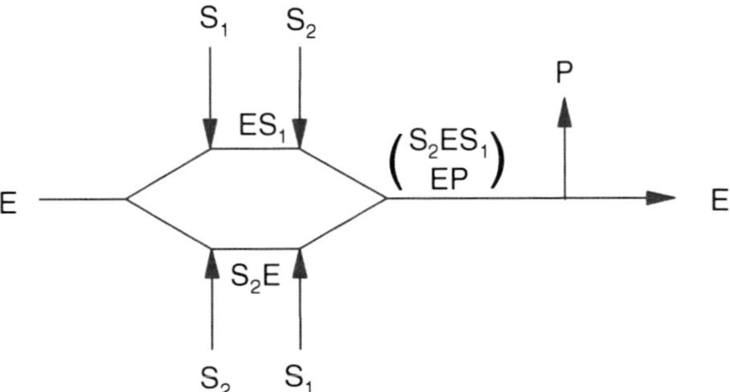

7

Haemoglobin and myoglobin: co-operativity

Blood is a very special juice (J.W. v. GOETHE: Faust I)

Haemoglobin is the most abundant protein in mammals, an adult male contains about 1 kg of it, equivalent to 3.5 g iron, about 80 % of the total iron in our body. Haemoglobin concentration in erythrocyte cytosol is 34 % by weight.

Haemoglobin and myoglobin are not only very important proteins, that attract funding, they are also available in large amounts and easily purified. Even human haemoglobin can be produced from outdated blood conserves without ethical problems, albeit with some risk of infection. For these reasons myoglobin and in particular haemoglobin are probably the best characterised proteins on this planet.

The 3D-structure has been solved by X-ray crystallography, and the structural changes that occur during binding of O_2, CO_2 and poisons like CO are known in great detail. The structural consequences of mutations in these proteins have also been investigated.

The kinetics of binding of ligands has been minutely studied, indeed the science of enzyme kinetics owes a lot to the progress made in studying oxygen binding to haemoglobin, which has been dubbed an "enzyme *honoris causa*" for this reason. Ironically, latest research shows that haemoglobins in some species indeed have enzymatic properties, they are involved in the destruction of nitric oxide and some xenobiotics. This surprise finding shows that our knowledge about these proteins is far from complete.

7.1 Structure

Both myoglobin, the oxygen storage protein in our tissues (and neuroglobin, which serves the same function in neuronal tissues), and haemoglobin, the

Table 7.1. Some important mile-stones in haemoglobin research.

Year	Author	discovery
1825	J.F. Engelhart	Iron makes 0.334 % of the weight of haemoglobin in all species. Thus haemoglobin is a defined molecule with a molecular weight of $n * 16700$.
1853	L. Teichmann	isolates haem by treating haemoglobin with NaCl/glacial acetic acid. The protein remains as colourless precipitate, the haem can by crystallised from the solution.
1864	F. Hoppe-Seyler	suggests the name haemoglobin and crystallises it (first crystallisation of a protein).
1904	C.H.L.P.E. Bohr	measures oxygen binding to haemoglobin as function of oxygen and carbon dioxide concentration.
1912	W. Küster	discovers the tetrapyrrol-ring structure of haem. In the following year R. Willstätter finds a similar structure in chlorophyll, but with Mg instead of Fe as central atom.
1914	J.S. Haldane	describes oxygen binding as equilibrium reaction with monodisperse haemoglobin, the binding constant depends on the animal species used.
1920	L.J. Henderson	investigates the influence of pH on the oxygen binding curve. The results are interpreted by J.B.S. Haldane (son of J.S. Haldane).
1925	G.S. Adair	determines the molecular weight of haemoglobin to be 65 kDa by measuring its osmotic pressure. He also discovers that oxygen binding to haemoglobin is a co-operative process.
1926	J.B. Conant	discovers the coordination of iron with amino acid side chains in haemoglobin.
1929	H. Fischer & K. Zeile	synthesise haem.
1934	T. Svedberg	measures the molecular weight of haemoglobin with his newly developed ultracentrifuge and finds haemoglobin to consist of a single, well defined molecule.
1935	F. Haurowitz	isolates HbF and compares oxygen binding curves of adult and fetal haemoglobin.
1940	R.R. Porter & F. Sanger	show that all haemoglobin molecules have the same end-groups.
1959	M. Perutz & J.C. Kendrews	solve the 3-dimensional structure of haemoglobin by X-ray crystallography.

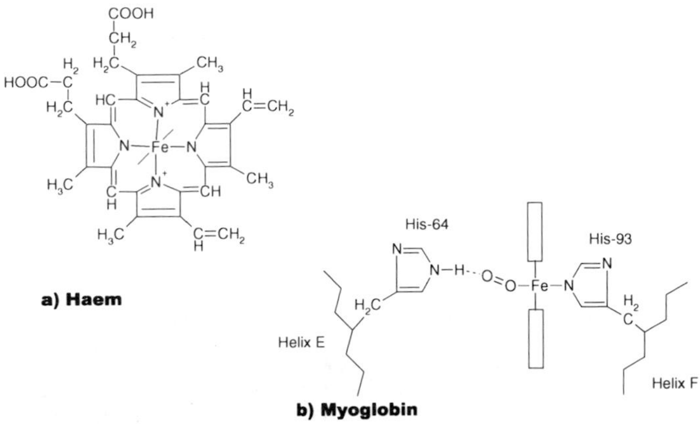

a) Haem

b) Myoglobin

Figure 7.1. Binding of haem to myoglobin. a) Haem is a flat molecule with iron at its centre. The iron is held in place by complex bonds to the 4 ring nitrogen atoms. The remaining 2 coordination sites of iron are oriented perpendicular to the haem ring system. b) One of these binds to a His residue of myoglobin (called the proximal His), the second binds the oxygen molecule. Binding of oxygen to haem is stabilised by a hydrogen bond to a second (distal) His residue. Note that the axis of the oxygen molecule forms an angle with the Fe-O bond, in case of the poison CO the system would be straight. The subunits of haemoglobin have a similar structure, the proximal histidines are His-87 in the α- and His-92 in the β-subunit.

oxygen transport protein in erythrocytes, contain haem as oxygen binding site. Haem is a porphyrin molecule with Fe^{2+} bound in its centre. Because it is buried deep in a small pocket inside the protein without direct access to the medium, haem can not leave the protein. Such firmly bound molecules in the active centre of a protein are called **prosthetic groups**.

Binding of haem into the protein has several functional consequences:

- Free iron can catalyse the formation of reactive oxygen species, which are dangerous to cells. Iron bound inside myoglobin or haemoglobin has a significantly reduced tendency to do that.
- Free haem has an extremely high affinity for some poisons, like nitric oxide and carbon monoxide. Carbon monoxide binds about 20 000 times better to free haem than oxygen, while in myoglobin it is only 200 fold better. The reason is the different orbital geometry of oxygen and carbon monoxide; while the axis of the O_2 molecule forms an angle with the Fe−O bond (see fig. 7.1), CO binds perpendicular to the haem ring. Steric hindrance by the distal His-residue reduces the affinity of haem for CO in myoglobin and haemoglobin.
- In free haem oxidation of the iron centre to Fe^{3+} destroys the oxygen binding site. The iron-histidine bond seems to protect the iron centre against oxidation.

- In a protein the affinity of haem for oxygen can be controlled by protein conformation. As we will see in a moment, this has important physiological consequences.

Myoglobin is a monomeric protein with a single haem group. Haemoglobin however is a heterotetrameric protein (2 α- and 2 β-subunits in normal adult haemoglobin), each of the subunits has its own haem. And this apparently small difference has important consequences.

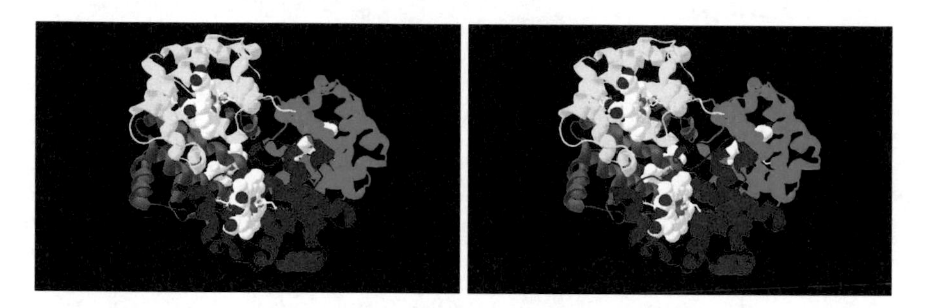

Figure 7.2. Three-dimensional structure of haemoglobin, in stereo (PDB-code 1HGA). The molecule consists of 2 protomers, each containing an α- and a β-subunit. The 4 haem molecules are in protected cavities inside the subunits.

7.2 Oxygen binding and co-operativity

In the enzyme section we have derived an equation for the binding of a ligand to a protein as a function of ligand concentration:

$$\Theta = \frac{[S]}{K_d + [S]} \tag{7.1}$$

Oxygen binding to myoglobin is well described by this equation. However, the binding curve of haemoglobin looks completely different (see fig. 7.3).

If oxygen binds to one of the 4 subunits in haemoglobin, the conformation of the whole protein changes, and with it the affinity of the other 3 binding sites for oxygen. This interaction is called **co-operativity**. There are three key points about co-operativity that you have to remember:

- Co-operativity requires the coexistence of several binding sites for a ligand in a protein, either on the same or (usually) on different subunits.
- Co-operativity results in sigmoidal (S-shaped) rather than hyperbolic binding curves.

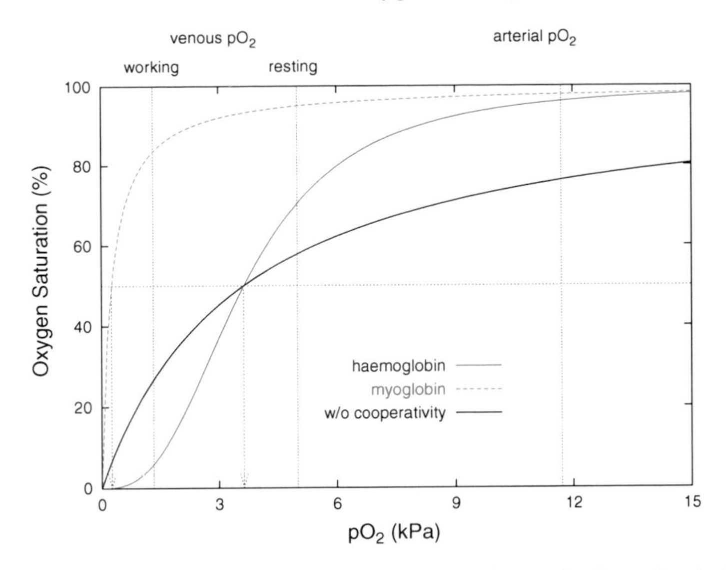

Figure 7.3. Oxygen binding to myoglobin, haemoglobin and a hypothetical protein with the same half-saturation point as haemoglobin, but a hyperbolic binding curve. Oxygen partial pressure in normal air is about 20 kPa, in the lung it is ≈ 13 kPa. Venous pO_2 depends on the exercise status (1–5 kPa).

- Co-operativity allows the body to regulate ligand binding and, in enzymes, substrate turnover to its needs in a given physiological situation. You will find co-operative kinetics in enzymes at key regulatory steps in metabolism (e.g. phosphofructokinase in glycolysis)

7.2.1 Functional significance of co-operativity

Let us look at the function of myoglobin and haemoglobin, so we can appreciate the importance of co-operativity:

Haemoglobin picks up oxygen in the lungs, and transfers it onto myoglobin in the tissues. In the lungs, oxygen partial pressure is about 13 kPa, somewhat lower than pO_2 in outside air (20 kPa). Arterial oxygen partial pressure is about 11 kPa, in tissues, pO_2 is between 1 and 5 kPa, depending on the physiological situation. Myoglobin is half-saturated at 0.26 kPa, so even in the low pO_2 of an exercising muscle myoglobin will be 85 % saturated with oxygen, and deliver it to oxygen-utilising enzymes.

Haemoglobin should be almost saturated with oxygen at the pO_2 of arteries (13 kPa), but should give off almost all bound oxygen in tissue, at 1 kPa. If you look at fig 7.3, you will find that a hypothetical protein with hyperbolic binding curve and a half-saturation point like haemoglobin can not do this. In the arteries, only 77 % of the molecules would bind oxygen, and in tissue 24–66 % of the molecules would retain their oxygen. In other words only 11–53 % of the transport capacity of such a protein would be used. Simply shifting the half-saturation point is no solution either: A lower affinity would improve oxygen release in the tissue but result in lower oxygen loading in the lung, a higher affinity would have opposite effects. Thus neither change would improve net oxygen transport.

Now look at the sigmoidal binding curve of haemoglobin. At arterial pO_2 haemoglobin is about 97 % saturated with oxygen. In a resting muscle, with $pO_2 = 5$ kPa haemoglobin is about 70 % saturated, in a working muscle with $pO_2 = 1$ kPa only 3 %. This corresponds to a transport efficiency of 27–95 % and makes haemoglobin a virtually ideal transport protein: almost completely filled at the source and, if required, virtually empty at the sink.

Note also how myoglobin buffers oxygen supply in muscle: A change of haemoglobin saturation from 3 to 75 % results in a 80–95 % saturation range for myoglobin.

7.2.2 Mechanism of co-operativity

How has evolution achieved this miracle of adaptation? As already mentioned the key is cross-talk between binding sites of haemoglobin. Each binding site can occur in a low affinity state (called the "tense" or T state) or in a high affinity ("relaxed" or R) state.

> The names "tense" and "relaxed" where originally chosen because in the T-state haemoglobin molecules are stabilised by additional ionic bonds between the α- and β-subunits, which do not exist in the R-state. The reason is that in the absence of oxygen the haem iron is pulled towards the proximal His, putting a strain on the haem molecule that results in a movement of the surrounding protein helices. Oxygen binding releases this strain.

If there are no oxygen molecules bound to haemoglobin, all 4 subunits will be in the T state, and their affinity to oxygen will be low.

If all 4 subunit have oxygen bound, then all will be in the R state and have a high affinity for oxygen.

You can already see how this works: With low oxygen saturation (in tissue) affinity will be low, and haemoglobin will tend to release oxygen completely. With high oxygen saturation (in the lungs) haemoglobin will have high affinity, and tend to bind as much oxygen as possible — just what is functionally required.

The interesting question is: What happens at intermediate oxygen pressures, how is the transition between all-T and all-S achieved? There are two models for this, the **concerted model** of J. MONOD *et al.* [54] and the **sequential model** suggested by D. KOSHLAND *et al.* [44] (see fig. 7.4).

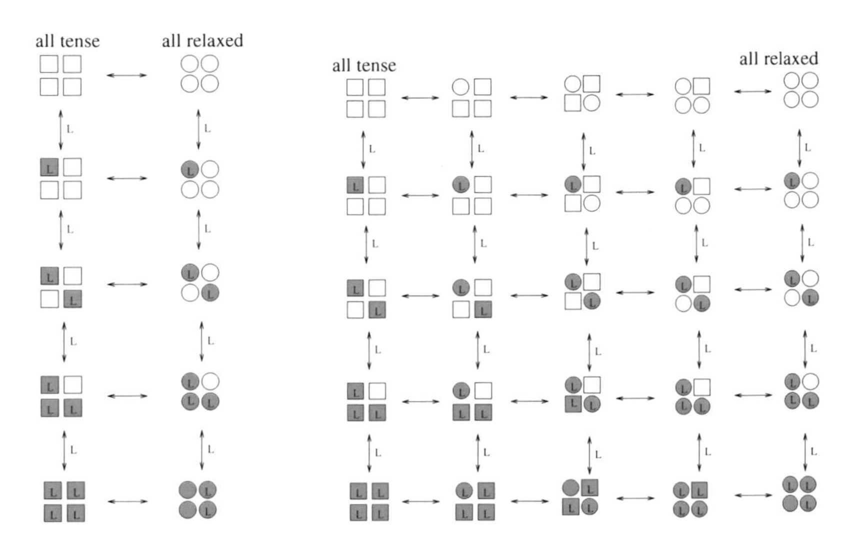

Figure 7.4. Two possible models for co-operativity. **Left**: In the concerted model all subunits in a protein are either in the T- (squares) or in the R-state (circles). The probability of the protein being in the R-state increases as more and more substrate is bound (filled vs empty symbols). **Right**: In the sequential model each subunit may individually switch between T- and R-state. The more subunits are in a given state, the higher the probability for the remaining subunits to follow suit. Note that the concerted model may be viewed as the limiting case of the sequential. Initial binding of oxygen is to the α-subunits, binding to the β-subunits occurs only after T to R transition.

The phenomenon that binding of a molecule at one site may influence binding at other sites on the same protein is called **allostery** (from the Greek words for "other shape"). When both ligands are identical, the interaction is called **homotropic**, if they are different, **heterotropic**. Some proteins have both homotropic and heterotropic interactions.

Equation for co-operative binding

A. HILL has proposed a mathematical model for co-operative binding in 1910 [36], long before the molecular causes of this phenomenon were discovered. He started with the assumption that several ligand molecules (L) bind to a protein (P):

$$P + nL \rightleftharpoons PL_n \tag{7.2}$$

Then the equations for the association constant K_a and binding Θ become:

$$K_d = \frac{[P] * [L]^n}{[PL_n]} \tag{7.3}$$

$$\Theta = \frac{[L]^n}{[L]^n + K_d} \tag{7.4}$$

The latter equation can be linearised by rearrangement followed by taking logarithms:

$$\log(\frac{\Theta}{1 - \Theta}) = n * \log([L]) - \log(K_d) \tag{7.5}$$

This is called the HILL-equation, and a diagram of $\log(\frac{\Theta}{1-\Theta})$ vs $\log([L])$ is called HILL-plot. The function $f(x) = \log(\frac{x}{x_{max} - x})$ is called logit(x).

When HILL plotted the oxygen binding data for haemoglobin in this way, he found that the line that best fitted the data had n_H=3, rather than the number of binding sites, which is of course 4. It has turned out that the number n_H is not the number of binding sites n, but reflects the degree of interaction between sites. n_H is called the **HILL-coefficient**, and can assume any value between 0 and the number of binding sites, it can even have (and frequently has) non-integer values.

$0 \leq n_H < 1$ signifies **negative co-operativity**, where binding of a ligand to one site impedes the binding at other sites.

n_H=1 means that there is no co-operation at all (binding follows HMM-kinetics).

$1 < n_H <$ n positive co-operativity, binding of one ligand to one site facilitates binding to the other sites.

n_H=n means complete co-operation (all sites are either filled or empty, no molecules with only some sites filled are allowed).

$n_H >$ n has never been observed, this situation is considered impossible on theoretical grounds.

7.2.3 Other factors involving oxygen affinity of haemoglobin

BOHR-effect

In tissues oxygen is used to burn food to carbon dioxide and water. The carbon dioxide needs to be removed, and this too is done by haemoglobin.

As a first step, carbonic anhydrase converts CO_2 into the much more soluble bicarbonate ion:

$$CO_2 + H_2O \rightleftharpoons H^+ + HCO_3^- \qquad (7.6)$$

As protons are produced in this reaction, venous blood has a somewhat lower pH (7.2) than arterial (7.6), even though the effect is mitigated by the buffering capacity of blood. This is important, because a drop in pH lowers the affinity of haemoglobin for oxygen, and increases that for bicarbonate. This is called the BOHR-**effect**[1]. Haemoglobin actually releases one proton for each molecule of oxygen bound, thus decreasing pH interferes with oxygen binding.

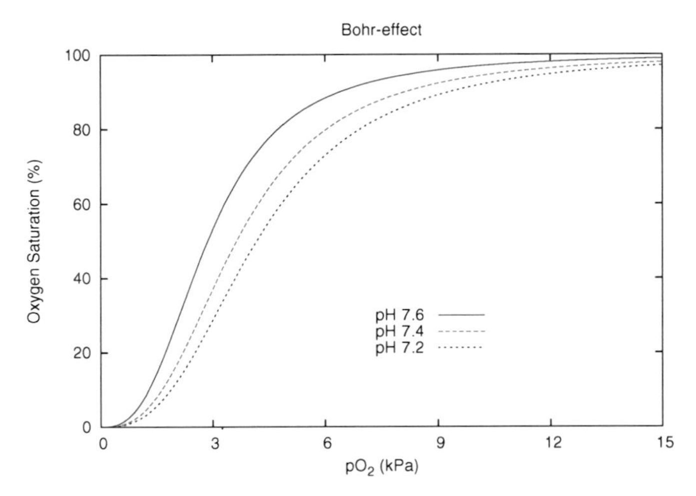

Figure 7.5. BOHR-effect. Oxygen binding to haemoglobin is pH-dependent, oxygen affinity is higher in the lungs than in muscle.

> One of the most important binding sites for protons is His-146 of the β-subunit. When protonated, this His can form an ionic bond to Asp-94, stabilising the T-state of haemoglobin. Because of that ionic interaction His-146 is stabilised in the protonated form, giving it a high pK_a in the T-state. However, several other amino acids beside His-146/Asp-94 are also responsible for the BOHR-effect.

CO_2 binds to the α-amino group of all 4 subunits, forming carbamino-haemoglobin. In this reaction too protons are produced, increasing the BOHR-

[1] named after CHRISTIAN BOHR, physiologist and father of the physicist NIELS BOHR

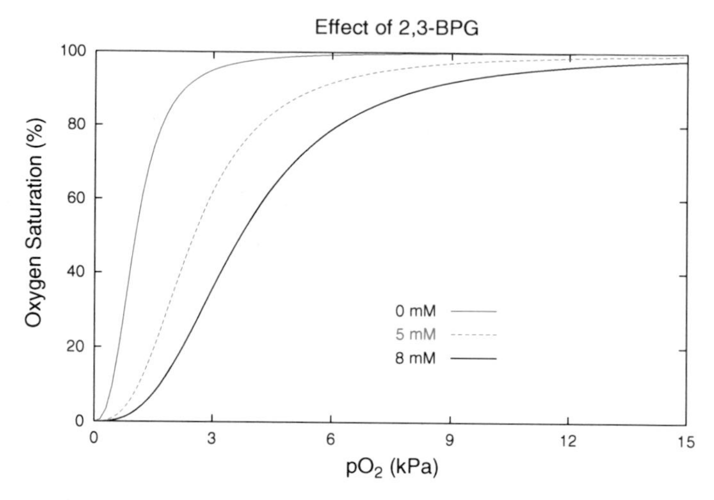

Figure 7.6. 2,3-BPG lowers the affinity of haemoglobin for oxygen. Increased 2,3-BPG in blood is an adaptation to lower environmental pO_2, for example in people living at high altitude or suffering from certain lung diseases.

effect. Additionally, this gives the amino-termini a negative charge, making them available for salt-bridge formation.

$$Protein - NH_2 + CO_2 \rightleftharpoons Protein - NH - COO^- + H^+ \tag{7.7}$$

2,3-Bisphosphoglycerate as heterotropic allosteric modulator

2,3-BPG binding reduces oxygen affinity of haemoglobin. 2,3-BPG binding and release allows our bodies to adapt quickly to different environmental oxygen partial pressures. For example, if we fly in a plane or climb onto a mountain, air pressure and therefore oxygen partial pressure is reduced. Under these conditions, 2,3-BPG concentration in blood will rise, this leads to a drop in the oxygen affinity of haemoglobin. This makes oxygen delivery to the tissues more efficient, but has little effect on oxygen saturation in the lung (see fig. 7.6). Hypoxia caused by lung damage or other pathological situations will also increase blood 2,3-BPG.

Each haemoglobin tetramer has one binding site for 2,3-BPG, a pocket formed between the β-subunits in the T-state, lined with positive charges that can interact with the negative charges of the phosphate groups of 2,3-BPG. In the R-state, this pocket is to small to accommodate 2,3-BPG, hence binding of 2,3-BPG into the pocket stabilises the T-state of haemoglobin.

There is a direct medical application for this knowledge: Erythrocytes from donated blood used to be stored in acid-citrate-dextrose medium ("ACD-blood"), where they are stable but loose much of their 2,3-BPG. As a result the haemoglobin in transfused erythrocytes had an abnormal high affinity for oxygen and could not participate in the oxygen supply of tissue until the 2,3-BPG level had been restored (about 24–48 h after transfusion). This deprived the patients of most of the benefit from transfusions.

Today **hypoxanthine** is added to the storage medium, which is split to inosine and ribose in the erythrocytes. Ribose can be converted to 2,3-BPG, maintaining its level in the cells.

7.3 Haemoglobin related diseases

Subunit composition of human haemoglobin

As already mentioned, normal adult haemoglobin (HbA) consists of 4 subunits, two α- and two β-subunits. One α- and one β-subunit are held together tightly by ionic and hydrophobic bonds, forming a protomer. Two such protomers then form the haemoglobin molecule. The protomers are held together mostly by hydrogen bonds. This allows the protomers to move with respect to each other during oxygen binding and -release.

However, HbA is only one member of the haemoglobin family. During embryonal development the fetus needs to get their oxygen supply from the blood of the mother. In order for this to work, their haemoglobin has a different composition with a higher affinity for oxygen.

During the first 3 months of development embryos produce **embryonal haemoglobin** (Hb Gower 1) in the yolk sac. Embryonal haemoglobin has the composition $\zeta_2\epsilon_2$.

As the bone marrow of the embryo develops, it starts producing **fetal haemoglobin (HbF)** ($\alpha_2\gamma_2$). The γ-subunits lacks some of the positive charges found in the β-subunit, resulting in weaker binding of 2,3-BPG and therefore higher oxygen affinity (oxygen affinity of HbF is similar to HbA striped of 2,3-BPG).

Production of haemoglobin β-subunits starts at about the 6th month of development, and HbA will have completely replaced HbF about 3 months *post partum*.

Normal adult blood contains not only HbA, but also about 2 % HbA_2 and 3–9 % glycated HbA (HbA_{1c}). HbA_2 has the composition $\alpha_2\delta_2$. Its production starts about 3 months *post partum*. The physiological significance of its presence is not known.

HbA_{1c} is produced from HbA by a reaction between the N-terminal valines of the β-subunits and blood glucose. This reaction is spontaneous and does not require enzymes. The rate of HbA_{1c}-formation, and therefore its accumulation during the 120 days life span of an erythrocyte and therefore its concentration in blood, is proportional to the blood glucose concentration. Thus the determination of $[HbA_{1c}]$ is valuable to check the long term average blood sugar concentrations in **diabetics**.

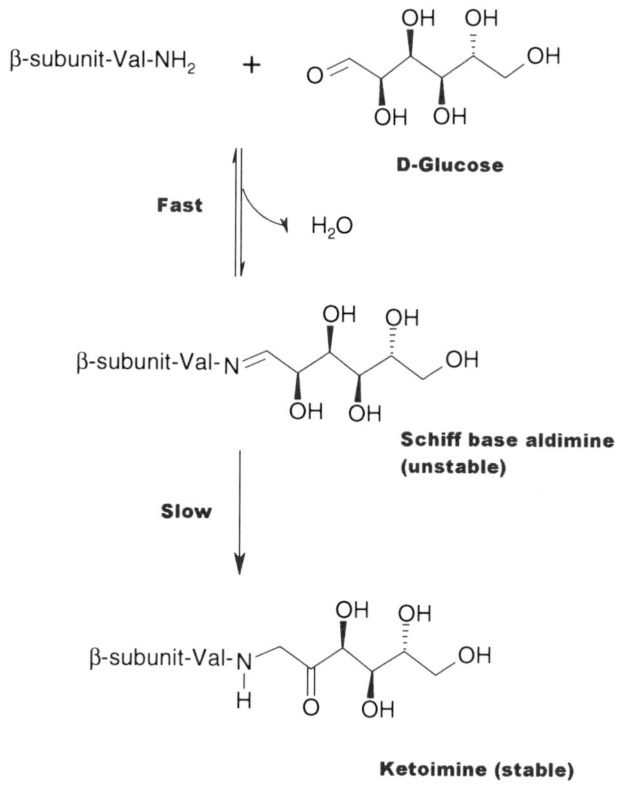

Figure 7.7. Formation of glycated haemoglobin (HbA_{1c}). The aldehyde group of glucose reacts with the N-terminal amino group of the β-subunit of haemoglobin, forming an unstable SCHIFF-base. This intermediate undergoes AMADORI- rearrangement to a stable ketoimine. Unlike glycoproteins glycated proteins are formed spontaneously, without enzymes.

In humans the information for the haemoglobin subunits $\alpha 1, \alpha 2$ and ζ is located on chromosome 16, for β, δ, γ and ϵ on chromosome 11. Each of these genes contains 3 coding segments (exons) and 2 non-coding segments (introns), which have to be spliced out of the mRNA before the proteins can be synthesised on the ribosomes. Synthesis of the haemoglobin subunits is normally synchronised, so that equal amounts of α- and β-subunits are produced.

Inherited diseases relating to haemoglobin

Sickle cell anaemia

Sickle cell anaemia is an autosomal recessive disorder (i.e. the gene needs to be inherited from both parents for the disease to become manifest) which affects mostly Africans and their descendants. It is caused by a mutation in the gene for the β-subunit of haemoglobin (β^S), leading to the substitution of Glu-6 with Val. The resulting haemoglobin is called HbS. The substitution of a charged amino acid by a hydrophobic one creates a hydrophobic patch, which is exposed in deoxy-haemoglobin, but hidden when oxygen is bound. [Haemoglobin] in erythrocytes is quite high (35 % w/v), and if a large proportion of the molecules have such a hydrophobic patch, they aggregate, forming lumps in the erythrocytes. The erythrocytes take on an unusual shape, which gives the disease its name.

As a consequence, the life time of the erythrocytes in sickle cell disease is reduced, leading to **anaemia** (lowered erythrocyte concentration in blood). This happens in particular when less oxygen is bound to the haemoglobin (exertion, pregnancy, high altitudes, increased CO_2-concentration, decreased blood pH). Sickled erythrocytes aggregate in the capillaries, blocking them and causing pain and tissue death from oxygen deprivation (**infarct**). Susceptibility for infections is increased due to low tissue oxygen.

It has been found that heterozygous individuals (one normal and one β^S gene), who have both HbA and HbS in their erythrocytes, are less prone to infection with *Plasmodium falciparum*, the causative agent for **malaria**. This parasite must spend part of its life cycle in erythrocytes, feeding on haemoglobin. Since the presence of HbS decreases the erythrocyte life span (even in heterozygotes, although less than in homozygotes), the parasite can not complete its life cycle in such individuals, giving resistance.

Haemoglobin C disease

The cause of HbC-disease is quite similar to HbS-disease, but Glu-6 is replaced by Lys rather than Val. In other words, a negatively charged amino acid is replaced by a positive one. Homozygous patients will show mild anaemia, but do not suffer from haemolytic crises and usually do not require treatment.

Haemoglobin SC disease

This disease occurs when a patient inherits the HbS mutation from one and the HbC mutation from the other parent. The course of the disease is much milder than in HbS disease, but a (potentially fatal) haemolytic crisis may occur in special situations (e.g. during child birth or surgery).

Thalassaemias

As mentioned before, the synthesis of the α- and β-subunits of haemoglobin is normally synchronised, so that equal amounts are produced. Thalassaemias are caused when this synchronisation no longer works and either not enough α-subunit (α-thalassaemia) or not enough β-subunit (β-thalassaemia) is produced. Production of this subunit can be **absent** (α^0- and β^0-thalassaemias) or **reduced** (α^+- and β^+-thalassaemias).

Since people carry the information for the α-subunit in 4 copies (2 on each of the chromosomes 16), α-thalassaemia can occur in a variety of degrees. An individual with one defect copy is called a **silent carrier** and will not show overt symptoms. Individuals with two defect copies are said to have the **thalassaemia trait** and with three defect copies **haemoglobin H disease** (mild to moderate anaemia). Individuals with 4 defect copies will die during embryonal development, when the embryonal haemoglobin is replaced by the fetal one. The abortus will have high levels of γ-tetrameres (Hb Bart) or β-tetrameres (HbH), which are soluble but show hyperbolic oxygen binding curves (no co-operativity). As discussed above, they are useless as oxygen carriers.

In β-thalassaemia the α-subunits are produced at a normal level, but in the absence of the β-subunits normal haemoglobin can not form. Instead the α-subunits precipitate in the cells destined to become erythrocytes, which die as a consequence. As there are only two copies of the β-globin gene, affected individuals either have the thalassaemia trait (**β-thalassaemia minor**, no treatment required) or **β-thalassaemia major** (COOLEY's anaemia, with both copies defect). The later individuals appear perfectly healthy at birth, where the fetal haemoglobin is still present, but become rapidly anaemic as HbF should be replaced by HbA. Originally these patients were treated with regular blood transfusions, but the resulting iron overload resulted in death from **haemochromatosis** (iron poisoning) in their mid-twenties. Today **bone marrow transplantation** offers a better prognosis.

Exercises

7.1. Which of the following statements is *false*? Allosteric enzymes

a) have sigmoidal v vs $[S]$ plots.

b) have binding site(s) for regulatory ligand(s), which are located apart from the substrate binding site.

c) have several substrate binding sites, binding of substrate to one site influences binding to the other.

d) are characterised by a HILL-coefficient which is identical to the number of substrate binding sites.

e) Often regulate pacemaker-reactions at the beginning of metabolic pathways.

7.2. Which of the following statements is *false*? 2,3-Bisphosphoglycerate

a) binds mainly to desoxy-haemoglobin.

b) binds to haemoglobin covalently.

c) binds less strongly to fetal than to adult haemoglobin.

d) concentration in blood increases as air oxygen partial pressure decreases.

e) reduces the HILL-coefficient of haemoglobin for oxygen binding.

7.3. Myoglobin

a) has a lower oxygen affinity than haemoglobin.

b) is a diprotomer containing 2 α- and 2 β-subunits.

c) has a hyperbolic oxygen binding curve.

d) contains Cu^{2+} as central atom.

e) contains Fe^{3+} as central atom.

7.4. Sickle cell anaemia

a) is caused by the inability to synthesise haemoglobin β-subunit.

b) is caused by a change in the amino acid sequence of the β-subunit of haemoglobin.

c) is caused by a failure of haemoglobin to bind iron.

d) increases the risk of the carrier to become infected with malaria.

e) symptoms are reduced at lower oxygen concentrations.

7.5. Carriers of the gene for sickle cell anaemia have increased resistance to malaria
because
aggregation of HbS reduces the the life span of the erythrocyte.

8

Enzyme kinetics: special cases

In an enzyme following HMM-kinetics, enzyme activity rises gradually with substrate concentrations. For many physiological situations this is appropriate, the enzyme activity adapts to the metabolic needs of the cell.

In this chapter we want to look at some cases, where such a gradual response would be inadequate, where instead a steep, switch-like, **all-or-nothing response** is required.

Binding of a few hormone molecules to the receptors on the membrane of a cell can affect considerable changes in the metabolism of the cell. For example, binding of **mitogens** (growth factors) can make a cell enter mitosis (see figure 8.1). A cell however should be either in mitosis, or not. A gradul transition between these two states is impossible. The switching between **mutually exclusive metabolic states** by hormones (e. g. **insulin** and **glucagon**) would be another example.

You have already learned how the association of subunits into oligomeric enzymes with co-operating binding sites can produce S-shaped control curves, which offer such switch-like characteristics. There are however other ways.

8.1 Activation cascades

The usual explanation given for such cascades is that they allow amplification of the signal. If each enzyme activated 100 substrates, than the Receptor/ G-protein/MAPKKK/MAPKK/MAPK/MAP cascade in fig. 8.2 would allow amplification by a factor of $100^5 = 1 \times 10^{10}$. This explanation is incorrect on two grounds:

- Such amplification factors are unmanageable by the cell and would lead to constant erroneous activation of MAPs due to noise.

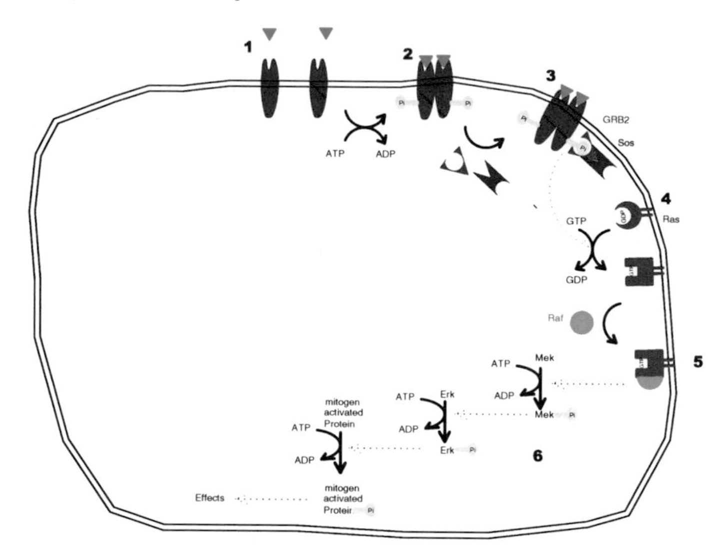

Figure 8.1. Binding of mitogens results in the phosphorylation of so called **mitogen activated proteins (MAP)**, which control entry of the cell into mitosis. Mitogens bind to cell surface receptors (1), which then dimerise and autophosphorylate each other on tyrosine residues (2). These phospho-tyrosine residues serve as binding sites for GRB2 and Sos (3). Bound Sos acts as a GDP/GTP exchange factor for the small G-protein Ras, converting the inactive GDP-Ras into active GTP-Ras (4). This in turn binds and activates Raf, which is a Ser/Thr kinase (**MAPKKK**) (5). Activated Raf will phosphorylate and thus activate MAP-kinase-kinases (**MAPKK**) like Mek, which will in turn phosphorylate MAP-kinases (**MAPK**) like Erk, which act on MAPs. Not shown are the GTP-hydrolysis stimulating factors that turn off Ras and the protein phosphatases that remove the activating phosphate groups from the other proteins. For more details see text books on molecular cell biology.

- Experimentally determined amplification factors for this system are in the order of $1 \times 10^1 - 1 \times 10^2$. Such factors could be achieved also by a single enzyme.

The reason for cascades is probably not amplification, but the shape of the resulting **response curves**. A cell should be either stimulated into mitosis, or not. Thus rather than the usual hyperbolic response curves something steeper, like a S-shaped HILL-curve, is required.

Assume we have a phosphorylatable substrate in the presence of a kinase and a phosphatase. The left column of the figure 8.2 shows this situation for low substrate concentrations, where the enzyme activity is **linearly** related to [S], the right column shows high concentrations with **hyperbolic** curves.

The curves were drawn under the assumption that the total substrate concentration (sum of phosphorylated and dephosphorylated substrate) does not change over time. Only the fraction of substrate phosphorylated changes.

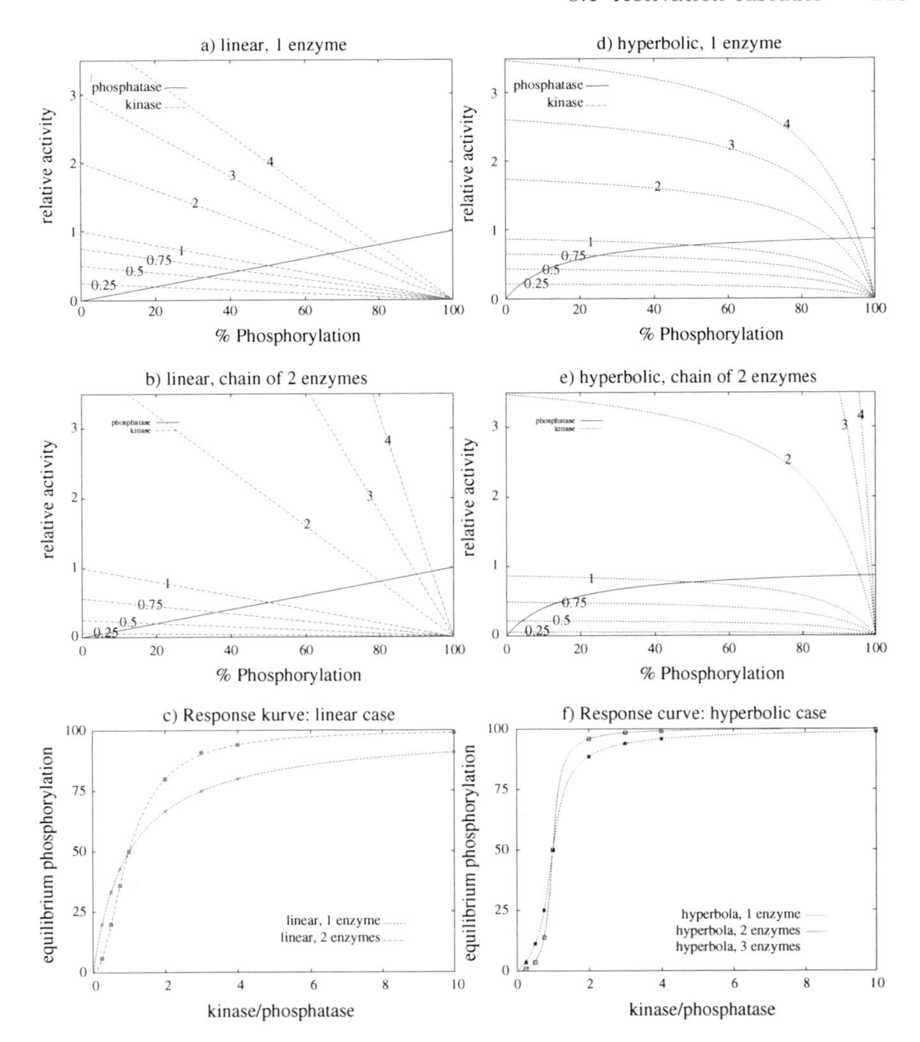

Figure 8.2. Analysis of the response curves for enzyme cascades

Under this condition if all the substrate is dephosphorylated, the kinases will be maximally active, whilst the phosphatases have no substrate to work on and therefore their activity will be zero. In the case of fully phosphorylated substrate the opposite is true.

It is easy to see that if phosphatase and kinase are present at the same activity, an **equilibrium** will be established were half the substrate is phosphorylated. At this x-coordinate the lines of phosphatase and kinase activity in figure 8.2a) intersect.

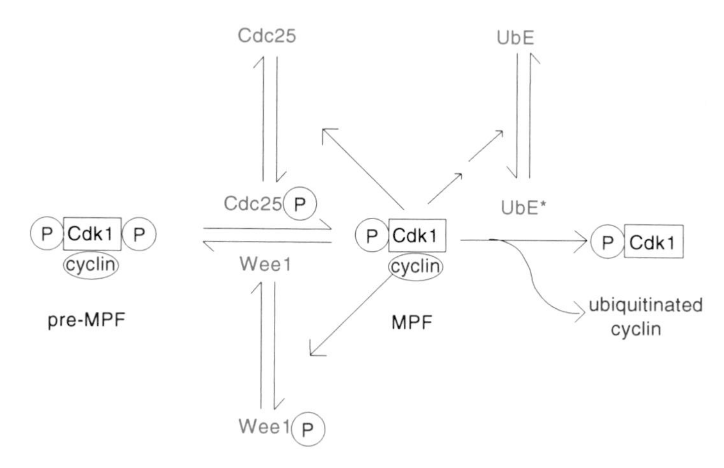

Figure 8.3. Part of the feedback-control of MPF-activity. For details see text.

If we keep the phosphatase activity constant but change the kinase activity (say, after hormone binding), the equilibrium between phosphorylated and dephosphorylated substrate, and as a result the intersection between the lines denoting phosphatase and kinase activity will also change.

Now look at the at fig. 8.2e). It shows how the equilibrium phosphorylation changes as a function of the ratio of kinase / phosphatase activity. If the kinase activity is regulated directly by some input stimulus we get a **hyperbolic response curve**.

Now let us assume that our kinase is regulated by a cascade: The stimulus acts on an enzyme that stimulates the kinase with an amplification factor of 2. Under this condition, a doubling of the input will lead to a 4-fold increase of kinase activity (plotted in the middle row of fig. 8.2). The resulting response curve is S-shaped, and it is the steeper the higher the number of enzymes in the cascade is.

In case of a linear v vs [S] relationship (left column of fig. 8.2) the result will be a HILL-curve. If the v vs [S] relationship is hyperbolic, the response curve will also be S-shaped, but steeper than predicted by the HILL-equation.

For further discussion of this topic see [25]

8.2 Feedback-networks

Many events of metaphase are controlled by **maturation promoting factor (MPF)** (see for example [48] for a detailed discussion on MPF). MPF is a complex of two proteins, the Ser/Thr-kinase **Cdk1 (Cyclin dependent kinase 1)** and its control protein **Cyclin B**. Cdk1 can be phosphorylated on two sites: Tyr15 and Thr161. Phosphorylation on Thr161 is required for activity, the enzymes involved are not shown here for reason of clarity. Phosphorylation on Tyr15 inhibits the kinase activity. This phosphorylation is caused by a protein kinase called **Wee1**, the dephosphorylation is accomplished by the phosphatase **Cdc25**.

The clou of this system is that both Wee1 and Cdc25 can be phosphorylated by MPF (or kinases activated by MPF). Wee1 becomes inactivated, Cdc25 activated by phosphorylation. Thus the presence of a small amount of active MPF will, by phosphorylating these two proteins, stimulate its own production. This is called **positive feed-back**.

MPF-activity at the end of the cell cycle is destroyed by ubiquitin-dependent proteolysis of Cyclin B (see fig. 10.14 on page 162). The ubiquitinating enzymes however are also activated by MPF, that is, high MPF activity results in its own destruction. This is called **negative feed-back**.

Positive and negative feed-back together explain the saw-tooth like activity pattern of MPF, as can be shown in computer simulations.

8.2.1 Advanced Material: Simulation of feed-back systems

What are the properties of a feed-back system?

It is frequently observed that equations for the rate of change of a parameter (that is, a differential equation) can be written down much easier than the integrated equation for the parameter itself (we have already encountered such systems of differential equations when we derived the HMM-equation (see page 72).

For example, the change in concentration of MPF in fig. 8.3 is given by its formation from preMPF ($v_1 = [preMPF] * k_{Cdc25}$) and its destruction by Wee1 and UbE ($v_2 = -[MPF] * k_{Wee1} - [MPF] * k_{UbE}$). Thus

$$\frac{d[MPF]}{dt} = [preMPF] * k_{Cdc25} - [MPF] * k_{Wee1} - [MPF] * k_{UbE} \quad (8.1)$$

Similar equations can be written for all other participants in this system (whether shown in fig. 8.3 or not). This results in a system of linked differential equations.

However, we are not really interested in the rate of change of the activities over time, but in the activities themselves. To get those, the system of differential equations needs to be integrated. With small systems, this can sometimes be done analytically, but systems as complex as MPF-regulation can be integrated only numerically on a computer.

$$\frac{dx}{dt} = k * x \approx \frac{\Delta x}{\Delta t} \tag{8.2}$$

$$x(t + \Delta t) \approx x(t) + \frac{\Delta x}{\Delta t} \tag{8.3}$$

In other words, if you know the concentrations of all components in a system at time t=0, you can calculate the concentrations after a small time increment ($t_1 = t_0 + \Delta t$) by calculating the change over the time period and adding this to the original value. The results can then be used to calculate concentrations at t_2 and so on. This is called EULERS algorithm.

There is an error associated with this procedure, because it assumes that the rate of change is constant during the period Δt. The error can be minimised by using the average rate of change during this period, rather than the rate of change at its beginning. This is the principle of the RUNGE-KUTTA algorithm (a good implementation of this method in C is found in [34]). Solving systems of differential equations in this way is central to areas as diverse as enzyme kinetics, weather prediction or airplane construction.

If these principles are applied to cell cycle control enzymes, the results mimic the laboratory observations closely, for example saw-tooth patterns of MPF-activity can be seen. For further details see [77]

8.3 Multiple phosphorylation

Another method used by cells to control response curves is exemplified by the *S. cerevisiae* Sic1 protein. This protein binds to the yeast S-phase cyclin/Cdk1 and inactivates it, thereby preventing DNA replication. At the G_1/S-transition Sic1 is phosphorylated (by another Cyclin/Cdk-complex), leaves the complex and becomes ubiquitinated. At least 6 of the 9 phosphorylation sites of Sic1 need to be phosphorylated for this to happen.

Why 9, wouldn't one do? Ubiquitination starts by binding of phosphorylated Sic1 to Cdc4, which is part of the ubiquitination machine. Other proteins can bind to Cdc4 after phosphorylation of only a single site. However, none of the phosphorylation site sequences in Sic1 matches the binding site of Cdc4 very closely. Thus multiple phosphorylation is required to increase the binding strength.

But why then does Sic1 not have a sequence that matches the binding site on Cdc4? By site-directed mutagenesis this question can be studied experimentally [55]. In short, such a mutated Sic1 does not work.

Assuming that each phosphorylation is accomplished by a single productive collision between Sic1 and its kinase, the first 5 events introduce a delay, before finally the 6th event marks Sic1 for destruction. Thus the requirement for multiple phosphorylation acts as a kind of timed fuse.

Additionally, the first 5 sites act as a noise filter. Random phosphorylation by unspecific kinases may happen on one site, may be even two. But all 6 – that is unlikely. If the rate of unspecific phosphorylation is ϵ, requirement for 6 phosphorylations reduces unwanted Sic1 destruction by a factor of ϵ^6. This is called **kinetic proof-reading**.

The same is true for phosphorylation by low levels of kinase. Only when the concentration of kinase has risen to a significant level will Sic1 be destroyed. A doubling of kinase concentration could then increase the rate of Sic1 destruction by $2^6 = 64$-fold, resulting in a step-like inactivation.

Part III

Special proteins

9

Prion proteins and prion diseases

The so called prion diseases (for more details see [40]) result in destruction of the brain, this is eventually fatal. The first such disease described (in the 18th century) was **scrapie** in sheeps. It was noted that this disease could be spread to other animal species by feeding them infected brain. One of the most remarkable features of the infectious agent of scrapie is its incredible resistance to destruction. A sheep head, buried for three years, proved still infectious when fed to mice. The agent can not be inactivated with formalin (methanal), with UV light or by heat sterilisation.

Similar observations were made with **kuru**, a disease prevalent in the Fore people of Papua New Guinea. Kuru was spread by cannibalism (ritual eating of infected human brain), until the colonial power (Australia) stomped out this habit in 1957. New cases are still reported from time to time, but only in older people who probably became infected while cannibalism was still rife. This points to a very long incubation period which can exceed 40 years. Victims first giggle and tremble uncontrollably (kuru actually means "the laughing death"), later they lose all awareness and control over body functions and finally die.

Natural prion diseases in humans

CREUTZFELDT-JAKOB-disease (CJD), GERSTMANN-STRÄUSSLER-SCHEINKER-disease (GSS) and **fatal familial insomnia** (FFI) are rare inherited diseases in humans. CREUTZFELDT-JAKOB-disease (CJD) for example starts with neurological symptoms (difficulties with coordination, tremor), patients loose speech, then the control over body and mind. Death occurs within 6 month of outbreak. Most victims are more than 60 years old.

Pathologically, all these diseases, collectively known as **spongiforme encephalopathies**, are very similar. They all result in the formation of intracellular vacuoles in brain tissue (at locations specific to a particular disease), which eventually cause death of the affected cell and fatal failure of brain function. These plaques contain an abnormal protein, PrP^{sc} (Prion protein scrapie, **prion** stands for proteinaceous infectious particle).

Pathomechanism

Researches who tried to isolate an infectious agent from these plaques so far invariably have ended up with preparations which contained only PrP^{sc}, but no nucleic acids. Methods usually employed to destroy nucleic acids (nuclease treatment, UV-irradiation at 250 nm) fail to destroy prion infectivity. This led TIKVAH ALPER to suggest that the causative agent is not a microorganism as in other diseases [2, 3]. S. PRUSINER has vigorously followed up on this idea, finally succeeding in the isolation of PrP^{sc} (Nobel Price in 1997) [63].

PrP^{sc} is derived from a protein normally present in brain, PrP^{c} (c for cellular, the gene is located on chromosome 20). Both forms are membrane bound via a GPI-anchor (see page 235), but whilst PrP^{c} is bound to the plasma membrane PrP^{sc} occurs mainly inside the cell on lysosomal membranes.

The function of PrP^{c} is unknown, knock-out mice strains without the genetic information for this protein are resistant to infection with PrP^{sc}. Otherwise they appear normal, but show subtle changes in electrophysiological experiments. If copper is present during the *in vitro* folding of PrP^{c} the protein acquires superoxide dismutase activity. Thus it may play a role in the protection of the brain against damage from **reactive oxygen species (ROS)**.

Apparently PrP^{sc} causes an autocatalytic conversion of PrP^{c} to PrP^{sc}. Because these two forms of the protein have different conformations, they show different fragments after treatment with protease. Indeed, different strains of PrP^{sc} also result in different proteolytic fragments, these fragments therefore can be used to identify them.

Inherited encephalopathies like CJD, FFI or GSS apparently are caused by mutations in the PrP^{c} gene, which make spontaneous conversion to PrP^{sc} more likely (for example P102L in GSS and E200K in fCJD). They are inherited in an autosomal dominant manner.

Apart from prion diseases the conversion of proteins into an insoluble form (**amyloid**) occurs also in other diseases like Morbus ALZHEIMER. Like in prion disease conversion of part of the protein into β-sheets seems to be the root of the problem. Recent evidence may indicate that the strands in these sheets are relatively short and have little twist [83]. As a result, amino acids in amyloid β-sheets have a much narrower range of ϕ and Ψ values than normal β-sheets, making them occupy a small region in the RAMACHANDRAN-plot near the diagonal. However, although this hypothesis is attractive, experimental evidence for it is currently limited.

It is currently unclear whether the precipitated amyloid proteins cause cell death, or whether these may be quite inert, with the damage done by soluble intermediates of amyloid formation.

BSE and vCJD

Scrapie infected sheep offal, converted to bone meal, had been used to feed cows in Britain without adequate precautions. This led to the appearance of **mad cow disease** or **bovine spongiforme encephalopathy (BSE)**, some humans eating infected beef (about 750 000 infected animals were converted to human food) have come down with "**new variant** CREUTZFELDT-JAKOB-**disease (vCJD)**". Apparently prions that have crossed a species barrier once are much more prone to do so again. Because the incubation period is very long (at least 15 years) we do not yet know how many people ultimately will fall victim to this new disease. Infected cow offal, converted to bone meal, has been exported from Britain all over the world and fed to other animals. Export to third world countries even continued after export to other EU countries was banned because of BSE. This may mean that BSE was spread in animals fed with bone meal containing food world wide (not only cows and sheep, but also pigs, chicken, cats, dogs...). One of the current concerns is that although Scrapie in sheep does not appear to be infectious to humans, BSE can be transmitted to sheep and is then difficult to distinguish from scrapie.

Prions, taken up in the gastrointestinal tract, apparently propagate first in the lymphatic tissues of the intestine, then spread to the nerves that supply the abdomen. From there they get into the spinal cord and the brain. Lymphatic (cecum, Peyer's patches, spleen, tonsils, bone marrow) and in particular neural tissue (spinal cord, brain, eyes) of infected people and animals contain the highest concentration of PrP^{sc} (1×10^6 and 1×10^9 infectious particles per g,

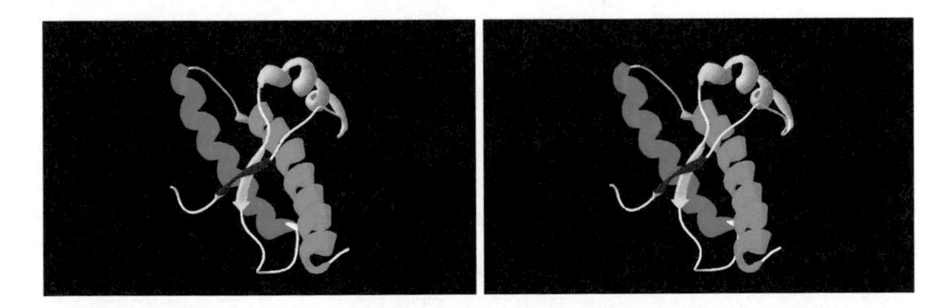

Figure 9.1. Stereo view of the normal PrP^c (bovine, residues 23–230, determined by NMR, PDB-code 1DX0). According to one hypothesis the entire N-terminal region (the two β-strands and the small α-helix turns into a sheet of 4 anti-parallel β-strands during conversion to PrP^{sc}. However, due to its tendency to aggregate no experimental structures are available for PrP^{sc} yet.

respectively), in animal experiments 0.1 g of infected cow brain can transmit the disease to other animals.

Compared to classical CJD, the patients with vCJD are much younger (29 compared to 65 a on average, youngest was 14), this may have to do with the more active immune system in children, which ironically may make them more susceptible to infection. Also, children tend to eat more junk food (hamburger, sausages). The course of the disease is protracted (14 months from outbreak to death compared to 4.5 months in classical CJD). First symptoms usually are psychiatric: depression, aggression, loss of memory. To this time (02/2005) 150 British, 6 French and one patient each from Canada, Ireland, Italy, Japan and the USA have been reported.

Vertical transmission of scrapie in sheep has been known to occur for some time, and there were some fears that it may occur for vCJD in cattle and humans. One case of suspected human vertical transmission was reported but a recent epidemiological study [84] found no evidence for it.

Several people who later succumbed to vCJD were blood donors, and infected blood can transmit the disease in animal experiments. As the blood was pooled with that from other donors for the production of blood products, several hundred people have been given potentially tainted medicines. Again newer studies suggest that transmission by transfusion, if it occurs at all, must be a rare event.

Another possible route for nosocomial infections are surgical instruments (1 case described) and transplants of neuronal tissue (e.g. *Dura mater* or corneas) or pharmaceuticals made from brain (e.g. growth hormone), as the infectious agent is difficult to destroy with sterilisation techniques currently in use.

As long as there are only relatively few cases, there will be no cure: The costs of developing a pharmaceutical are so high (according to industry figures 800 000 000 € on average, independent watchdog organisations have claimed that "only" 200 000 000 € are required) that the industry will do this only for widespread diseases, which allow these costs to be recovered.

It has been observed that in vCJD all victims so far had a mutation of Val-129 to Met in both copies of the PrP^c gene (MM genotype). However, kuru victims with VV and VM genotype are also observed, thus the MM genotype may simply reduce the incubation period (or kuru and vCJD may behave differently in this respect).

Exercises

9.1. Which of the following statements is *false* about prion diseases?

a) Scrapie is a prion disease in sheep and not infectious to humans.

b) Kuru is a prion disease in humans, spread by cannibalism.

c) Prion diseases are caused by a virus infection.

d) Mutations in PrP^{c} can increase the risk of prion disease in affected patients.

e) In new variant CREUTZFELDT-JAKOB-disease (vCJD) the infectious agent can spread from the gastrointestinal tract through the lymphatic tissue and peripheral nervous system into the brain.

10

Immunoproteins

If you wish for peace, prepare for war.
(Latin proverb)

10.1 Overview

The human body is a nutrient rich environment and would be colonised rapidly by viruses, bacteria and other parasites if it were not protected by an immune system. This system not only recognises and destroys invading parasites (and even cancer cells), but does not attack the bodies own cells. Both properties are equally important for our survival. An attack of our immune system against our own cells is called **autoimmune disease**. Such diseases tend to be debilitating, protracted and eventually fatal. Examples are *myasthenia gravis* (antibody production against the muscle acetylcholine receptor), systemic *lupus erythematosus* (SLE) (reaction against nuclear antigens) or *diabetes mellitus* type 1 (destruction of the β-cells in the islets of LANGERHANS in the pancreas).

We will look at the immune system from the biochemical point of view, and focus on the proteins involved. Later in your studies you will take immunology courses that take a broader perspective. For self-study, [42] is suitable.

There are two different defense systems in our body, the **innate** and the **acquired** immunity.

The **innate immunity** is our first line of defense. It consists of phagocytic cells that devour any foreign material. Because this system is inborn and does not require training, it can respond quickly to a new antigen, but its efficiency is much lower than that of the acquired immune system.

Within about two weeks of encountering a new antigen the body mounts a specific immune response against it. This **acquired immunity** is highly specific and very effective, it also results in a permanent memory of that antigen. If the antigen is encountered a second time, the specific immunity can be mounted much faster. For that reason we can get many diseases only once ("childhood diseases").

10.1.1 Cells of the immune system

The cells that make up our immune system are originally produced in the bone marrow, they are derived from the same multipotent stem cells that give rise to erythrocytes and platelets. Some of these cells however migrate to other lymphoid organs during their development for final differentiation.

Primary **lymphoid organs** are bone marrow (and liver during early embryonal development), thymus and in birds the *bursa fabricii* (an organ associated with the end gut). Secondary (peripheral) lymphoid organs are tonsils, spleen, lymph nodes, PEYER's patches and the appendix.

Apart from the erythroblast and megakaryoblast cell lines, which are not involved in the immune system, the multipotent stem cell of the bone marrow gives rise to the following cell lines:

- The lymphoblast cell line develops into **lymphocytes**, the principal carriers of the acquired immune system:
  - **T-lymphocytes** migrate from the bone marrow to the thymus for final differentiation, before they are released into the blood stream. We distinguish the $CD8^+$ T-killer- (T_k) and the $CD4^+$ T-helper (T_h) cells.
  - In birds the **B-lymphocytes** differentiate in the *bursa fabricii*. A lot of effort was spend on the search for the bursa-equivalent in mammals, it now appears that in mammals the B-cells do not move to another organ, but develop completely in the bone marrow, before they are released into the blood stream. Once stimulated with antigen, a B-cell can develop either into an antibody-producing **plasma cell** or into a long lived **memory cell**, which enables our body to mount a specific immune response much faster when it encounters an antigen a second time.
  - **Natural killer (NK-) cells** kill virus infected and cancer cells marked by antibodies (IgG1 and IgG3) on their surface. They are also important in innate immunity, killing cells with low level of MHC-I, or cells expressing foreign carbohydrates.
- The myeloblast cell line leads to granulocytes, which are responsible for innate immunity, although some of them help in acquired immunity as well. All granulocytes have lobed nuclei.

- **Basophils** have granules that interact with basic dyes like methylene blue, they will appear blue in blood smears stained with the usual GIEMSA-type of stain. These granules contain heparin (an anticoagulant) and mediators of inflammation like histamine, which increase the permeability of blood vessels and attract other cells to a site of infection. These mediators are responsible for the typical signs of inflammation: *calor* (heat), *rubor* (redness), *tumor* (swelling) and *dolor* (pain). In connective tissue **mast cells** are found instead, which have similar function as basophils.
- **Neutrophils** contain many lysosomes (primary granules) and smaller, secondary granules. Both stain only weakly with GIEMSA-stain. Neutrophils are phagocytic cells that engulf and digest foreign material. They have few mitochondria and therefore an anaerobic metabolism, an adaptation to the low oxygen environment in tissue. They have a limited capability for regeneration and die after a single burst of activity, forming the principal component of puss. Phagocytosis by neutrophils is increased when pathogenes are covered with antibodies.
- **Eosinophils** are packed with large, uniformly sized granules that bind acidic dyes like eosin, giving them a dark pink colour in blood smears. Eosinophils are highly phagocytic for antigens covered with IgE, they are thought to be involved in defense against large parasites like worms. They are also involved in **allergic** responses.

- The monoblast cell line gives rise to **monocytes**, which spend only a short time in blood before settling in tissues as **macrophages** (in connective tissue, often abbreviated MΦ), KUPFFER-cells (in liver), microglia (in brain), dendritic cells of the reticuloendothelial system etc. All these cells are phagocytic and form part of the innate immune response, they also present peptides from digested pathogens on their surface, allowing the acquired immunity to develop (see the section on MHC, page 159ff.). Some have additional specific functions, for example microglia cells which are part of the blood-brain barrier.

10.2 Humoral immunity: immunoglobulins

Immunoglobulins are highly soluble proteins found in blood, lymph and interstitial fluid. They are Y-shaped molecules, the two short ends of the Y interact with the antigen, the long F_c-end with effector cells (see fig. 10.1).

Antibodies can protect against infection in four different ways:

Neutralisation occurs when antibody binding to a pathogen makes this pathogen unable to complete its life cycle. This process is of particular importance in virus infections. Viruses bind to and enter cells of their host in

Table 10.1. Properties of blood cells

	Erythrocyte	Platelet	Neutrophil	Eosinophil	Basophil	Lymphocyte	Monocyte
size (µm)	6–8	2–3	10–12	10–12	9–10	7–8	14–17
number per ml	4–6 Mio	150 000-400 000	2800–5300	70–420	0–70	1400–3200	140–700
differential leucocyte count (%)	–	–	40–75	1–6	< 1	20–45	2–6
duration of development (d)	5–7	4–5	6–9	6–9	3–7	1–2	2–3
half life period (d)	120	7	0.3	0.5	0.25	variable	month to years

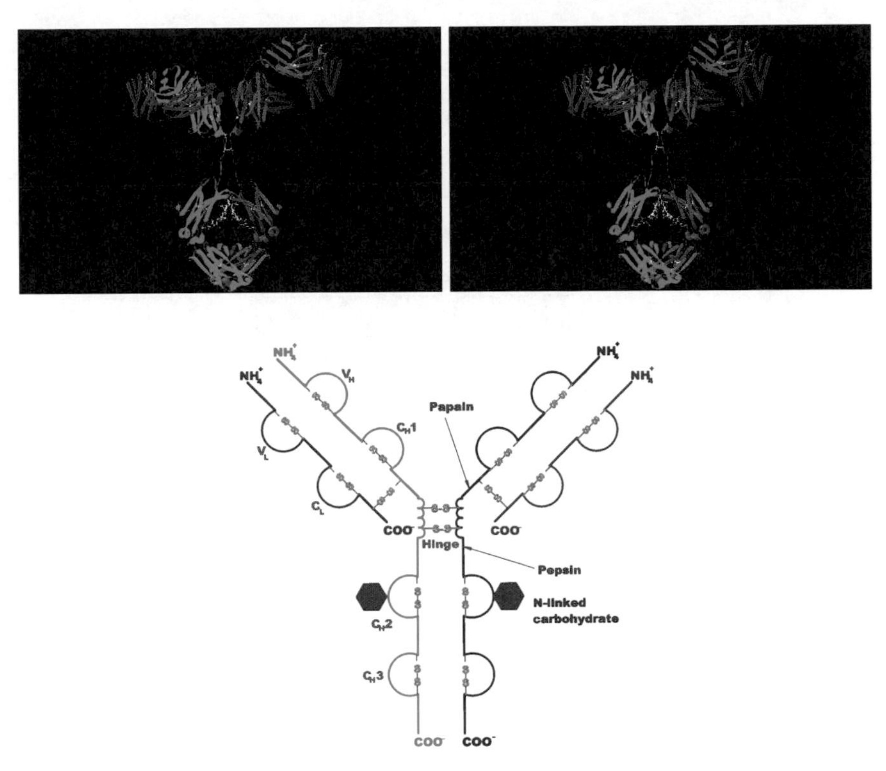

Figure 10.1. IgG (PDB-code 1FC2) is the archetypical immunoglobulin. It consists of 2 heavy and 2 light chains, connected by disulphide bonds. The light chains consists of 2 immunoglobulin motives each, the heavy chains of 4. The latter are decorated with N-linked oligosaccharides. **Top**: Stereo view of the crystal structure of IgG. **Bottom**: Pepsin can cleave IgG so that the antigen-binding site is still dimeric, but the F_c-end is removed. Papain on the other hand cleaves IgG above the disulphide bonds holding the heavy chains together. This results in F_{ab}-fragments with only one antigen binding site. Thus the effects of antigen crosslinking and of interactions of the F_c-end with other molecules can be experimentally dissociated.

order to be replicated by them. Special surface proteins on the virus bind to proteins expressed on the surface of the host cell, to start this process. If antibodies bind to the viral proteins, the virus can no longer multiply.

Opsonisation means that pathogens covered with antibodies are devoured more efficiently by phagocytic cells like macrophages and neutrophils. These have special receptors for the F_c-ends of antibodies.

Complement activation occurs when antibodies bound to pathogens start a chemical reaction of soluble proteins in our blood, that lyse pathogens. The function of complement will be described in detail later.

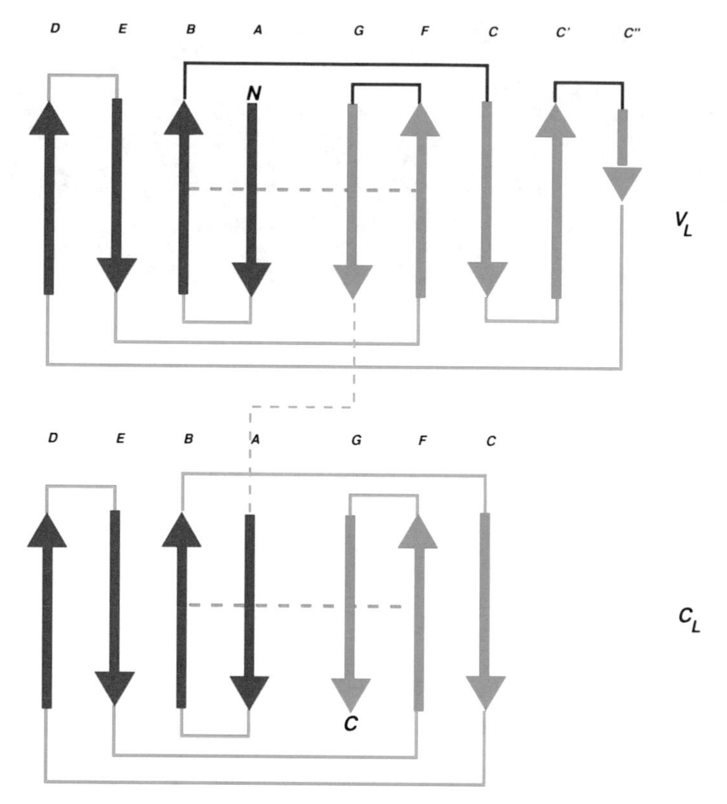

Figure 10.2. The segments of the immunoglobulins all have the same structure, which is known as Ig-fold and also found in other molecules: It consists of 2 antiparallel β-plaited sheets stacked on top of each other (purple and green). The loop segments connecting the β-sheets are shown in cyan, the hypervariable loops that form the antigen-binding side in red. The disulphide bonds are shown in yellow.

Production of reactive oxygen species catalysed by antibodies may lead to direct killing of pathogenes. This enzymatic activity of antibodies (which appears to be independent of antibody specificity) has been discovered only recently [81] and needs further investigation.

10.2.1 Structure of immunoglobulins

Antibodies consist of several copies of a characteristic motive, two anti-parallel β-pleated sheets on top of each other and connected by a disulphide bond (see fig. 10.2). This **Ig-fold** is found in many proteins in the immune system and some other proteins with different function.

There are some differences in the sequences even of the constant regions of antibodies between individuals. About 20 such **allotypes** have been identified in human heavy, a few more in the light chain. Some such allotypes are enriched in specific populations, they may be a cause of different disease susceptibility in different individuals, but this has not been confirmed.

Interactions between antigen and antibody may be based on ionic, hydrogen and hydrophobic bonds.

There are 5 immunoglobulin isoforms

Apart from IgG, there are 4 other immunoglobulins produced in our body (see fig. 10.3), IgA, IgD, IgE and IgM, some of which occur in several subclasses.The reason for this diversity in molecular structure is the specialisation for different functions, as shown in table 10.2.

From the clinical point of view, IgE deserves special mentioning because it is this antibody isoform that is involved in **allergic reactions**. IgE helps

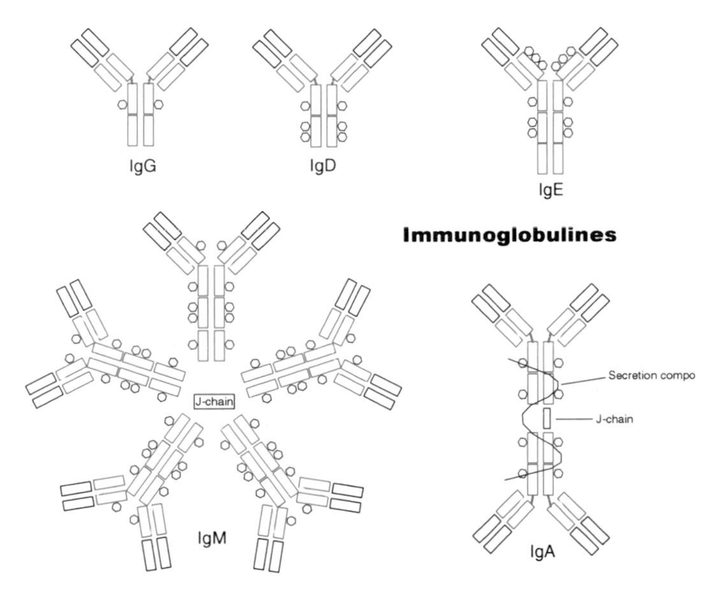

Figure 10.3. The 5 different immunoglobulins are composed of similar units: variable (blue) and constant (green) domains, forming heavy and light chains. These are held together by disulphide bonds (yellow). The proteins are decorated by sugar side chains in N-glycosidic bond (purple hexagons). IgA and IgM additionally contain a small J-chain (purple), IgA also a secretion component.

Table 10.2. Structure and function of immunoglobulin isoforms. Protein A is isolated from the cell walls of pathogenic strains of *Staphylococcus aureus*, Protein G from *Streptococcus ssp*. These proteins bind to the F_c-end of antibodies, resulting in a coat of host protein surrounding the bacteria. This makes them less prone to phagocytosis by macrophages and neutrophils. These proteins have become very useful in the laboratory for the isolation of antibodies and antibody-antigen complexes

	IgA1	IgA2	IgD	IgE	IgG1	IgG2	IgG3	IgG4	IgM
heavy chain	α_1	α_2	δ	ϵ	γ_1	γ_2	γ_3	γ_4	μ
light chain	κ, μ	κ, μ	κ, μ	κ, μ	κ, μ	κ, μ	κ, μ	κ, μ	κ, μ
# of protomers	1 or 2	1 or 2	1	1	1	1	1	1	5
molecular weight (kDa)	160	160	184	188	146	146	165	146	970
serum level (mg/ml)	3.0	0.5	0.03	5×10^{-5}	9	3	1	0.5	1.5
half life (d)	6	6	3	2	21	20	7	21	10
virus neutralisation	strong	strong	no	no	strong	strong	strong	strong	weak
classical complement	no	no	no	no	moderate	weak	strong	no	strong
alternative complement	weak	no	no	no	no	no	no	no	no
MΦ + Neutrophils	yes	yes	no	yes	yes	no	yes	no	no
Mast cells + basophile	no	no	no	yes	no	no	no	no	no
Natural killer cells	no	no	no	no	strong	no	strong	no	no
passes placenta	no	no	no	no	strong	weak	moderate	very weak	no
secretable	yes	yes	no	no	no	no	no	no	weak
Binds to Protein A	no	no	no	no	yes	yes	weak	yes	no
Binds to Protein G					strong	strong	strong	strong	

protecting our body against large parasites, for example worms. A person with an allergy produces IgE antibodies against an inappropriate target, for example against a pharmaceutical (penicillin being a notorious example). The ability of antigen-bound IgE to interact with eosinophils and in particular mast cells leads to the rapid release of inflammatory mediators, in the worst case resulting in an **anaphylactic shock**.

IgA is unusual in that it can be secreted across mucus membranes. This can be significant against pathogens that invade through mucus membranes like HIV. Early attempts to find an AIDS-vaccine were made by injecting proteins from simian immunodeficiency virus (SIV) into apes. These vaccines protected the animals against infection with SIV, but only if the virus was injected. If the virus was smeared onto the vagina, no protection was observed. The monkeys had only mounted an IgG, but not an IgA response against the vaccine.

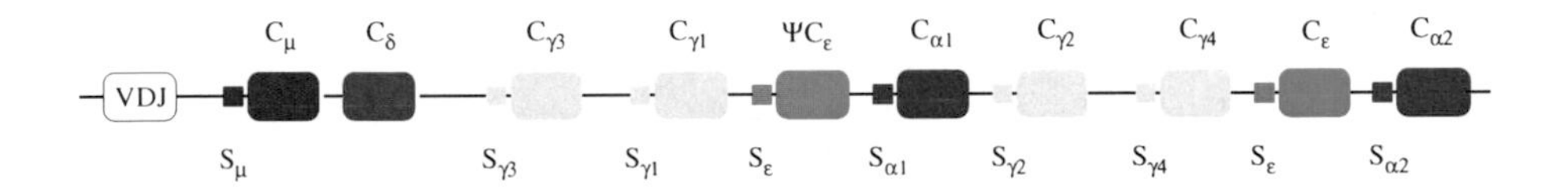

Figure 10.4. All B-cells start their life producing IgM and IgD, which are made by alternative splicing of mRNA. Once stimulated by antigen binding to the membrane-bound antibody, a B-cell may perform an **isoform switch** by recombination between the switch regions in front of the heavy chain genes. The sequence of the S_μ-region is $[(GAGCT)_nGGGGGT]_m$ with n = 3-7 and m $\approx$ 150. The other switch regions have similar sequences. Note that there is no S_δ-region. There appears to have been a gene duplication event, of a unit with two γ-, an ϵ and an α-gene. One of the ϵ-genes is a pseudo-gene (ΨC_ϵ), thus only one IgE isoform is expressed in humans.

Additionally, there are two light-chain isoforms (κ and λ), which can occur in any of the Ig-classes. No functional differences between these isoforms has been found to date.

10.2.2 How is the large number of Ig-molecules obtained?

It is estimated that our body can produce about 1×10^{15} different antibodies. This is a very large figure, in particular as our genome contains only about $3 - 4 \times 10^4$ different genes. Clearly it is impossible to have one separate gene for each antibody. How then are antibodies encoded?

The genetic information for antibodies is contained not in a single chunk, but in small building blocks (see fig. 10.5). Each of these building blocks is

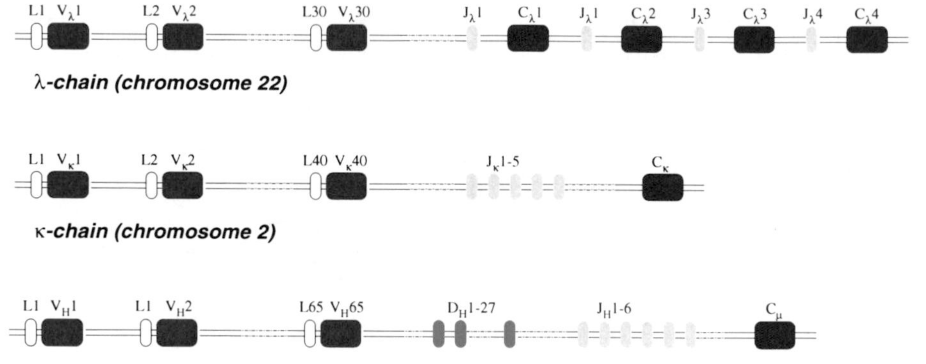

Figure 10.5. Organisation of Ig-genes on human chromosomes. The genetic information for the λ-light chain is on chromosome 22. There are about 30 different L+V and 4 different J+K regions, resulting in 120 different combinations. The information for the κ-light chain is on chromosome 2. About 40 L+V regions, 5 J-regions and one K-region result in 200 possible combinations. Since each antibody contains either λ- or κ-chains, 120+200=320 combinations are possible for the light chains. The heavy chain is encoded on chromosome 14. About 65 L+V, 27 D, 6 J and one C region result in 10 530 possible combination for the heavy chain. Thus, the genetic information in our germ line can directly encode 320*11 000 = 3 369 600 different variable domains.

polygenic, and as a B-cell matures, these building blocks are brought together by random **somatic recombination**, resulting in a unique sequence.

The resulting antibodies are presented on the surface of the cell via a short transmembrane segment (removed from soluble antibodies by alternative splicing of mRNA). Antigen binding is signalled into the cell by tyrosine kinases and the cell starts to proliferate. During proliferation mutations are introduced into the sequence of the variable domains, adding to the number of different antibody molecules. Those cells whose mutations lead to higher affinity antibodies have a higher chance of being stimulated into further division by renewed antigen contact, resulting in the selection of antibodies with higher and higher affinities. This is called **affinity maturation**.

These two processes together lead to the large number of antibodies in our body. Joining the various regions together is done by a combination of somatic recombination, RNA-splicing and protein-processing (see fig. 10.6 and 10.7).

Somatic recombination is done in such a way that a variable number of random bases (P and N nucleotides) are added between the regions, greatly increasing the number of possible antibodies (see fig. 10.8).

A B-cell is, like all other mammalian cells except gametes, diploid. Thus there are 2 heavy chain and 4 light chain genes (2 κ and 2 λ) present, but only

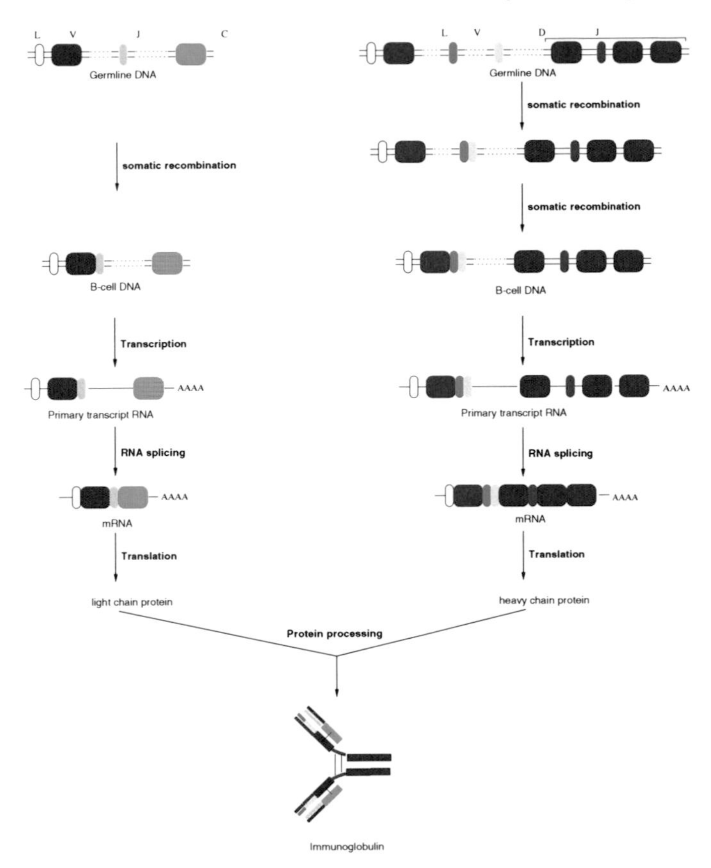

Figure 10.6. Somatic recombination joins the V, D and J-regions. mRNA-splicing then joins the L- and C-regions. The final antibody is made by joining heavy and light subunits during protein processing.

one of each gets expressed. This selection is called **allele exclusion**. Recombination occurs first in the heavy chain locus, D- and J-segments recombine on both chromosomes, but only one of them recombines also the V-segment. Recombination of the light chains starts in one κ-locus, if that does not lead to a functional light chain, the second κ-locus will recombine, then the λ-loci one after the other.

Once a functional IgM molecule is produced and expressed on the cell surface, this makes neighbouring stroma cells send out a signal that lead to suppression of Rag1-Rag2 expression in the B-cell, and therefore to an end of recombination. Thus B-cells express only one antibody idiotype. If neither of the recombination events leads to a functional antibody, the cell dies.

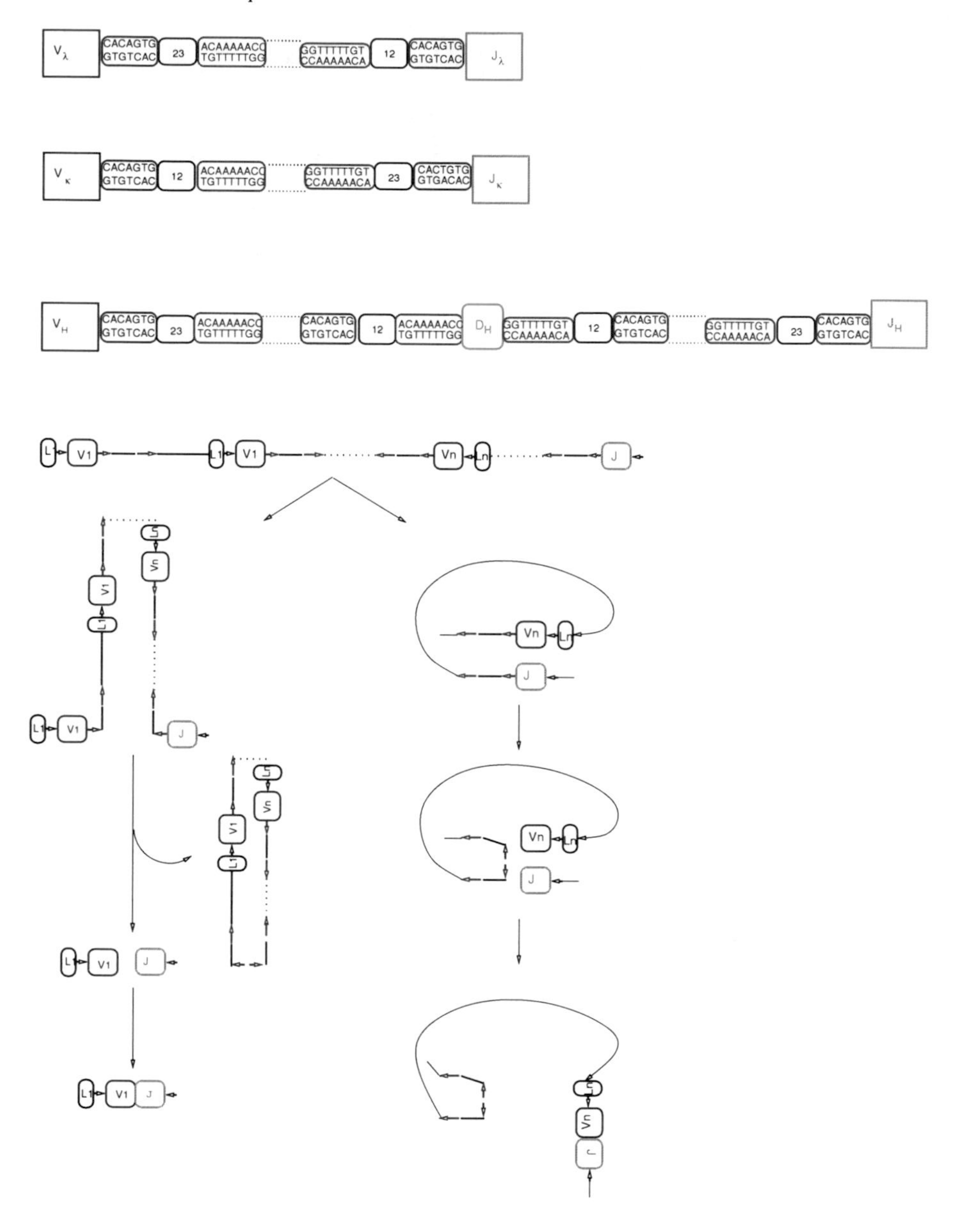

Figure 10.7. Top: The variable regions are flanked by conserved heptamer (7 bases) and nonamer (9 bases) regions, which are needed for recombination. Between those are regions of either 12 or 23 bases. A segment with a 12 base region is joined with a region with 23 bases during recombination. **Bottom:** There are two possible mechanisms of recombination, one results in the loss of the intervening sequences, the other does not.

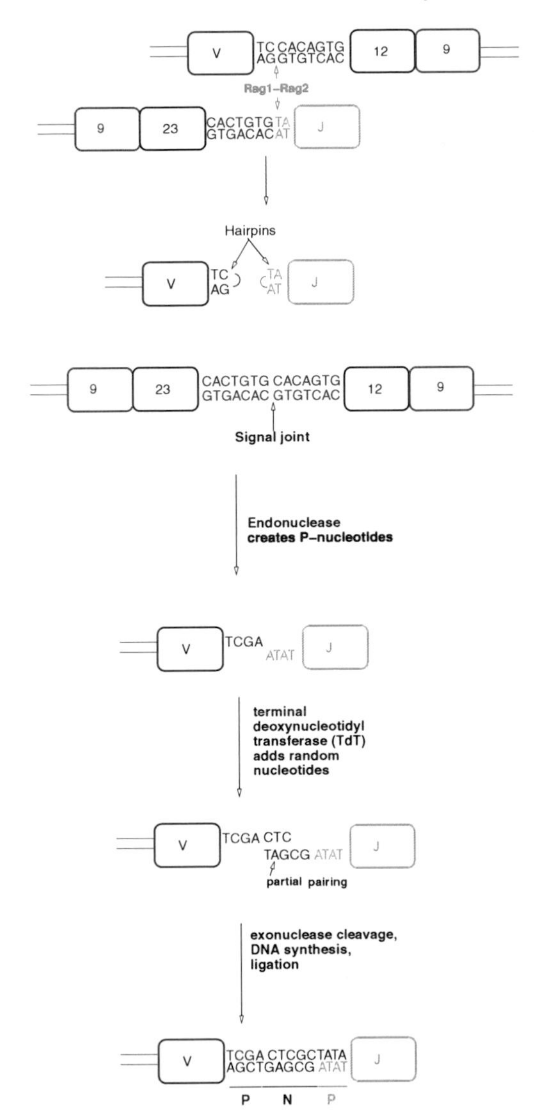

Figure 10.8. Somatic recombination of Ig-genes requires a set of specialised enzymes. 1) First Rag1-Rag2 (Recombination Activating Gene) recognises the recombination signal and cuts one strand of the DNA at the end of the heptamer sequence. 2) The cut ends then react with the uncut strand, cutting it and forming a hairpin. The two heptamers are ligated into a signal joint. 3) The hairpins are cleaved by endonuclease at a random site, to generate the P- (palindromic) nucleotides. 4) Deoxynucleotidyl-transferase (TdT) adds a variable number of N- (non-template) nucleotides, until partial pairing of the ends is possible. 5) Exonuclease trims of any non-paired ends. 6) The ends are ligated together.

Each clone of B-cells has unique sequences of P- and N-nucleotides, which can be used to follow its descendants through affinity maturation and isoform-switch in experimental studies.

10.2.3 Time course of antibody response

If an antigen enters our body for the first time, it will encounter only few B-cells that by chance express on their surface antibodies against this antigen. Binding of antigen to these surface antibodies stimulates these cells to proliferate. At the same time, the cells will start producing soluble rather than membrane bound antibodies (by alternative splicing), which they secrete into the blood stream. Such **plasma cells** are protein factories, filled with rough ER. Each plasma cell produces about 2000 antibody molecules per second. This is probably the reason for the short life span of such cells.

These antibodies will invariably be of the **IgM** isoform, their affinity to the antigen will probably be low. However, IgM is pentameric, and there are 10 antigen binding sites per molecule. Most pathogens have highly repetitive antigens on their surface, for example the coat proteins of viruses or the glycosaminoglycan subunits in the bacterial wall. Once several of the binding sites of an IgM molecule have bound to a pathogen, release of the antibody requires simultaneous dissociation of all binding sites, which is an unlikely event. Thus the apparent binding strength of an antibody, called its **avidity**, is higher by several orders of magnitude than its affinity would suggest (up to 10 000-fold).

As the B-cells proliferate, somatic hypermutation in the variable region of the immunoglobulin gene creates a large number of different antibodies, those cells that produce higher affinity antibody (on their surface) have a higher chance to be activated by antigen binding and to proliferate further. Thus antibodies with higher and higher affinity are created. After affinity maturation cells switch from producing IgM to other isoforms, in particular **IgG**.

IgG has only 2 binding sites, its avidity is only about 10-fold higher than its affinity. However, because of its higher affinity, this is sufficient.

Antibody-antigen complexes have dissociation constants of between about 10 µM and 1 pM.

If the antigen is no longer present, the antibody-producing plasma cells die within a few days, and the serum antibody concentration drops quickly to undetectable levels. However, not all descendants of the stimulated B-cells developed into plasma cells. Others became **memory cells**, which have a life span of many years. If these encounter the same antigen again, they can

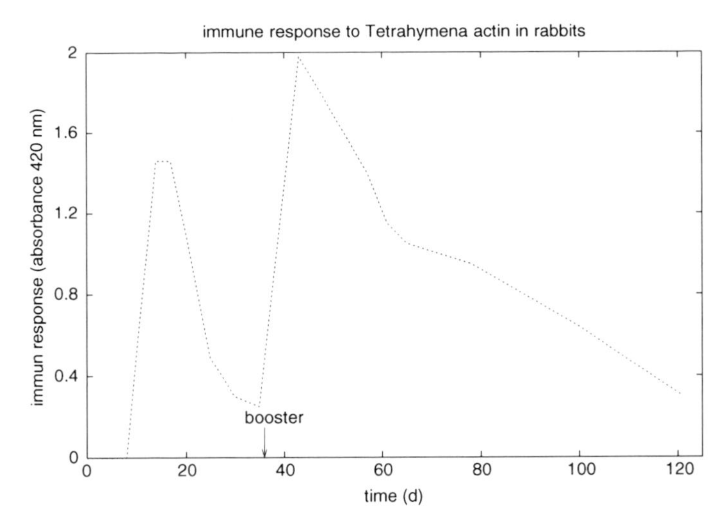

Figure 10.9. Time course of the antibody response. A rabbit was injected with antigen (here actin from *Tetrahymena ssp.* in complete FREUND's adjuvant) and the concentration of antibodies against actin in the rabbits blood was measured by ELISA. A booster injection (with incomplete FREUND's adjuvant) was given on day 36. Note that the response in ELISA is not linear, the change in antibody concentration is much larger than the change in absorbance would suggest. After first contact it takes about 2 weeks for the rabbit to develop a measurable antibody titre against the antigen, and the concentration of antibodies drops rapidly afterwards. However, memory cells remain in the rabbits blood, a second contact with the antigen results in a quick, long lasting, strong response. Repeated booster injections could maintain the high antibody concentration (not shown).

mount a rapid, strong, high affinity and long lasting antibody response (see fig. 10.9). For this reason we can get many diseases only once, if we overcome the initial infection, we are protected from re-infection for the rest of our life. Some pathogens however, in particular viruses like influenza, can overcome this protection by rapidly changing their surface proteins.

Note that the variable parts of the antibody are considered by our immune system as foreign material, and that therefore after a while an immune response against them is mounted. These anti-antibodies themselves are antigenic and so on. This complicated network is involved in the regulation of the antibody production rate and in particular in switching off the production of antibodies no longer needed. This "network theory" of the antibody response was developed by N.K. JERNE (NOBEL-Price 1984).

10.2.4 Immunisation

The observation that one contact with many pathogens is enough to result in long-term protection against them is used for the purpose of **immunisation**. In principle, there are three ways to accomplish immunisation:

- **Passive** immunisation is achieved by transferring antibodies from one individual to another. Since only the proteins, but not the memory cells are transferred, protection lasts only a few days. This method is used to intervene with an acute crisis (for example by injecting **anti-snake venom** antibodies raised in animals like horse after a snake bite). Another example for this technique is the transfer of serum from a patient who just survived an infection to another patient suffering from the same infection (**hyperimmune serum**).
- **Active** immunisation is achieved by injecting the antigen so that a lasting (in ideal cases life long) protection is achieved. Dead pathogens or their purified components can be used as **vaccines**, however even better results are often achieved by pathogens which are still alive, but have lost the ability to cause disease (**attenuated** pathogens). Genetic engineering allows the production of vaccines were surface markers of very dangerous pathogens, which are not available in attenuated form, have been inserted into relatively harmless organisms like **vaccinia viruses**[1]. Antigens are often injected together with an **adjuvant**, a substance that increases the immune response of our body to the antigen.
- **Adoptive** immunisation requires the transfer of spleen cells from an immunised individual to another. So far, this procedure has been used only for experimental purposes.

10.2.5 Monoclonal antibodies

The high affinity and specificity of antibodies makes them an ideal tool in research, diagnostics and even treatment of disease. It would therefore be useful if a particular antibody could be produced in large quantities.

As we have seen, antibodies are produced by plasma cells, which have a life span of only a few days. They are also **terminally differentiated** and do no longer divide. At any given time, there are many different plasma cells active in our bodies, the immunoglobulins in our plasma are directed against many different antigens.

[1] Vaccinia is the causative agent of cow pox, and is relatively harmless in humans. It was used for immunisation against **small pox** until this terrible disease was declared extinct, one of the greatest successes of mans fight against disease.

Milstein & Köhler have found a way around these problems (Nobel-price 1984 together with Jerne for the network theory of the immune system). They isolated spleen cells from a mouse immunised against a particular antigen. These cells, which included antibody producing cells, they fused with cells from a **myeloma** of a mouse from the same strain. Some of the fusion products had both the ability to produce antibodies against the antigen in question (from the spleen cells) and to divide indefinitely (from the tumor cells). Such cells were isolated and cloned. Because each B-cell produces only antibodies with a particular variable region (that is a particular **idiotype**), the antibodies produced by such clones are homogeneous. They can be isolated from the culture medium. Such antibodies are called **monoclonal**, because they are produced from a single cell clone.

It is of course more difficult to get spleen cells and matching myeloma cells from humans. However, using techniques of molecular biology, it is possible to replace the genetic information for the constant region of IgG in those cloned mouse cells by that for human IgG, so that the resulting cells will produce **humanised** antibodies. Such antibodies can be injected into patients repeatedly without causing an immune-reaction against mouse IgG, and are therefore useful as **therapeutic agents**.

By the same method it is also possible to change the isoform of an antibody, which can be useful in allergy research.

Indeed modern molecular biology techniques allow the whole process of somatic recombination, hypermutation and selection to be done entirely *in vitro* producing the DNA for a perfectly good antibody (from any species) against any antigen without the use of research animals. This DNA is then integrated into a cultured cell line for antibody production, or even into bacteria like *E. coli*. This allows researchers to raise antibodies for example against self-antigens, that could not be raised in animals.

10.2.6 Laboratory uses of antibodies

Since antibodies bind specifically and with high affinity to antigens, the antibody-antigen reaction can be used to detect either antibodies or antigens, both in research and in clinical laboratories.

Fig. 10.10 shows the basic principles involved in immunological assays. The exact procedure followed depends on the sort of information required. Assays that detect only the presence of an antigen are usually simple, fast and cheap, however, if additional information like the molecular weight of the antigen is required, costs can increase significantly. It is therefore not uncommon to use simple assays for screening purposes and the more expensive ones for closer investigation of the samples only that were identified in the screen.

The following main assay types are used:

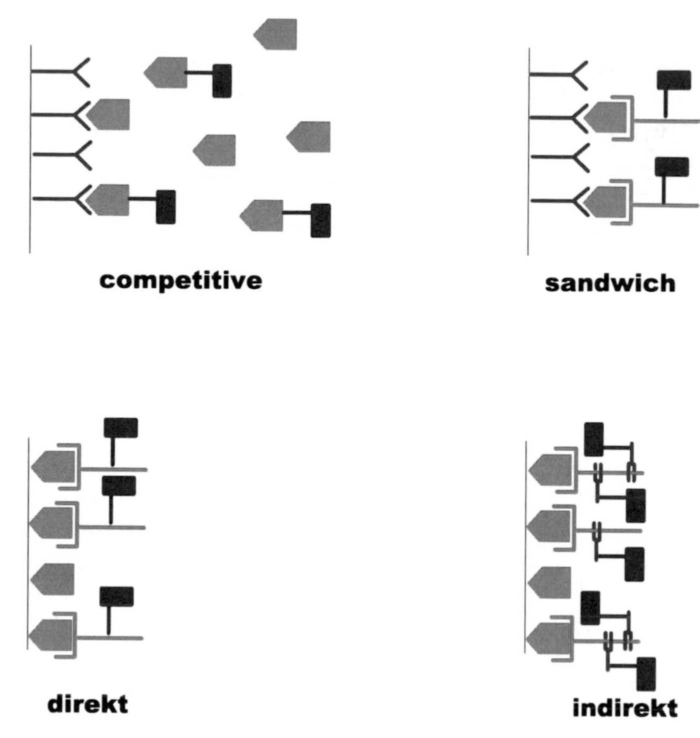

Figure 10.10. Basic principles of immunological assays. For details see text.

competitive: The sample is added together with a known amount of labelled antigen to a known amount of antibodies. The labelled antigen binds to the antibody, resulting in a defined signal. If the sample also contains (unlabelled) antigen, this would compete with binding of the labelled antigen molecules, reducing the signal measured in a concentration dependent manner. Competitive assays are often used for the detection of small molecules, like drugs. The problem with that type of assay is to label the antigen without destroying antigen-antibody interaction.

sandwich: Antibodies are fixed on a solid support, then incubated with the antigen. Antigen binding creates binding sites for a second antibody, directed against a different epitope of the antigen, thus the amount of second antibody bound is proportional to the concentration of antigen. This procedure works only with fairly large antigen (with several epitopes), in addition it requires antibody pairs, which are not always available.

direct: The labelled antibody binds to the antigen, which is fixed on a solid support. The amount of antibody bound and hence the signal depends on the amount of antigen present. This type of assay is often used against large antigens like proteins, viruses, bacteria or even whole cells.

indirect: assays work similar to the direct ones, however, the primary antibody is unlabelled. After it has bound to the antigen, a secondary antibody directed against the constant part of the primary antibody bears the label. Since each primary antibody can bind several secondary ones this leads to an increase in sensitivity. Additionally, the complicated labelling procedure needs to be performed only once in the secondary antibody (obtained from large animals like sheep or horse by manufacturers), which can then be used to detect several primary antibodies (obtained in the lab from small animals like rats).

Various different labels may be used to detect the antibody, the most important are radioactive isotopes, fluorescent molecules and enzymes. Note that the assays described above can easily be set up to detect antibodies, rather than the antigens.

Development of immunological assays for drugs, disease marker proteins or for pathogens is a multi-billion € industry, and has revolutionised especially clinical laboratories. In the last couple of years it has been complemented by the detection of nucleic acids, for example by polymerase chain reaction (PCR).

10.3 Destroying invaders: the complement system

Blood and interstitial fluid contain a group of proteins, that can destroy cell membranes. Collectively, they are known as **complement system**. The name indicates that these proteins complement and enhance the disease-preventing activity of antibodies and immune cells. There are about 20 different proteins in the complement system, they are produced mostly in the **liver**.

Obviously, such proteins are very dangerous, and their activity needs to be tightly controlled. The early components of the complement system are highly specific serin-proteases, activation of the complement system is a chain of protein cleavage reactions. Apart from activating another protease, each cleavage results in the production of a small peptide, which acts as **inflammatory mediator (anaphylatoxin)**.

Complement

- **opsonises** the cell for destruction by macrophages and neutrophils
- catalyses the formation of the **membrane attack complex** from components C5–C9. A complex of C5–C8 catalyses the oligomerisation of C9, which forms pores of about 10 nm diameter in the cell membrane. As a result the affected cell is osmotically lysed.

10.3.1 How is complement activated?

There are three principle sequences of complement activation, known as **classical, alternative** and **lectin** pathway respectively.

The **classical pathway** (see fig. 10.11) is started by IgM or IgG molecules bound to a pathogen.

In the **lectin pathway** of complement activation the mannan-binding lectin MBL replaces the antibody-antigen complex as starting point. This binds MASP (MBL associated serin protease), which cleaves C4 and C2 just like C1s does in the classical pathway.

The **alternative pathway** starts with spontaneous cleavage of C3 into C3a and C3b, the latter inserts into membranes. Our own cell membranes have C3b-inactivating components, but some pathogens are lacking those. Thus C3b can hydrolyse B and form the C3bBb-complex, which leads to activation of further C3-molecules. The C3b,Bb complex is the functional homologue of C2b,4b; C2 and B are encoded next to each other in the MHC-locus. Both complexes can bind C3b to form a C5-convertase (C2b,3b,4b or $C3b_2Bb$).

Note that the conversion of complement molecules works as a catalytic cascade quite similar to what is seen in hormone signalling (see fig. 8.1 on page 122). A small number of C1-molecules bound to antigen-antibody-complexes leads to the rapid production of large amounts of C3b and C5b, flooding the surface of the invader.

Cleavage of complement molecules converts a water-soluble protein without enzymatic activity into a membrane-bound Ser-protease and a soluble peptide. This protective mechanism is also observed for example with digestive enzymes.

Once the C5-convertases have been formed, alternative and lectin pathway proceed like the classical. In evolution the alternative pathway probably developed first as part of our innate defense. The lectin- and classical pathway are later additions to direct complement more specifically and with higher activity to foreign material.

10.3.2 What does complement do?

C4a, C3a and C5a are potent activators of inflammation (**anaphylatoxins**). They induce smooth muscle contraction, increase vascular permeability and recruit granulocytes and macrophages to the site of infection.

Macrophages and other phagocytic cells have complement receptors for C3b (CR1, CR2, CR3 and CR4) and the collagen-like region of C1q, thus complement not only lyses pathogens, but also opsonises them. C5a (and to a

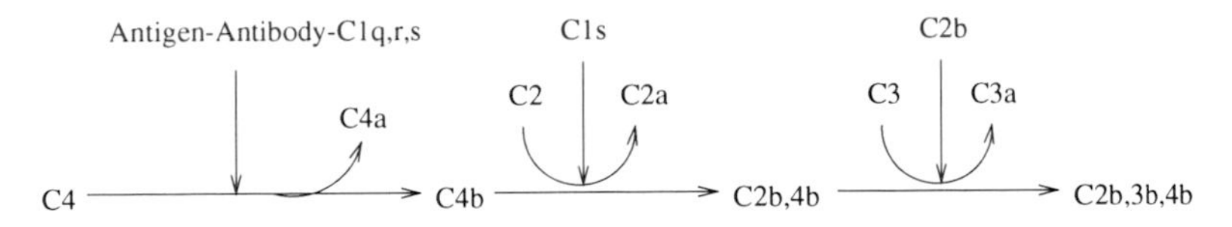

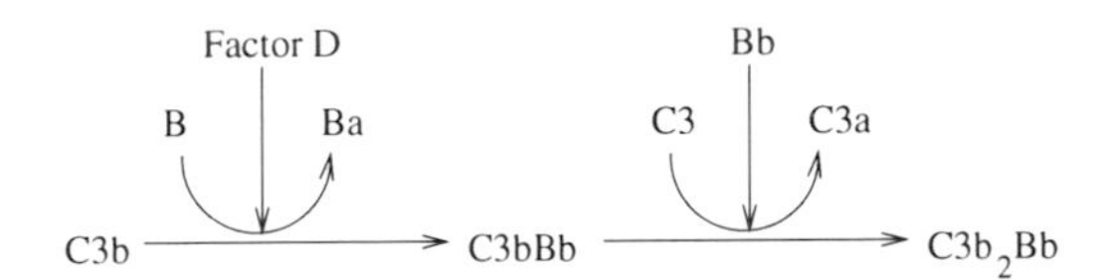

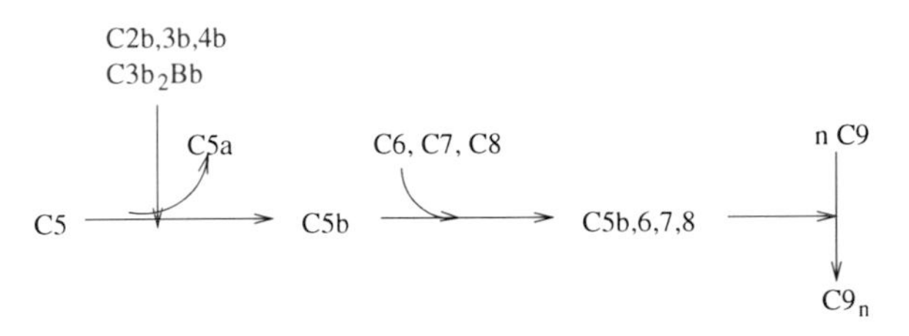

Figure 10.11. Classical pathway of complement activation. **Top:** The C1q,r,s complex can bind to immunoglobulins bound to antigen. At least 1 IgM or 2 IgG are required. Binding of the hexameric C1q protein to Ig results in C1r cleaving and activating C1s. Activated C1s can cleave C4 into C4a (a weak inflammatory peptide) and C4b, which binds to the membrane. This binding is covalent: Cleavage of C4 exposes a reactive **thioester** bond (between Glu and Cys) in C4b, which reacts with OH-groups in proteins and carbohydrates forming ester and thiol. C4b then recruits C2, which in the C4b-bound state is also cleaved by C1s into C2a (inflammatory peptide with moderate activity) and C2b. The C2b,4b complex recruits C3, which is cleaved to C3a (inflammatory peptide) and C3b. C3b can stay bound as C2b,3b,4b complex, but as each C2b,4b complex cleaves up to 1000 C3 molecules, non-complexed C3b is also formed. **Middle:** C3b binds factor B, which is then cleaved by factor D, a soluble protease. This results in the formation of a C3b,Bb complex, which also has C3-convertase activity (i.e. splits C3 into C3a and C3b). This leads to a rapid, autocatalytic flooding of the pathogens surface with C5-convertase. **Bottom:** C2b,3b,4b and $C3b_2Bb$ both have C5-convertase activity, they recruit C5 to the membrane, where it is split into C5a (a highly active inflammatory mediator) and C5b. C5b then binds C6, C7 and C8, the latter inserts into the membrane and catalyses the oligomerisation of 10-16 C9 molecules into a large transmembrane pore.

lesser extent C3a) also stimulates the phagocytic response of macrophages by binding to a 7 membrane-spanning domain type of receptor (G-protein coupled). This is particularly important for IgM, as there is no F_c-receptor for this isoform.

CR1 also occurs on **erythrocytes**. Immune complexes left over after an infection trigger the complement system, covering them with C3b and C4b. This complex binds to the erythrocyte, in liver and spleen the immune complexes are then abstracted from the erythrocytes by specialised macrophages (without destroying the erythrocytes). By this process potentially dangerous immune complexes (which can cause rheumatic diseases, hence the name "rheuma factor") are removed from the blood stream.

10.3.3 How is complement degraded?

Activated component is unstable and degraded rapidly.

There is a complement inhibitor circulating in the blood (C1 inhibitor), which dissociates C1r and C1s from C1q. C1q remains bound to the immunoglobulins on the pathogens, but can no longer activate C4. Inability to form C1Inh leads to **hereditary angioneurotic oedema**, where uncontrolled complement activation leads to swelling (for example in the trachea).

If C4b does not form a covalent bond with the pathogen's membrane, its thioester is hydrolysed by water and C4b is inactivated.

The C2b,4b complex is dissociated by binding of C4-binding protein (C4BP). The complex between C3 and Bb is dissociated by complement receptor 1, factor H, factor I and decay-accelerating factor (DAF). Factor H is recruited to our cells by binding to sialic acid, which is absent from most pathogens.

CD59 is widely expressed on our cells and prevents the formation of a membrane attack complex, protecting our cells from "stray bullets". **Paroxysmal nocturnal haemoglobinuria** is caused by the inability to synthesise the glycolipid tail that binds CD59 (and decay-accelerating factor (DAF)) to the cell membrane. This leads to the destruction of erythrocytes by complement, explaining the haemoglobinuria.

10.4 Cellular immunity

Antigen-presenting cells present antigens to T-cells bound to specific cell-surface proteins, called MHC-I [2] and MHC-II (Major Histocompatibility Complex, also known under their German name *Hauptlymphozytenantigen*, HLA).

[2] According to standard nomenclature in human genetics all-caps names like MHC-I are reserved for genes, for the corresponding proteins Mhc-I should be used. How-

T-cells have a receptor for the MHC-antigen complex, the **T-cell receptor**.

10.4.1 The major histocompatibility complex

Sampling proteins produced in our cells: MHC-1

Our cells constantly present samples of the proteins produced by them on their surface, this status is read by T-killer cells. Cancer cells (characterised by expression of embryonal antigens) and virus-infected cells are killed, hopefully before more damage occurs.

Ubiquitin and the proteasome

Proteins no longer needed in our cells are marked by linking several copies of **ubiquitin** to it, a 8.6 kDa protein. Transfer of ubiquitin is performed by a group of ubiquitin-ligases, of which 3 classes exist. **E1-ligase** (UbA1) (see fig. 10.12) forms a thioester-bond with the C-terminal glycine-residue of ubiquitin in an ATP-dependent reaction. This activated ubiquitin is then transferred to an **E2-ligase** (UbCs) and from there to an **E3-ligase**. All three ligases bind ubiquitin as thioester. There is only one E1 (UbA1) but several UbCs and many E3-ligases, which transfer the ubiquitin to the ϵ-amino-group of a lysine in the proteins to be destroyed, forming an isopeptide bond. Ubiquitin contains several Lys-residues, whose ϵ-amino groups form iso-peptide bonds with the C-terminal Gly of other ubiquitin molecules, resulting in long chains of poly-ubiquitin. For the discovery of ubiquitin A. CIECHANOVER, A. HERSHKO & I. ROSE received the Nobel-Price for Chemistry in 2004.

Ubiquitin is the most well known member of a whole family of proteins with ubiquitin-fold, which are transferred to proteins by a similar mechanism as ubiquitin, except that transfer is often by an E2-ligase directly. E3-ligases are required for ubiquitin presumably because of the large number of different proteins labelled with this marker. These **ubiquitin-like modifiers** (UbLs) are involved in the regulation of endocytosis, apoptosis, cell cycle, DNA repair and other processes. There are even ubiquitin like proteins (Isg15 and Fat10), which are regulated by interferon and modulate immune response. The mechanisms involved in these regulatory pathways is however poorly understood. Poly-ubiquitins formed via Lys-48 are recognised by the proteasome, special **isopeptidases** are able to remove ubiquitin from iso-peptide bonds and recycle it. Formation of poly-ubiquitin via Lys-63 seems to be involved in regulatory processes like receptor-mediated endocytosis.

---

ever, in the literature the all caps spelling is found for many proteins, including MHC. I will follow the common use here and use italics (*MHC-1*) to indicate genes. May be one day we will get a universally accepted, perhaps even species independent, nomenclature.

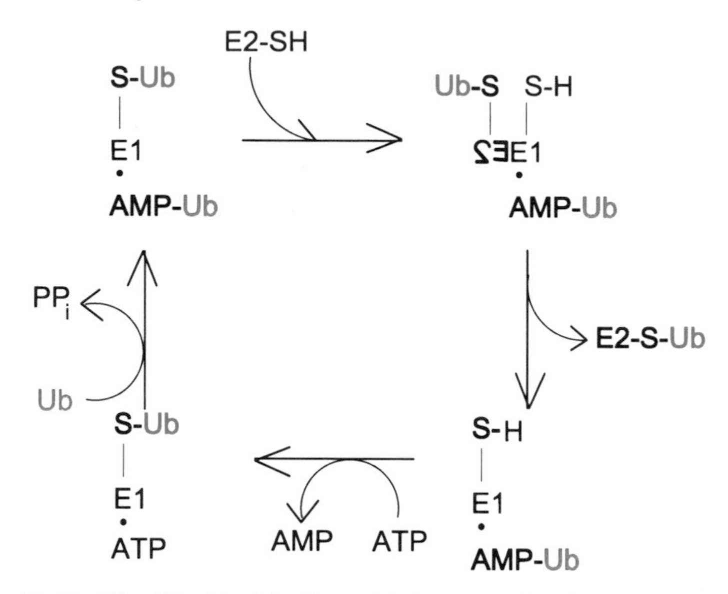

Figure 10.12. The E1-ubiquitin ligase binds two molecules of ubiquitin, one is covalently bound to a thiol-residue of the enzyme, the second is bound to AMP. The first ubiquitin molecule is transferred to the E2-ubiquitin ligase, then the second ubiquitin is moved to the SH-group, the AMP is exchanged for ATP. Then another ubiquitin is bound to the nucleotide, pyrophosphate is released in the process and the enzyme is ready for a new cycle.

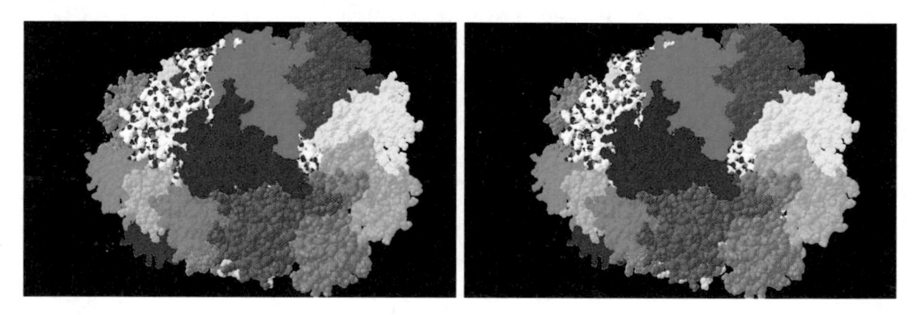

Figure 10.13. Stereo view of the proteasome (PDB-code 1RYP) from archaebacteria. This complex consists of 4 rings with 7 proteins each. It has ATP-dependent protease activity. Presumably unfolded proteins are threaded through the hollow core, where they are degraded. There are 2 isoforms of the proteasome in mammals, a constitutively expressed and an interferon-inducible one.

The 26 S proteasome (see fig. 10.13) consists of the 20 S subunit, a stack of 4 rings made of 7 proteins each, which acts as an ATP-dependent protease. Both ends of that stack are covered with a 19 S cap[3], which recognises ubiquitinated

[3] The SVEDBERG (S) is the unit of the sedimentation velocity of a particle in an ultracentrifuge. It depends not only on the size, but also on the shape of the

proteins and acts as an anti-chaperone (unfolding proteins so that the amino-acid thread can pass through the pore of the 20 S-subunit).

For the cells, protein production and destruction is a big number game. About 4×10^6 protein molecules are produced each minute on the 6×10^6 ribosomes of each cell, resulting in a total of 2.6×10^9 protein molecules per cell. Of the 4×10^6 protein molecules produced per minute, 2×10^6 are immediately destroyed again by its 30 000 proteasomes, resulting in 50 000 peptides principally capable of binding to MHC-1 (see fig. 10.14). Of those about 1000 actually bind and are displayed on the cell surface. About 10 foreign peptides on the surface of a cell are required to mount an immune response [62].

One of the functions of the proteasome is the destruction of **cyclins**, proteins required for the control of cell division. For this reason inhibition of the proteasome with **Bortezomib (Velcade®)** can prevent cell division in cancer cells. The substance was recently introduced as a second line drug to treat multiple myeloma and is in clinical tests for some other forms of cancer. Given the many other functions of the proteasome it is not surprising that Bortezomib has considerable side effects.

Some pathogens have evolved mechanisms to inhibit antigen presentation to the cell surface, the following mechanisms have been identified so far:

Inhibition of proteasomal degradation of proteins is found for example in EPSTEIN-BARR-virus, the causative agent of "kissing disease". The exact mechanism is unclear.

Prevention of peptide transport into the ER can occur by several mechanisms

- Blocking of peptide binding to Tap1/Tap2 by Herpes simplex virus 1 (HSV-1) ICP47, protein which associates to the ER membrane on the cytosolic side and then binds to Tap1/Tap2 with a dissociation constant of 50 nM. ATP-binding to Tap is still possible afterwards, but not its hydrolysis.
- Blocking of ATP binding by human cytomegalovirus US6, a type 1 membrane glycoprotein. It attacks Tap from the ER lumen, but prevents ATP binding and hydrolysis on the cytosolic side. Peptide binding is still possible, but transport is prevented.
- Inhibition of transport without inhibition of peptide or ATP binding is caused by the luminal and transmembrane part of bovine herpes virus type 1 glycoprotein N.
- Destruction of Tap1/Tap2 by the cytosolic domain of bovine herpes virus type 1 glycoprotein N occurs by stimulation of proteasomal digestion. Glycoprotein N will be digested in this process as well.

---

particle. Thus S-values are not additive, as in the proteasome, where 20 S + 19 S + 19 S = 26 S)

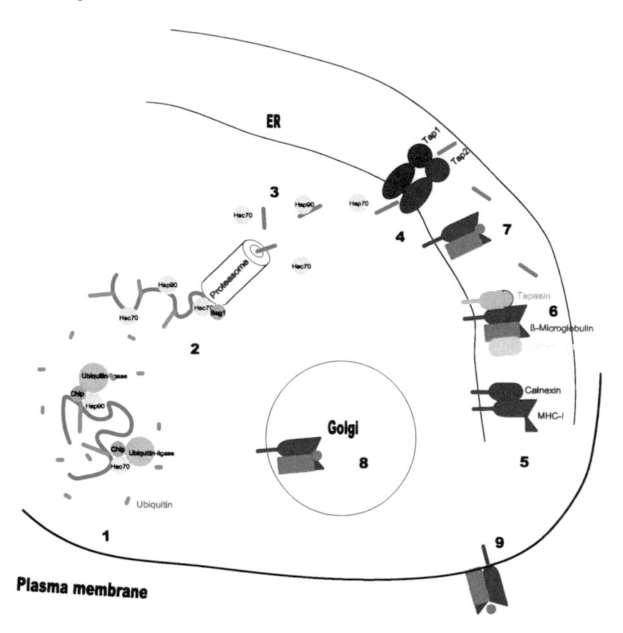

Figure 10.14. **1**: Damaged proteins are recognised by molecular chaperones like Hsp70, Hsc70 and Hsp90 (for a detailed discussion of chaperones see page 211ff), together with some co-chaperones like Chip. This leads to binding of ubiquitin-ligases, which transfer a destruction-marker peptide, ubiquitin, to the damaged protein (see also fig. 10.12). **2**: Ubiquitinated proteins are recognised and digested by the proteasome, an ATP-dependent protease. **3**: Peptides released from the proteasome are shuttled to the ER by chaperones and **4**: transported into the ER by the Tap1/Tap2 system, an ABC-type transport ATPase (see section 17.2.1 on page 280 for a fuller discussion of Tap). **5**: Newly synthesised MHC-1 molecules are associated with the chaperone **calnexin** , until β_2-microglobulin has bound to them. **6**: Then they interact with **calreticulin, tapasin** and Tap, until the peptide binding site has been filled. **7**: This stabilises MHC-1 enough that from now on it can make do without chaperones. It is transported via the GOLGI-apparatus (**8**) to the cell surface (**9**), where the peptide can be read by $CD8^+$ T_k-cells, which destroy cancer cells and cells expressing foreign antigens (after infection with viruses).

Destruction of the loaded MHC-I/peptide complex by stimulation of the quality control machine of the ER (see section 16.3.4 on page 243). The complex is retro-transported into the cytosol, deglycosylated, ubiquitinated and degraded by the proteasome. Examples include human cytomegalovirus US1 and US2 proteins.

Prevention of transport of the MHC-I/peptide complex to the plasma membrane. Human cytomegalovirus US3 and US10 proteins lead to retention in the ER, murine cytomegalovirus m152 to retention in *cis*-GOLGI and

HSV-7 (cause of infectious mononucleosis) U21 protein re-routs them to the lysosome for degradation.

Degradation of MHC-I/peptide complex after arrival at the plasma membrane is for example caused by the HSV-8 (causes KARPOSI-sarcoma in AIDS-patients) K3 and K5 proteins. These type 1 membrane proteins reside in the ER and ubiquitinate the MHC-I protein. Once the latter arrives at the plasma membrane, it becomes internalised and destroyed along the endosomal/lysosomal pathway. The Nef-protein (negative factor) of HIV is membrane associated via a myristoyl-tail and causes MHC-I to be incorporated into endocytic vesicles by interacting with adaptor protein (see section 16.4 on page 243).

As seen in the section on complement (10.3 on page 155), our immune system is capable of detecting such a situation and destroying cells that do not express MHC-I on their surface. Some viruses have reacted to this by encoding proteins that look like MHC-I to our bodies immune system (e.g. human cytomegalovirus UL18 or murine cytomegalovirus m144). This is a beautiful example of co-evolution of parasite and host.

Phagocytic cells present foreign material: MHC-II

Professional antigen presenting cells like macrophages present on their surface peptides from the material they ingested to T-helper cells. These in turn can stimulate the production of reactive oxygen species and hydrolytic enzymes in the phagosomes of macrophages (see fig. 10.15).

B-cells can bind foreign material via their surface expressed antibodies, this material is phagocytosed, degraded and presented on the surface. $CD4^+$-T-helper cells then stimulate the production of soluble antibodies.

Dendritic cells, which stimulate T_k-cells, have a special adaptation that allows them to present materials endocytosed or pinocytosed on MHC-I instead of MHC-II. By an unclear mechanism such proteins are first transported into the ER, then retro-transported into the cytosol by the ER quality control machine, where they are ubiquitinated and then degraded by the proteasome. The resulting peptides are then treated like those resulting from the breakdown of indigenous material.

Structure of MHC-I and -II

MHC-I (see fig. 10.16) is a heterodimer, consisting of a membrane spanning α-chain (43 kDa), and a smaller protein, β_2-microglobulin (12 kDa). The α-chain has 3 domains, called α_1, α_2 (which form the peptide binding site) and

Antigen presentation by MHC class-II

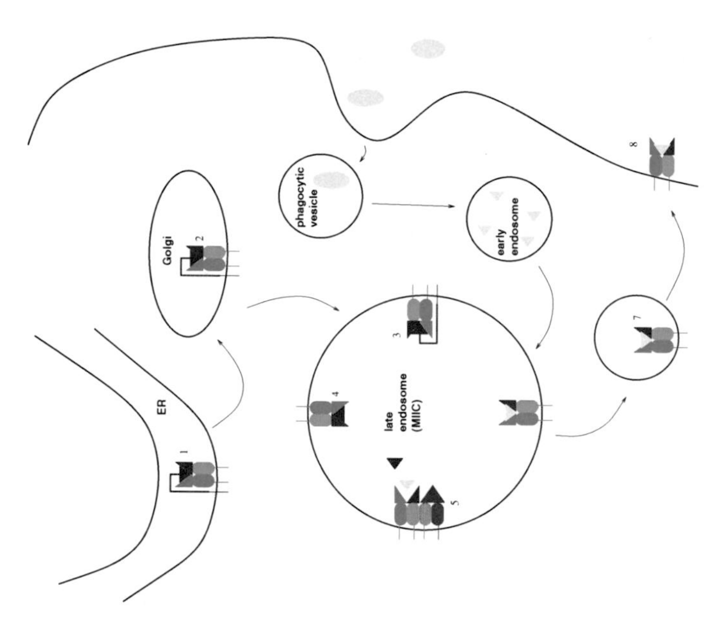

Figure 10.15. 1: Newly synthesised MHC-II is assembled in the ER, the peptide binding site is blocked by the MHC-II invariant chain (**Ii**) (brown), so that no self-peptides can bind. Ii also serves as a target signal for transport to the endosomes. **2,3**: MHC-II is transported to the MHC-II compartment (**MIIC**) of the endosomes, where the Ii chain is degraded (**5**), leaving only the **CLIP**-fragment (Class II associated Invariant chain Peptide). **6**: CLIP is exchanged for peptides from phagocytosed material. For this process a chaperone, **HLA-DM** (pink) is required. Once stable peptide binding has occurred, the MHC-II molecules are transferred to the cell membranes (**7,8**), where they are read by $CD4^+$-T_H-cells.

α_3, the membrane spanning domain which is structurally similar to the β_2-microglobulin. Both α_3 and the β_2-microglobulin have an immunoglobulin fold.

MHC-II is a heterodimer consisting of two similar membrane spanning glycoproteins, α and β. Each of these has two domains, α_2 and β_2 are membrane spanning and have immunoglobulin fold, α_1 and β_1 together form the peptide binding site.

Both MHC-I and MHC-II have a peptide binding cleft with a bottom formed by a β-sheet and side walls made from two α-helices. Peptides bind into the valley between the helices.

The peptide binding sites of MHC-I and -II are similar, except that in MHC-I the peptide is tightly bound at both ends, while in MHC-II binding occurs

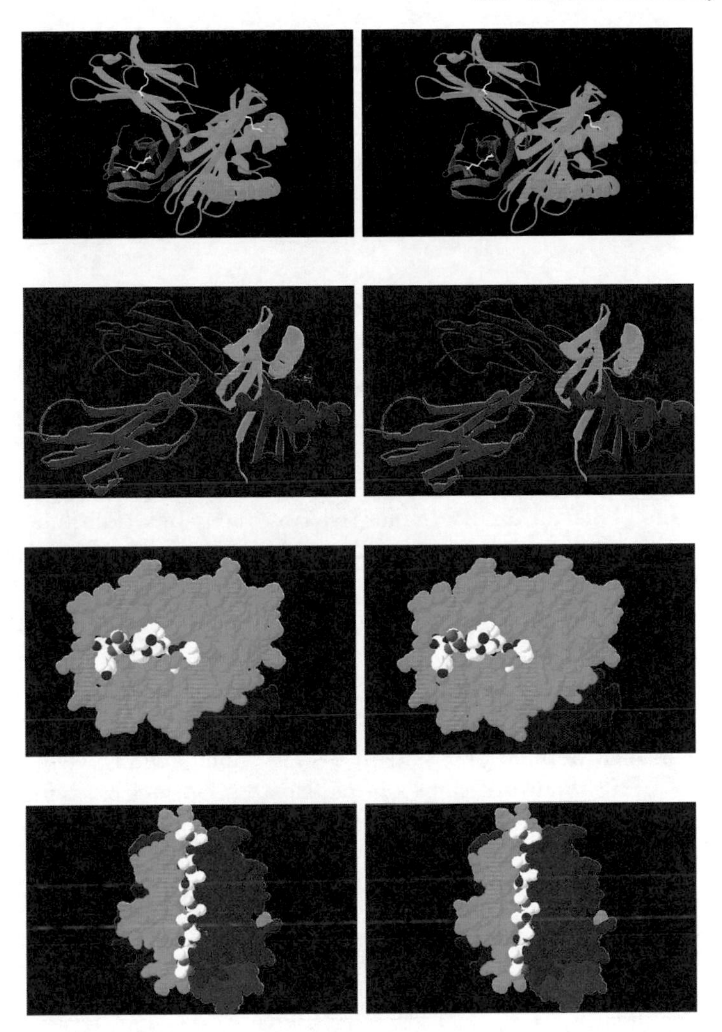

Figure 10.16. Top: Ribbon-representation of the MHC-I molecule (PDB-code 1QSE). α cyan, β_2-microglobulin blue. **Second:** MHC-II (PDB-code 1SEB). α blue, β red. In both molecules disulphide bridges are marked in yellow, the antigenic peptide is also drawn. Note the immunoglobulin domains in both molecules. The peptide is bound in both molecules in a valley with a bottom formed by a β-sheet and two α-helices as side walls. In MHC-I peptide is bound only by the α-subunit, in MHC-II by both α and β. **Bottom**: Space filling models of the peptide binding clefts of MHC-I and MHC-II, as seen by the T-cell receptor. The peptide is in an extended conformation. In MHC-I the peptide is short and both ends are surrounded by the binding side. In MHC-II the peptide is longer, the ends stick out of the binding site. All diagrams are stereo views.

more centrally, the ends of the peptide are free. Thus peptides bound to MHC-I tend to have the same length, those bound to MHC-II are more variable, and usually longer than those binding to MHC-I.

Peptides are bound to MHC at only a few amino acids, the **anchor residues**. Once binding of peptide to MHC has occurred, the complex tends to be stable for several days.

MHC variability

Both MHC-I and -II genes occur in several copies in the MHC-gene cluster (they are **polygenic**), which in addition are highly **polymorphic**, with several 100 alleles in some cases. Thus each person has a unique set of MHC-proteins, being heterozygous at most loci. A full match is usually only seen in identical twins. This variability is increased by recombination and gene conversion between misaligned chromosomes during meiosis, which constantly creates new variants of MHC.

The reason for the high variability is obvious: If all people had the same MHC-alleles, a pathogen could avoid detection by our immune system by not expressing proteins that have the anchor residues to bind to that MHC. This is indeed observed with small, isolated populations for example on remote islands. Such populations tend to be very homogenous with respect to MHC-alleles, and the viruses (for example Epstein-Barr virus) infecting such populations have adapted to this situation by avoiding epitopes that bind to such a MHC-molecule.

Several cases have been described in recent years where persons with a particular MHC-type are more resistant against a certain disease than other people. If a certain disease, say malaria, is endemic in an area, an MHC-type offering resistance to that disease will be selected for, explaining the uneven distribution of MHC-alleles in different populations (see for example [68]).

Allelic variation tends to occur particularly often in those amino acids of the MHC-molecules that interact with the bound peptides.

The MHC gene complex contains more than 200 genes in humans, occupying about 4 centimorgans on chromosome 6 (except the β_2-microglobulin gene, which is on chromosome 15). The complex is divided into 3 areas:

I HLA-A, -B and -C (MHC-I genes) and a variety of so called class-Ib genes

II HLA-DR, -DP and -DQ (MHC-II genes), an additional β-chain in the HLA-DR gene that can pair with any MHC-allele, Tap, LMP (interferon inducible proteasome subunit), DMα and β, DOα and β, tapasin

III C4, C2 and B (complement proteins), TNF-α and -β (cytokines), also 21-hydroxylase for steroid biosynthesis

Many of the MHC-II genes (except the negative regulator of peptide binding, HLA-DO) are induced by interferon-γ via the **MHC Class-II transactivator (CIITA)**. Patients lacking this transactivator suffer from an inherited form of immunodeficiency: **bare lymphocyte syndrome**.

MHC-1b genes encode many proteins with immune-related function, their number is variable between species and even individuals of the same species. One example is **HLA-G**, which apparently prevents fetal placenta-cells from being killed by the mothers immune system. **HLA-E** is encoded together with MHC-I, an imbalance between HLA-E and MHC-I protein on the surface informs T-killer cells that this cell has been infected with a virus that suppresses peptide presentation by MHC-1. **CD1** specialises in the presentation of bacterial cell wall components like mycolic acid.

Transplantation

As the name Major Histocompatibility Complex indicates, the high variability creates a problem in organ transplantation, as foreign MHC-molecules are powerful antigens. Not only would a recipient of a non-matched organ destroy it (**rejection**), but lymphocytes invariably transplanted with the organ would also attack the recipient (**graft-versus-host response**). About 1–10 % of our T-cells actually recognise foreign (**allogeneic**) MHC-molecules, these are called **alloreactive**.

One might ask now why there are so many T-cells against transplants. After all, we are not normally exposed to foreign MHC-molecules. It appears that T-cells recognise MHC in general. However, T-cells interacting with our own MHC-molecules die during their early development by apoptosis.

Each potential organ recipient is **typed** according to which MHC-variants they have, these data are stored in a central computer in a country or cooperating group of countries (like **Eurotransplant** in Amsterdam for Europe and the **Organ Procurement and Transplantation Network** in Richmond, Virginia for the USA). If an organ becomes available, the MHC-type is compared with those stored in the computer. The organ is given to the recipient with the closest match. Medical urgency is also taken into account. This procedure ensures that valuable organs go to those recipients who benefit most.

10.4.2 The T-cell receptor

Peptides bound to MHC-molecules on the surface of our cells are recognised by a specific protein expressed on the cell membrane of T-cells. This protein is called the **T-cell receptor**. Each T-cell has about 30 000 receptor molecules on its surface.

Structure of T-cell receptor

The T-cell receptor resembles the F_{ab}-fragment of an antibody. It is a heterodimer consisting of an α- and a β-chain, each of which consists of two **immunoglobulin domains** (one variable and one constant) and an additional **tail domain** at the C-terminus. The tails are held together by a disulphide bond just outside of the cell membrane. The tail can be divided into a **hinge-domain** (extracellular), a **transmembrane domain** and a **cytoplasmic domain**.

Because the function of the T-cell receptor is determined by cell-cell interactions, it does not need a F_c-domain. There is an alternative receptor isoform with γ and δ-chains instead of α and β, but its function is unknown.

Thus T-cell receptors are **monovalent**, unlike the immunoglobulins which have 2, 4 or 10 binding sites. Also, T-cell receptors are **never secreted**, they only exist membrane bound.

The α-chain is homologous to the immunoglobulin light chain and has V- and J-regions in its variable domain, the β-chain is homologous to the immunoglobulin heavy chain, with V-, D- and J-segments.

However, there are some differences between T-cell-receptor and immunoglobulin structure. The C_α-domain has only one β-sheet, were most immunoglobulins have the second β-sheet there are loosely packed strands and even an α-helix in C_α. α- and β-strand are held together in part by hydrogen bonds with carbohydrate site chains.

Origin of T-cell receptor variability

The gene for the α-chain has 70–80 V-regions, 61 J-regions and one constant region, giving 4270 – 4880 possible combinations.

The gene for the β-chain has 52 V-, 2 D-, 13 J- and 2 C-regions, resulting in 2704 possible combinations.

Thus, recombination can account for about 13×10^6 T-cell receptors. The mechanism of recombination is very similar to that seen in immunoglobulins.

T-cell receptors do not change by somatic hypermutation, this is an adaptive specialisation of B-, not T-cells. The reason might be that hypermutation would create receptors that either can not recognise MHC-I or that become self-reactive.

Specificity of T-cell receptors

If a mouse with MHC-I of a particular type (lets call it MHC-I^a) is infected with a virus, proteins from that virus will be presented on the surface of infected cells and lead to the stimulation of killer cells, which destroy the infected cells.

If such T_k-cells are isolated and brought into contact with mouse cells of another MHC-I type (say, MHC-I^b), which are infected with the same virus, no killing will occur. In other words: The T_k-cells are able to recognise the viral peptide only if presented by a particular MHC-I. This is called **MHC-restriction**.

If those same T_k-cells are brought into contact with mouse cells that have MHC-I^a genotype, but are not infected with virus, no killing will occur. Also, if the cells were infected with a virus unrelated to the first one, no killing will occur.

Thus T-cells are specific not only for an antigenic peptide, but for the combination of peptide and presenting MHC-I.

T-cell receptor interactions with MHC

The T-cell receptor, as we have seen, does not only interact with an antigenic peptide, but also with the MHC-molecules. Indeed, the T-cell receptor binds diagonally across the peptide binding site of MHC-1, with CDR-3 making contact with the peptide (see fig. 10.17). Not surprisingly, it is this region that contains most of the T-cell receptor variability. CDR-1 and -2 are much less variable, and interact with the MHC-surface.

V_α binds to the amino-terminus of the peptide, V_β to the carboxy-terminus.

The MHC-II/TCR complex has a quite similar structure to that of the MHC-I/TCR complex. Contact between T_h-lymphocytes and antigen presenting cells occurs in the peripheral lymphoid organs.

Superantigens

Some bacteria and viruses produce proteins that bind not to the peptide grove of MHC-II, but to the outer surface of MHC-II and the TCR-β-subunit. Such a superantigen can stimulate T_h-cells with one or a few T-cell receptor types. Rather than stimulating a specific immune response this type of binding results in suppression of adaptive immunity and systemic toxicity, caused by a massive release of cytokines by CD4-T_h-cells. Food poisoning by enterotoxins and toxic shock syndrome, both caused by *Staphylococcus ssp.*, are clinically important examples.

Figure 10.17. Binding of the T-cell receptor to the MHC-1/peptide complex, stereo view (PDB-code 1QSE). The T-cell receptor straddles the MHC-I/peptide complex diagonally, with the CDR-3s of the α- and β-subunit interacting with N- and C-terminal end of the peptide, respectively.

CD4 and CD8 co-receptors

CD4 and CD8 proteins have been used as immunological marker for T_h- and T_k-cells respectively for some time. They were discovered because they reacted with certain monoclonal antibodies. It was also discovered that T-cells express either the one or the other, but not both, and that expression of either surface marker correlated with a particular function of the cell (stimulation of antibody production and cell killing, respectively).

Only recently it has become clear that they act as co-receptor by binding to both the T-cell receptor and MHC. Thus they stabilise MHC-TCR interactions and reduce the concentration of antigen required for T-cell stimulation by about a factor 100.

CD4 has gained some notoriety by being the binding site for the **human immunodeficiency virus (HIV)**. It is a rod-like molecule made from 4 immunoglobulin domains and a peptide stalk in extended conformation, which goes though the cell membrane of the T_h-cell. The intracellular part can interact with a tyrosine-kinase, **Lck**. How this contributes to T-cell activation is unclear at present.

CD8 looks quite different, being a heterodimer of an α- and a β-chain. These two proteins are quite similar, each consists of an extracellular immunoglobulin domain and a stalk, which passes the cell membrane. The stalks are crosslinked by a disulphide bridge just outside of the membrane, and are heavily glycosylated. Like CD4 however CD8 interacts with the tyrosine kinase Lck.

In addition to the CD4 and CD8 co-receptors, cell adhesion molecules are involved in the contact of T-cells with the antigen presenting cells. This unspecific contact precedes MHC-TCR interactions, which result in the paracrin release of mediators by the T-cell.

Function of T_h- and T_k-cells

T_k-cells induce **apoptosis** (controlled cell death) in cells presenting foreign peptides on their MHC-I. These may come from a virus infection, or they may be embryonal proteins expressed in cancer cells. Lytic granules inside the T_k-cell contain cytotoxins (which kill the offending cell by inducing apoptosis or by forming pores in their membrane) and cytokines (which signal neighbouring tissue to activate defense mechanisms). The main cytokine is interferon-γ, which inhibits viral growth.

T_h1 cells activate macrophages. Macrophages can become infected by intracellular bacteria (for example *Mycobacterium tuberculosis* and *M. leprae*), which actually live in the phagosomes. This is possible because mycobacteria have a particularly resistant cell wall, and because they prevent the release of lytic enzymes into the phagosome. Activation of the macrophage by T_h1-cells overcomes at least the latter problem. Major cytokines of T_h1-cells are IFN-γ and TNF-β.

T_h2 cells activate antibody production in B-cells by producing interleukines 4 and -5. They also inhibit macrophages by producing IL-10.

TCR-MHC interactions lead to the clustering of TCR-molecules on the side of the T-cell closest to the antigen-presenting cell, and as a result to a reorientation of its cell skeleton. Thus the T-cell becomes **polarised** and can direct its mediators toward the antigen presenting cell.

Exercises

10.1. The proteasome

1) is a protease involved in the breakdown of ubiquitinated proteins.
2) makes peptides, which are displayed on the cell surface by MHC-I.
3) receives some of its substrate proteins from the chaperone Hsp70.
4) delivers its peptide products to Hsp70.

10.2. Critically discuss the physiological significance of the various immunoglobulin isoforms. What are the consequences for immunisation?

10.3. Which of the following statements about immunoglobulins is *false*?

a) The main function of IgA is the protection of mucous membranes.
b) IgD is very efficient in activating complement.

c) IgE is involved in allergic reactions.

d) IgG can cross the placenta.

e) The high avidity of pentameric IgM allows an effective immune response even before affinity maturation.

10.4. Immunoglobulin E

1) releases histamine from macrophages.

2) has an F_c-end which is bound by mast cells.

3) is involved in transplant rejection.

4) is involved in the defence against large parasites like worms.

10.5. The variable domain of the heavy chain (V_H) of IgG

1) is contained in the F_{ab}-fragment.

2) is O-glycosylated.

3) is part of the antigen binding site.

4) contains the hinge domain.

10.6. Disulphide bonds in antibodies

1) join the two heavy chains.

2) join one heavy with one light chain.

3) stabilise the domain structure.

4) join the two light chains.

10.7. Which of the following statements is/are *true* about antibodies?

1) The antigen-antibody interaction is reversible.

2) The large number of different antibody variable domains in our body is encoded by the huge Ig_{var}-cluster on chromosome 5, which contains about 1×10^6 different immunoglobulin genes.

3) Antisera contain antibodies against different epitopes.

4) Heavy and light chain of a given antibody have the same amino acid sequence in their variable part.

10.8. A hapten alone can not induce an immune response
because
a hapten alone can not bind to an antibody.

10.9. Antigen-presenting cells (APCs)

a) present peptides with a length of more than 50 amino acids.

b) express class I and II major histocompatibility complex.

c) recognise antigens with high specificity.

d) are present mainly in blood.

e) present antigens in phosphorylated form.

10.10. MHC-II

1) is an integral membrane protein which presents antigens from phagocytosed material.

2) is loaded with peptides in the cytosol.

3) is highly polymorphic.

4) bound peptides are recognised by the T-cell receptor together with the CD8 coreceptor.

10.11. The T-cell receptor (TCR)

1) occurs in the plasma membrane of both T_h- and T_k-cells.

2) is highly variable because of somatic hypermutation.

3) binds superantigens, resulting in immune suppression and systemic toxicity.

4) recognises antigens in solution.

10.12. Which of the following statements is *false*? The complement system

a) can be activated by immune complexes.

b) can be activated by antibodies bound to bacteria.

c) can form pores in cell membranes.

d) — when activated — increases the permeability of vascular walls.

e) is synthesised by B-lymphocytes.

10.13. Which of the following symptoms is/are related to inflammation?

1) Pain

2) Swelling

3) Redness

4) Heat

11

Cell skeleton

The shape of a cell is stabilised by a protein scaffold, known as **cell skeleton**. We distinguish the **microfilament** (7 – 9 nm diameter), the **intermediate filament** (10 nm diameter) and the **microtubules** (24 nm diameter).

It had long been assumed that the cell skeleton is a specific development of **eukaryotes** and not present in **prokaryotes.** In recent years **eubacterial** homeologues of the protein involved in the formation of all 3 types of cell skeleton types have been found. They seem to serve similar functions as in eukaryotes: stabilization of cell shape, mobility and all division. However, our knowledge about the bacterial cell skeleton is limited; its presence in **archaea** is still controversial. In the following we will limit the discussion to eukaryotes.

11.1 The microfilament

11.1.1 Basic actin structure

The microfilament is formed by **actin** molecules. Actin is a very abundant protein inside cells. About 1 – 10 % of all cellular protein is actin, leading to a concentration of several hundred µM. Actin is a clover-leaf shaped protein of approximately 40 kDa (see fig. 11.1).

Humans have 6 actin-isoforms, 4 α-actins are expressed in muscle, β- and γ-actin in non-muscle cells. They are highly conserved. α-actin is required for muscle contraction (see section 12.1.2 on page 187), β-actin for amoeboid movement of cells.

Free actin molecules (**G-actin**) tend to have MgATP bound to them. Actin is able to polymerise to long, helical filaments, called **F-actin** (see fig. 11.2). F-actin will have MgADP bound most of the time. Note however that ATP-hydrolysis only facilitates actin polymerisation, it is not strictly required and

Figure 11.1. Stereo view of actin (PDB-code 1YAG) complexed with gelsolin (purple). Note the MgATP-molecule in the actin and the Ca-ion associated with gelsolin.

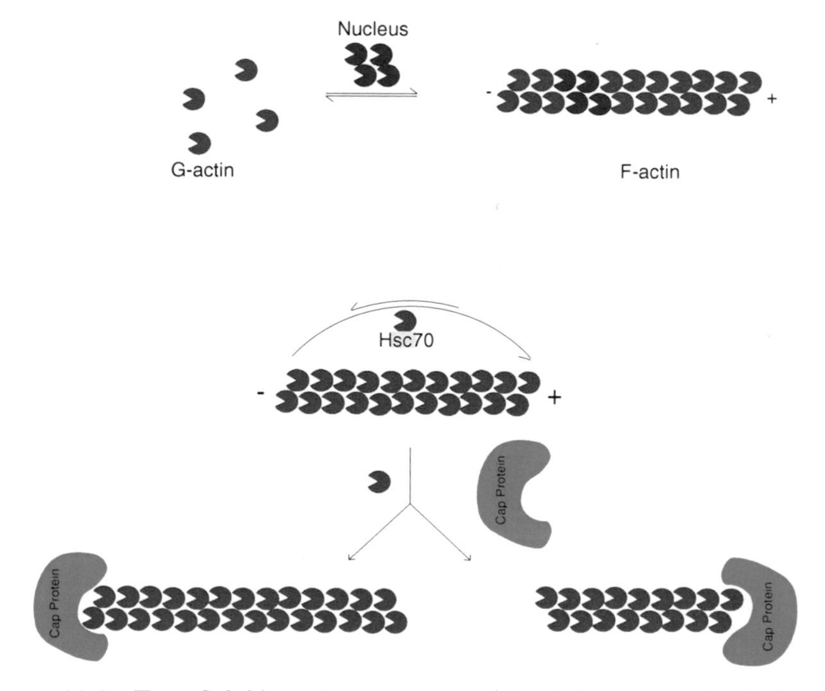

Figure 11.2. Top: Soluble actin monomeres (G-actin) can *in vitro* polymerise to form filaments of F-actin. This process is aided by the presence of nucleï of polymerised actin (but these are not absolutely required). The two ends of the actin filament are not identical, polymerisation occurs preferentially at the positive end. **Bottom**: *In vivo* G-actin is chaperoned by Hsc70, the cognate form of the 70 kDa heat shock protein (see page 212). Polymerisation and depolymerisation can occur on both ends, but polymerisation occurs preferentially at the positive, depolymerisation at the negative end. As a result the actin filament "moves" from – to +. Cap proteins can block either end of the actin filament. If the positive end is blocked, net dissociation from the negative end will lead to shortening of the filament. If the negative end is blocked, net association at the positive end results in filament elongation.

both G- and F-actin can occur with either ATP or ADP bound. Removal of nucleotide leads to rapid denaturation of the protein.

Polymerisation of G-actin to F-actin is spontaneous and occurs whenever the actin concentration is above a critical value (which depends on ion concentration, temperature and other factors). In this respect the polymerisation resembles crystallisation. Like crystal formation it is aided by the presence of nucleation sites, that is small aggregates of actin already assembled. Once three to four actin molecules have come together, they form a stable nucleus, on which further molecules can assemble.

Actin filaments have two different ends, designated plus (+) and minus (-). Assembly is preferentially to the **plus-end**, dissociation preferentially from the **minus-end**. At the minus-end the ATP-binding site of actin is exposed. Branching is possible when the **Arp2/3**-complex binds to an actin filament and serves as an anchor for the minus-end of an additional filament. This branching is tightly regulated and essential for the formation of filopodia. WISKOTT-ALDRICH-syndrom protein (WASP) is a regulator of the Arp2/3-binding to actin filaments, encoded on the X-chromosome. Failure to make the functional WASP results in an inherited immunodeficiency and blood clotting disorder, which is usually fatal. WASP itself is activated by a small G-protein, Cdc42. In plants the Arp2/3 pathway is responsible for correct cell morphology (root hair formation and interlocking of epidermal cells), but not essential.

Cells move by extending their β-actin filaments in one direction, the cytoplasm follows, forming **lamellipodia** (leading edges) and **filopodia** (leading spikes). These will constantly form **focal adhesions**, points of attachment to the substrate. The trailing end of the cell then simply follows, after its attachments to the substrate have been dissolved.

Actin polymerisation and depolymerisation are regulated in the cell by actin binding proteins. **Profilin** acts as an ATP/ADP exchange factor and is in turn regulated by the second messenger **phosphatidylinositol bisphosphate (PIP_2)**. **Gelsolin** and **profilin** break actin filaments into smaller units by binding to an actin molecule in the chain and twisting it so much that its interactions with other actin molecules break. This binding too is regulated by PIP_2 and Ca.

Several intracellular parasites (*Listeria monocytogenes* , Vaccinia virus) can stimulate actin polymerisation, thus pushing themselves through the cell.

One important function of actin polymerisation is the **acrosome reaction** during fertilisation. Once a sperm has made contact with an egg, actin bundles form in the head of the sperm (**acrosomal process**), which pierces the jelly-layer of the egg and allows the sperm nucleus to move into the egg.

Two fungal toxins interfere with the actin polymerisation/depolymerisation equilibrium: **Cytochalasin D** and **phalloidin**. Cytochalasin D caps the plus-end of F-actin, preventing further polymerisation. Phalloidin, the poison from the "angel of death mushroom" *Amanita phalloides*, binds to the minus end of F-actin and prevents its depolymerisation. The nasty bit about poisoning with *A. phalloides* is that symptoms appear only about three days after ingestion of the mushrooms, when it is to late to empty the GI-tract of the patients to minimise absorbtion. On the bright side, fluorescently labelled phalloidin is invaluable as a specific stain for actin filaments in microscopy.

11.1.2 Actin-networks

Actin filaments in the cell are cross-linked into a 2-dimensional network below the cell membrane and a 3-dimensional network throughout the cytoplasm. Cross-linking is performed by specialised proteins with 2 actin binding sites separated by coiled-coils or immunoglobulin domains. The actin binding sites usually have homology to the muscle protein **calponin** and are called CH-domains (calponin homology). Many of these proteins also have calmodulin-like Ca binding sites.

Filamin, ankyrin and **dystrophin** associate with transmembrane proteins like Na/K-ATPase and anchor actin filaments with respect to the plasma membrane. Actin filaments form a 2-dimensional network below the membrane, in the **cortex** of the cell. This may be further supported by a **spectrin** network.

Some of the transmembrane proteins transmit signals into the cell, for example **Gp1b-IX** in platelets. This protein binds to blood clots, and signals such binding to the actin filaments via **filamin**. As a result, the actin network (and therefore the platelet) contracts, closing the wound.

In muscle cells the actin network is connected to transmembrane glycoproteins via **dystrophin**. The glycoproteins in turn link the muscle cell to the extracellular matrix. Mutations in the dystrophin gene lead to DUCHENNE **muscular dystrophy**, an X-linked genetic disease with onset around age 6, which is eventually fatal during adolescence.

Actin bundles, cross-linked to the membrane, support **microvilli** and **filopodia**. Many cross-links between the actin filaments in the bundles make them stiff.

11.2 Microtubules

Microtubules are involved in:

- Beating of **cilia** and **flagella** (see section. 12.3 on page 192).
- **vesicle transport** in the cell and phagocytosis (see section 16.4 on page 231).
- Formation of the **mitotic spindle** (see section 12.4 on page 194).
- Stabilisation of **cell shape**, elasticity of erythrocytes and the stability of the long axons in nerve cells.
- Keeping of **cell organelles** like GOLGI-apparatus , ER and mitochondria in their proper place. If microtubules depolymerise (during mitosis or with colchicine), ER and GOLGI-apparatus fragment into many vesicles.

11.2.1 Microtubule structure

α- and β-Tubulin

Microtubules are made by the ordered polymerisation of **tubulin**, they have a diameter of 24 nm, their length can reach several hundred µm.

There are two isoforms of tubulin, called α and β, which form a heterodimer (see fig. 11.3). α-Tubulin contains **GTP** bound in a pocket which is covered by the β-tubulin, and which therefore can not be exchanged. β-Tubulin has an open binding pocket, which binds GTP. After polymerisation, GTP is hydrolysed to GDP, exchange of this GDP for GTP occurs during depolymerisation. The GTP in the α-tubulin is not hydrolysed at all.

These heterodimers polymerise into long chains, called **protofilaments**. 13 such protofilaments form a microtubule (see fig. 11.4).

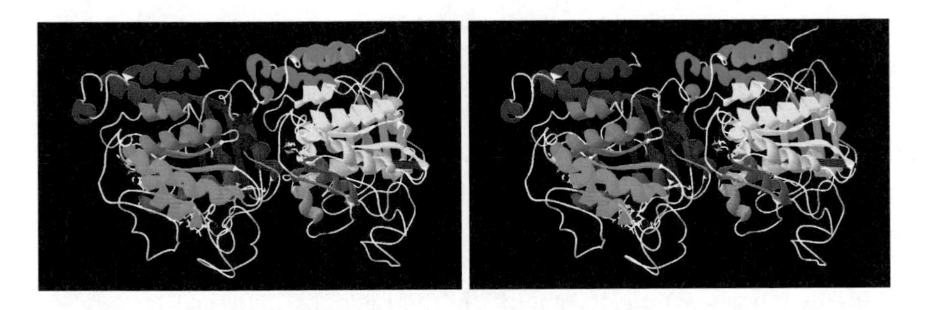

Figure 11.3. Stereo view of the α- (red-green) and β-tubulin (blue-purple) dimer (PDB-code 1TUB). α-Tubulin has GTP bound to it, the binding site is occluded after dimerisation. β-Tubulin contains bound GDP, in addition this subunit carries one molecule of paclitaxel, an anti-cancer drug, to stabilise subunit-interactions.

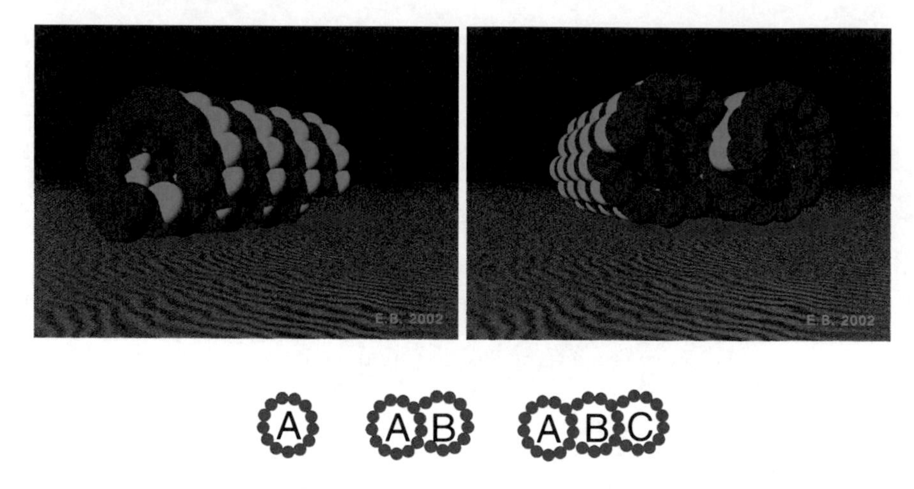

Figure 11.4. Microtubules consist of α- and β-tubulin. The α,β-tubulin-heterodimers polymerise into tubes, which consist of 13 **protofilaments** (long rows of dimers). This is called an A-tubule. 10 further protofilaments can be added to form a doublet, found in cilia and flagella. The second tubule is called a B-tubule. A further 10 protofilaments make a triplet, found in basal bodies and centrioles. This third tubule is called C-tubule, and distinguished from a B-tubule by the fact that it is connected to a B- rather than an A-tubule.

Because all tubulin-dimers in a microtubule have the same orientation, one end (called **minus-end**) is capped with α-, the other (called **plus-end**) with β-tubulin.

Mechanism of tubulin assembly and disassembly

The minus ends of all microtubules in the cell start from a central hub, the **microtubule organising centre (MTOC)**. These may be visible in electron microscopy as **basal body** or **centrosome**. Most cells have one MTOC, the centrosome which also contains the centrioles. Epithelial cells (and plant cells) have many MTOCs, which organise a network of tubules in the cell cortex. Cell polarity in these cells is linked to tubule orientation. Part of the MTOC is the 25 S-ring complex, which consists mainly of γ-tubulin (γTu-RC). The mechanism of microtubule organisation is not yet known, however.

During depolymerisation of tubulin the interactions between the protofilaments break first, so that under the electron microscope the ends of a shrinking microtubule look frayed, brush-like. Growth on the other hand must occur by the even addition of small units to the ends, because they look smooth in the EM. Also, disassembly is a much faster process than assembly (7 vs 1 µm/s).

Other proteins associated with microtubules

A number of proteins have been found to co-purify with microtubules in cell fractionation studies, fluorescent antibodies against these proteins label microtubules in fluorescence microscopy. These are called **microtubule associated proteins (MAP)**. Based on sequence similarity, these are grouped into type-I and type-II MAPs. These proteins can change the rate of tubulin polymerisation and depolymerisation, or cross-link microtubules to intermediate filament, membranes or with each other.

11.3 Intermediate filament

Intermediate filaments IF have a structural role, no transport functions seem to be associated with them, nor are they involved in cell motility. Because they are bound to transmembrane proteins, which link cells together, they are responsible for the stability of tissues.

Claws, nails, horns and hairs are composed mainly of keratin, an intermediate filament protein.

Intermediate filaments are quite stable and survive extraction of the cells with detergents and high salt concentrations, which remove the membrane, microfilament and microtubules. They can be depolymerised with urea, purified and re-polymerised by urea removal. Intermediate filament proteins polymerise into helical rods (see fig. 11.5). Assembly does not require ATP.

11.3.1 IF-proteins are cell-type specific

Six classes of IF-proteins can be distinguished:

Type I Acidic keratins are found in epithelial cells.

Type II Basic keratins are also found in epithelia. They form heterodimers with the acidic keratins, which then polymerise into fibres (see fig. 11.5). Many isoforms of both types of keratin have been described, we distinguish hard keratins (hair, horn, claws) and cytokeratins that are found in internal epithelia.

Type III Vimentin (leucocytes, blood vessel endothelia, fibroblasts), Desmin (stabilises sarcomeres in muscle cells), peripherin (found in peripheral neurons, function unknown) and glial fibrilary acidic protein (glial cells and astrocytes) can form homo- and heterodimers.

Type IV Neurofilaments NF-L, -H, -M and internectin are found in the nervous system. They are responsible for radial growth of the axon and determine its diameter.

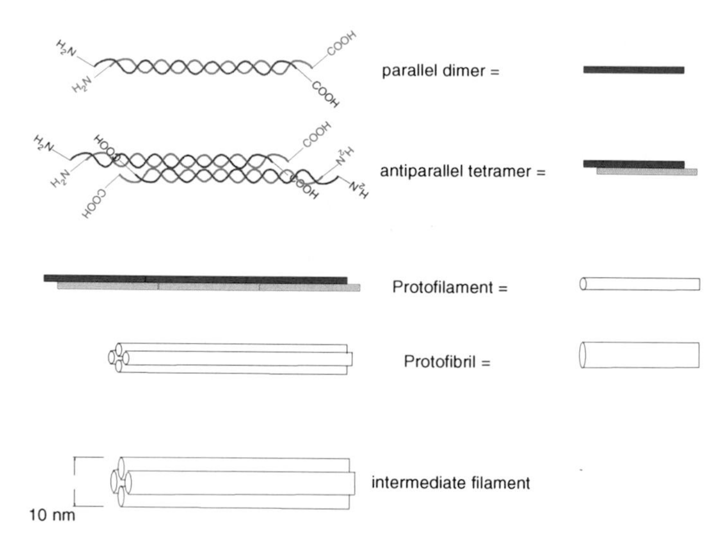

Figure 11.5. IF-proteins contain long helical segments, with more unordered heads and tails. 2 IF-proteins dimerise in parallel with their helical segments winding around each other. 2 such dimers then polymerise into an anti-parallel tetramer. These form protofilaments, 4 of those form protofibrils and four of those intermediary filaments. The head-parts of the protein stick out from the filaments like the fibres from a lamp-brush, which can be seen in EM-pictures of the IF.

Non-standard type IV Philensin and phakinin are found in the lens of the eye.

Type V Lamin[1] A, -B and -C are found in the nuclear lamina. Lamin-A and -B are produced from the same gene by different splicing, lamin-C is encoded by a separate gene. They maintain the nucleus and have an essential role during cell division (see for example [48]).

IF-proteins as tumour markers

Because IF-protein are cell type specific, they can be used to identify the origin of tumor cells in biopsies. For example epithelial cells, and tumours derived form them, express keratins while mesenchymal cells express vimentin. Antibodies against these proteins can be used to distinguish tumours of these two origins. This is important for the selection of the right treatment for gastrointestinal or breast cancers.

[1] Do not mix up lamin and laminin, a protein in the extracellular matrix (see section 13.1 on page 189)

11.3.2 Structure of intermediate filaments

It appears that the tetrameres of IF-proteins form a **soluble pool**, which is in dynamic equilibrium with the filaments, similar to what you have learned about actin and tubulin. Tetrameres can be isolated from cytosol. If labelled IF-proteins like keratin are microinjected into a cell, they become part of the IF after a couple of hours.

11.3.3 Intermediate filaments and cell cycle

IF are fairly stable structures, much more so than microfilaments and microtubules. However, during mitosis the IF must be reorganised. IF protein head domains can be phosphorylated by **Cdk2** (cell cycle dependent protein kinase) on serine residues, which leads to their depolymerisation.

11.3.4 Other proteins associated with intermediate filaments

The IF are cross-linked with each other, with microtubules and with the cell membrane by specialised proteins, collectively called **intermediate filament associated proteins (IFAP)**. The function of the known IFAPs is only to increase the stability of the cell skeleton, they do not influence the polymerisation/depolymerisation equilibrium nor do they act as motor proteins.

Treatment of cells with colchicine (see fig. 12.8 on 197) causes the microtubules to depolymerise. IF remain intact, but lump together in an unordered pile near the nucleus. Thus microtubule structure is required to organise IF.

Plectin and **ankyrin** are examples of IFAPs. Plectin crosslinks IF to microtubules and to the lamin network of the nucleus. Ankyrin connects them to plasma membrane proteins like Na/K-ATPase, and to microtubules. They also bind to cell/cell- and cell/matrix-junctions (see section 13.3 on page 208).

In muscle, where the forces created in a sarcomere are distributed to other sarcomeres, the cell membranes of neighbouring cells are connected by a **desmin** network.

Medical aspects

Mutated IF proteins like keratin do not kill a cell, but weakens its shape and its interactions with its neighbours. Thus forces can no longer be distributed between cells and tissues are weakened. An example for this is *epidermolysis bullosa simplex*, where skin keratin is mutated. As a result dermis and epidermis easily separate, forming blisters.

Exercises

11.1. Which of the following statements is/are *correct* about intermediate filaments?

1) Serine phosphorylation in the heads of IF proteins leads to intermediate filament breakdown during mitosis.
2) Ankyrin cross-links the intermediate filament to proteins of the plasma membrane.
3) Some intermediate filament proteins form a lining inside the nuclear membrane.
4) Intermediate filaments form the routes for retrograde transport of vesicles.

11.2. Cancer cells of mesenchymal and epidermal origin can be distinguished in biopsies
because
they contain vimentin and keratin, respectively.

11.3. Which of the following statements is *false* about the cell skeleton?

a) Conversion of microfilament F-actin to soluble G-actin requires the energy of ATP-hydrolysis.
b) DUCHENNE muscular dystrophy is caused by a failure to cross-link actin filaments to plasma membrane proteins.
c) Depolymerisation of microtubules leads to fragmentation of the GOLGI-apparatus.
d) Hair, nails and claws are made from intermediate filament protein.
e) Colchicine causes the intermediate filament to lump together near the nucleus.

12

Motor proteins and movement

The cell skeleton forms a scaffold, along which motor proteins can move. These proteins convert the chemical energy of ATP-hydrolysis into mechanical energy. Movement is uni-directional, either from minus to plus or *vice versa*. The most important systems are microfilament/myosin and microtubule/kinesin and -dynamin.

12.1 Myosin moves along actin filaments

Humans have 13 myosin genes, but only the functions of myosin-I, -II and -V are known. Myosin-II is required for muscle contraction and cytokinesis, myosin-V for the transport of intracellular vesicles along the microfilament, myosin-I on the plasma membrane interacts with actin-bundles in microvilli.

12.1.1 Myosin structure

All myosin isoforms consist of an N-terminal head and a C-terminal tail of different length (see fig. 12.1). In myosin-II and -V α-helical sections in the tails interact to form dimers, myosin-I molecules do not have this section and stay single. The tails of myosin-I and -V bind to **membranes**. Aggregation of tails in myosin-II forms a large supermolecular structure, the **thick filaments**.

Several regulatory **light chain** molecules bind to the neck regions of myosins, they have regulatory function and bind Ca. In myosin-I and -V these include **calmodulin**.

The head is the business-end of myosin, it contains the actin- and ATP-binding sites. The ATPase-activity of myosin is actin-activated.

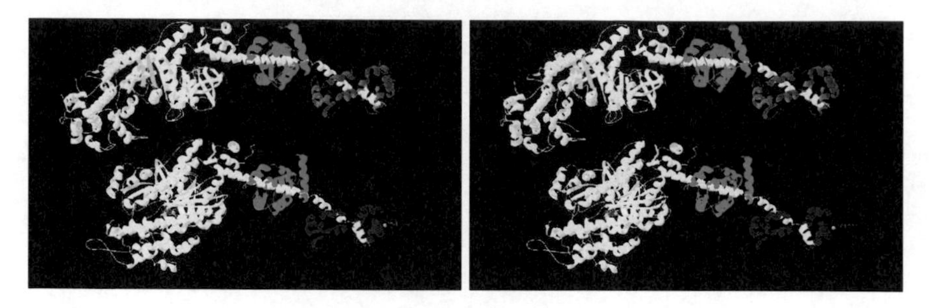

Figure 12.1. Stereo view of X-ray crystallographic structures of the head and neck regions of scallop muscle **myosin-II**, without (top, PDB-code 1DFK) and with (bottom, PDB-code 1DFL) bound ADP+Vanadate (ATP-analogue). The tail region was removed by protease treatment to allow crystallisation. The molecules have about the same orientation. Note the extensive conformational changes upon ATP-binding.

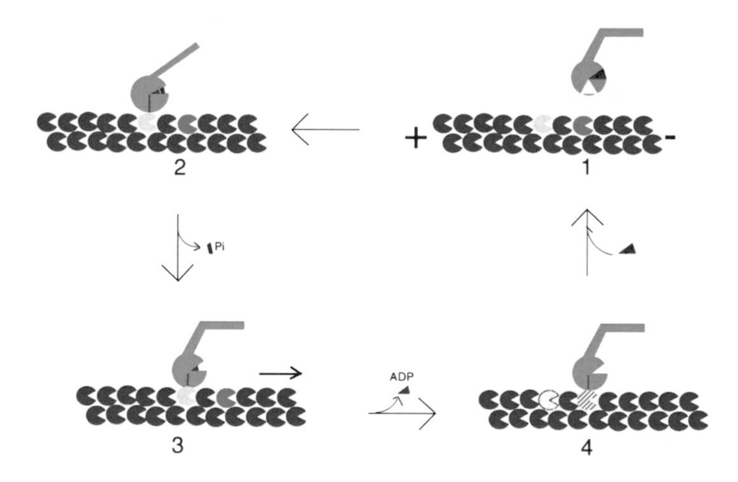

Figure 12.2. Actin-myosin interactions. 1) ATP-binding to myosin results in opening of its actin-binding site. Actin is released. 2) The bound ATP is hydrolysed to ADP and P_i, the joint between stem and head of myosin extends, a new actin molecule is bound. 3) P_i is released, the myosin joint flexes. As a result, the actin chain is pushed towards the minus end. This is the actual power stroke. 4) ADP is released, closing the cycle. Such interleaving of mechanical and chemical reactions ensures **tight coupling** between both, preventing unnecessary hydrolysis of ATP (a myosin molecule alone hydrolyses about 1 molecule of ATP per minute, in the presence of actin-filament that rate increases to 1000/min). This drawing shows monomeric myosin-I; myosin-II and -V are dimers, but work in a similar way.

12.1.2 Myosin-II

The power/weight ratio of muscle is about 1.5 times lower than that of a car engine and about 10-fold lower than that of an jet engine. We have three classes of muscle, **skeletal, smooth** and **heart** muscle. Skeletal muscle can contract rapidly but also tiers quickly. Smooth muscle contractions tend to be slower, but more prolonged. Heart muscle is specialised for the rhythmical contraction of this organ. Myosin-I in intestinal brush border membranes covers about 40 nm/s, myosin-II in skeletal muscle about 4.5 µm/s. Each individual cycle results in movement of 5 – 10 nm (the diameter of actin is about 5 nm), with a force of 1 - 5 pN (this is equivalent to the force of gravity acting on an *E. coli* cell).

Muscle activity may be **isotonical** (contraction at constant force) or **isometrical** (force is increased while length stays the same).

Skeletal muscle

Actin and myosin form a highly organised actomyosin complex , which results in a microscopically visible **stria** (see fig. 12.4).

Regulation of muscle contraction

Muscle contraction is stimulated by Ca-release from the **sarcoplasmic reticulum (SR)** (a specialised ER), where it is stored. Depolarisation of the cell after a nerve impulse leads to the opening of Ca-channels in the SR. Once the muscle has been re-polarised, Ca is moved into the SR by the **Ca-ATPase**, a P-type membrane transporter (see section 17.2.1 on page 265). Around the actin-filaments is wound a complex of **troponin** (a multisubunit protein) and **tropomyosin**. Tropomyosin molecules form a long continuous chain, which winds around the actin microfilament in the helical grove and hold the troponin in place. A rise in Ca-concentration changes the conformation of this complex and exposes myosin-binding sites on actin. Additionally, Ca binds to the regulatory light chain of myosin. This changes the conformation in the neck of myosin and allows myosin-actin interactions.

Smooth muscle

In smooth muscle the actomyosin complexes do not form ordered arrays as in skeletal muscle. Instead, they end in **dense bodies** and **attachment plaques**, the latter linking them to the cell membrane. Attachments plaques contain both α-actinin (like Z-disks) and **vinculin**. Smooth muscle contains

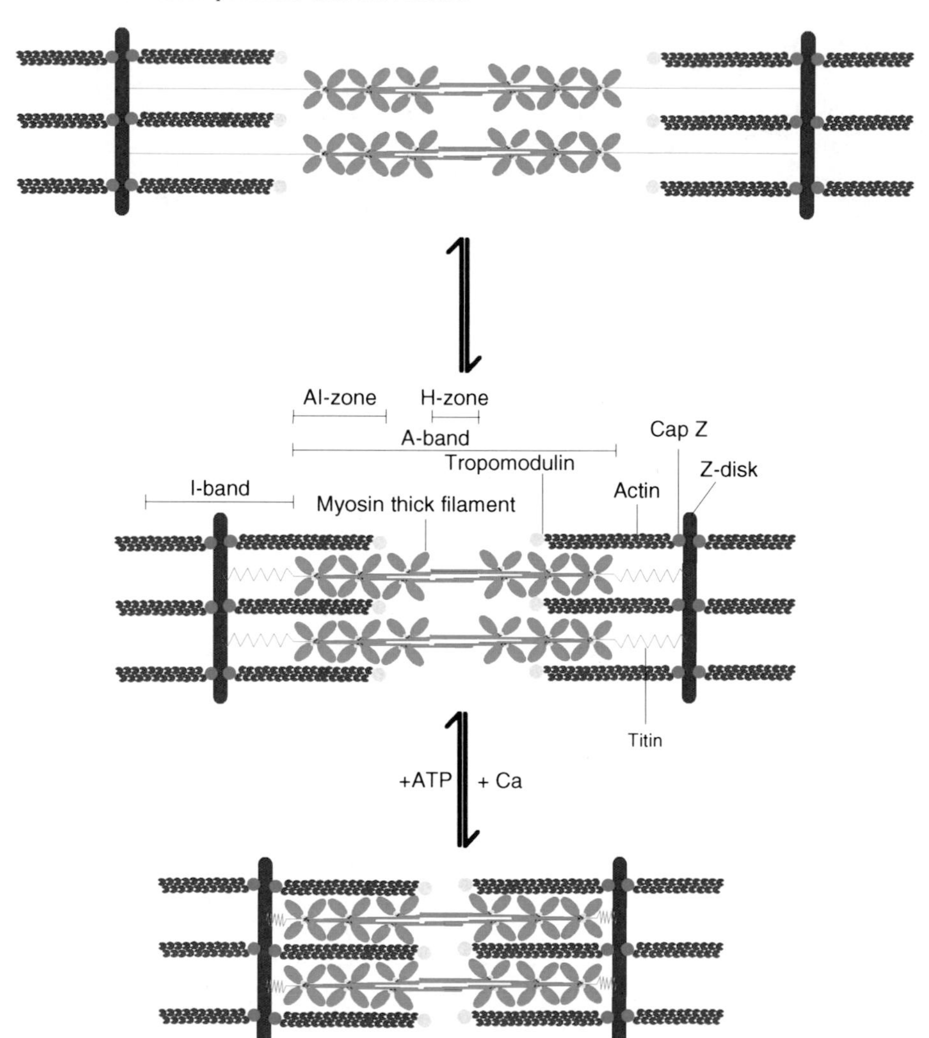

Figure 12.3. Sliding filament model of muscle action: The **sarcomere** is the smallest repetitive unit in skeletal muscle. Many sarcomeres in series make a **myofibril**, many myofibrils in parallel fill a muscle cell (**myofibre**), many myofibres form a muscle. Repeated binding of myosin heads to the actin filaments leads to muscle contraction, under the influence of ATP actin is released from myosin and the muscle relaxes. Muscle strength is the result of the summed binding strength of millions of myosin heads to actin microtubules. Passive stretching of muscle is possible, but elastically resisted by **titin** molecule that link the thick filaments to the Z-disks. Actin microfilaments are stabilised by capping proteins, **tropomodulin** at the minus and **Cap Z** at the plus end. **Nebulin** (not shown) binds along the entire length of the microfilaments and stabilises them. The association of actin and myosin in muscle results in an **actomyosin complex**. The Z-disk is composed mainly of α-actinin.

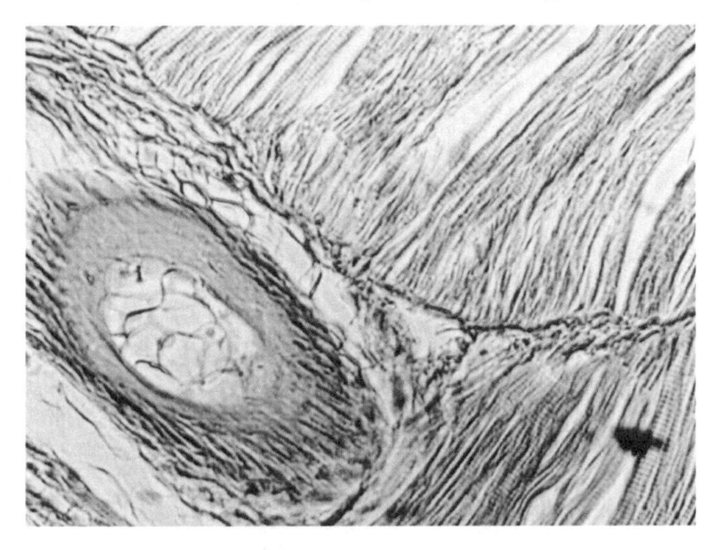

Figure 12.4. As a result of its structure, skeletal muscle appears striated in histological preparations (Zebrafish, HE staining).

caldesmon instead of troponin. Ca-dependent phosphorylation of caldesmon and of myosin light chains is required for smooth muscle contraction. Thus smooth muscles contract much slower than skeletal muscle. Ca-concentration in smooth muscle is regulated not only by nerve impulses, but also by hormones and growth factors.

Rigor mortis

Muscle relaxation occurs when ATP-binding to myosin leads to release of bound actin. After death the ATP-stores of the cell are rapidly depleted, but actin-myosin-interactions are still possible. This results in ***rigor mortis***, the stiffness of the dead. Only after lysosomal proteases have partially digested the muscle does it become soft again. Understanding of these processes is important for determination of the time of death in forensic medicine.

There is also an everyday application of this knowledge: Meat prepared immediately after slaughter is tough and hard, "hanging" it in a cool place for a couple of days makes it softer. Alternatively, the meat may be marinated in protease-containing fruit juices (pineapple, papaya). Meat tenderizers contain such proteases in powdered form.

Cytokinesis

Myosin-II and actin are found also in the cleavage furrow of mitotic cells. They form a **contractile ring** of parallel molecules around the cell. These molecules slide past each other and thereby reduce the diameter of the contractile ring, until the daughter cells are separated. If the expression of myosin-II is prevented, cells will still undergo mitosis, but because cytokinesis is prevented, multinuclear cells (**syncytia**) are formed.

12.1.3 Myosin-I

Myosin-I stabilises microvilli by linking their plasma membrane to the actin bundles in their core. Myosin-I co-purifies with certain GOLGI-derived **vesicles** and is apparently responsible for their movement along actin microfilaments in the cell. In single cell organisms like amoeba and cilliata it is required for contraction of the **contractile vacuole**, and therefore control of osmolarity.

Cell movement requires sliding of myosin-I on growing actin-filaments. **Profilin** on the leading edge membrane of the cell stimulate the extension of microfilaments in the direction of cell movement. At the same time **cofilin** stimulates depolymerisation of actin at the minus end. **Actin cross-linking protein** causes the formation of bundles from actin-filaments. These form the highways along which myosin-I slides, moving intracellular membrane systems and therefore also the cytoplasm. An extended lamellipodium will form a new focal adhesion to the substrate, at the same time the trailing edge detaches from old adhesions, leaving them behind. The retraction of the trailing edge requires myosin-II, which can be localised there by immunofluorescence microscopy, while myosin-I is found at the leading edge.

12.1.4 Myosin-V

Myosin-V is required for the transport of exocytotic vesicles to the plasma membrane.

12.2 Kinesin and dynein move along microtubules

If one observes living cells under the microscope, one can see organelles and vesicles moving over long distances on straight paths. This movement is clearly different from random, BROWNIAN motion. Such directed transport can cover large distances, for example in nerve cells, where proteins and membranes — synthesised on ribosomes and ER in the cell body — need to be moved along the axon into the synapse to replace material lost during exocytosis. An axon can be several meters long (whale, elephant). Thus if one incubates nerve cell bodies of the sciatic nerve (found in a dorsal root ganglion) with radioactive amino acids and then measures the concentration of radioactivity in different parts along the nerve at various time points, one can measure the speed of transport. This is called **anterograde transport**. Transport from the synapse to the lysosomes in the cell centre (**retrograde transport**) also exists. Interestingly, transport speed varies. Vesicles move with 1-3 μm/s, cytoskeletal components with about 1/100 of this speed. Organelles like mitochondria move at an intermediate velocity. We know today that such movement proceeds along microtubules which extend from the microtubule organising centre (near the nucleus) to the periphery. Movement is ATP-dependent.

12.2.1 Kinesin is responsible for anterograde transport

There are many isoforms of kinesin, with different structure and function. A typical kinesin consists of 2 heavy and 2 light chains, with a total molecular weight of 380 kDa. Each of the heavy chains carries an N-terminal head, which has an ATPase-domain and a tubulin binding site. The C-terminal ends wrap around each other in a coiled-coil structure (see fig. 12.5), these and the light chains together interact with the "cargo". Different kinesin isoforms are responsible for transporting different cargos.

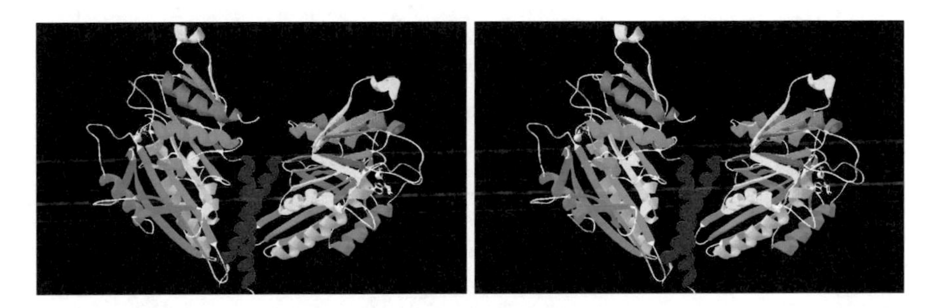

Figure 12.5. Stereo view of two kinesin heavy chain fragments from the fruit fly *Drosophila melanogaster*, with ADP bound to them (PDB-code 1CZ7). The light chains were removed to ease crystallisation of the protein, it would bind somewhere at the end of the coiled coil.

Incubation with the non-hydrolysable ATP analogue AMP-PNP leads to tight binding of kinesin to microtubules, the protein is released only after the AMP-PNP has been replaced with ATP.

Kinesin moves from the **minus to the plus** end of microtubules (**anterograde**).

12.2.2 Dynein is responsible for retrograde movement

Dynein complements kinesin in that it moves along the microtubules **from the plus- to the minus-end**, i.e. in **retrograde** direction. There are several dynein isoforms, their structure has not been elucidated to date, 2 – 3 heavy chains (about 500 kDa each) associate with several intermediate and light chains. We distinguish **cytosolic** dyneins (responsible for vesicle and organelle transport) from the **axonemal** dyneins found in cilia and flagella.

Cargos of dynein carry a number of ill-characterised so-called **microtubule binding proteins (MBP)**s, which actually bind dynein, which in turn interacts with the microtubules.

Table 12.1. Summary of motor proteins

Motor protein	Substrate	moves toward
Myosin	Actin	plus
Kinesin	Tubulin	plus
Dynein	Tubulin	minus

12.3 Cilia and flagella

12.3.1 Generic structure of cilia and flagella

Many protozoans, but also **sperm cells**, are propelled by cilia and flagella. Additionally, cilia on the **epithelial cells** of our air ways and in the ventricles of the brain remove contaminants. **Hair cells** in the inner ear carry one kinetocilium in addition to the sensory stereocilia (which have a different structure).

Cilia and flagella have the same basic structure, but cilia are shorter and tend to occur in greater number on a cell than flagella.

The travelling speeds of protozoans can be up to 1 mm/s, about 10 – 100 body lengths. Flagella from some insect sperm cells can reach a length of 2 mm.

The movement of cilia and flagella is complicated and can be studied with a high speed video camera mounted onto a microscope. These patterns ensure maximum resistance against the medium during the forward and minimum resistance during the backward stroke.

The basic structure of most flagella and cilia is a "**9+2**" arrangement of microtubules. 2 central tubulin singlets are surrounded by 9 peripheral tubulin doublets (see fig. 12.6).

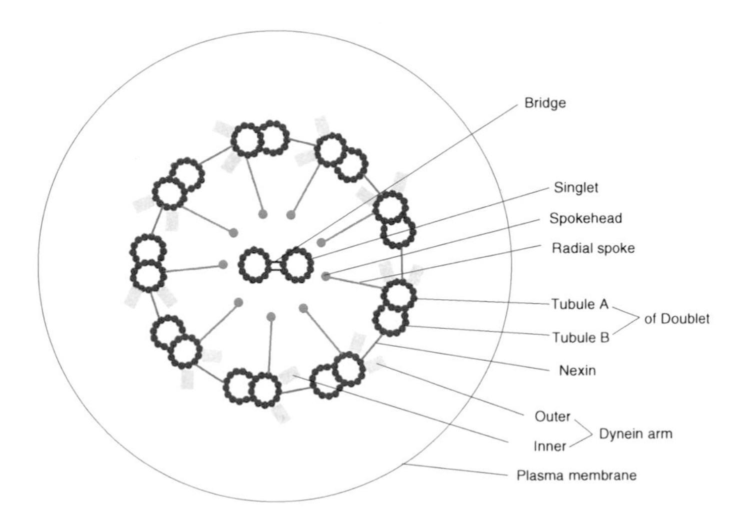

Figure 12.6. Structure of flagella and cilia. 2 singlet and 9 doublet microtubules (purple) form the core of the structure (hence "9+2"-pattern). Each doublet is composed of an A-tubule with 13 and a B-tubule with 10 protofilaments. They are linked by nexin-bridges (green) and carry a radial spoke (cyan) and dynein arms (yellow). The two singlet-tubules in the centre are connected by a bridge (blue). The entire flagellum is about 250 nm in diameter in all species, but the length can vary between 1 µm in ciliates to 2 mm in insect sperm. The plus-end of the microtubules points to the distal end of the flagellum.

Inside the cell each flagellum connects to a basal body, a structure resembling a centriole. They contain 9 triplet microtubules. In a triplet microtubule we have A-, B- and C-tubules (see fig. 11.4). A- and B-tubules continue into the flagellum, but the C-tubule terminates in a transition zone between basal body and flagellum.

12.3.2 Mechanism of movement

There are atypical flagella in some organisms with 3+0, 6+0 or 9+0 organisation, all of which are motile. At the very least this proves that the central pair of microtubules is not necessary for movement. Experimentally it is possible to remove the **plasma membrane** with detergent, the remaining structure is still motile, so the plasma membrane is not required for movement either. If such preparations are treated with proteases to remove the structural linkages between microtubules, bending is no longer possible, but upon addition of ATP the microtubules slide past each other instead. Removal of **dynein** by washing with high salt prevents any movement. Dynein can be added back to such stripped preparations, this restores motility. Thus the bending of flagella is caused by the dynein molecules on one A-tubule walking toward the minus-end (proximal) of the neighbouring B-tubule, while protein cross-links keep the structure intact.

No crystal structures are available for dynein, but upon negative staining dynein molecules appear under the EM as "blossoms" with two (inner dynein) or three (outer dynein in some species) heads, which are connected by smaller stalks. These heads probably represent the ATP- and tubulin binding unit, analogous to myosin.

Mutated *Chlamydomonas* algae lacking the **outer dyneins** are still motile, but the flagella beat slower. If the **inner dyneins** are mutated, no movement is possible. Thus it is the inner dyneins that are most important for flagellum beating.

12.3.3 Cilia and flagella start growing at the basal body

If the flagella of cells are removed and then allowed to re-grow in the presence of radioactive tubulin, it can be shown by autoradiography that tubulin is only incorporated at the distal end of the growing flagellum. Also, no radioactive tubulin is incorporated into a flagellar microtubule when moving cells are incubated with radioactive tubulin. Thus, these microtubules are fairly stable structure.

12.4 The mitotic spindle

The mitotic apparatus, which is responsible for chromosome separation during cell division, is also made from microtubules (see fig. 12.7). It consists of

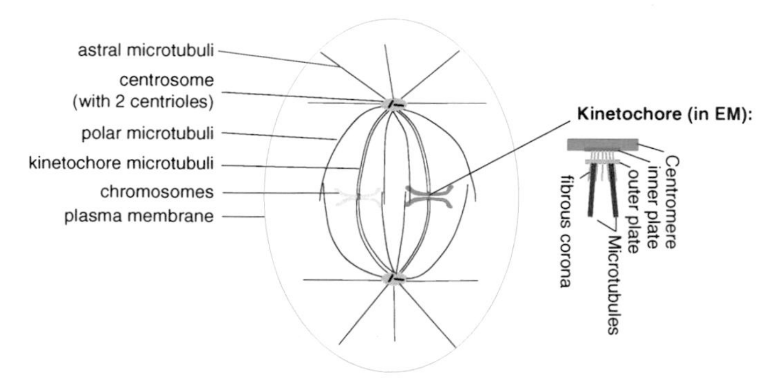

Figure 12.7. The mitotic spindle in animals.

- two centrosomes at opposing poles of the cell.
- a pair of "asters", tufts of astral microtubules which anchor the apparatus at the cell cortex.
- the central mitotic spindle, two symmetric bundles of microtubules shaped like a rugby ball. It consists of
  - kinetochore microtubules, which bind to the chromosome near their centromere[1] at a specialised structure, the kinetochore. The centromere is visible under the microscope as a constriction in the chromosome, it is also the point where the two sister chromatids are held together.
  - polar microtubules, which do not bind to chromosomes, but interdigitate at the equatorial plate with polar microtubules coming from the opposite centrosome.

As usual, all minus-ends of the microtubules originate from the centrosomes.

A proteomic screen has identified more than 700 proteins involved in the spindle apparatus, this number is enormous if you consider that from the 30 000 proteins encoded by the human genome only about 2000 are expressed in a particular cell at a particular time.

Many proteins required for spindle function are stored in the nucleus and become cytosolic only once the nuclear membrane breaks down. In yeasts on the other hand the spindle is located inside the nucleus (spindle pole body, **closed mitosis** as oposed to the **open mitosis** of higher animals) and spindle proteins are specifically imported into the nucleus during mitosis. The switch from closed to open mitosis probably became necessary because of the greater chromosome number and mass in higher animals, which required an increase in the microtubuli number by some two orders of magnitude. Thus the spindle apparatus no longer fit inside the nucleus.

Microtubules attach to the chromosomes at specialised structures, called kinetochores (see insert in fig. 12.7). In ultra-thin sections 2 plates can be seen, which are connected by a fibrous corona. This structure binds to specific DNA-sequences, which are identified by their ability to turn plasmids into artificial chromosomes. The precise structure of the kinetochore has yet to be elucidated.

In prophase kinesin-like motor-proteins bound to the microtubules of one centrosome will move toward the plus-end of microtubules from the other centrosome. As a result, the centrosomes are pushed to opposing sides of the cell. At the same time minus-directed kinesins are responsible for parallel orientation

[1] Do not confuse centromere, centrosome and centriole! The centrosome (or central body) contains the centriole(s) and is a MTOC. The centromere is that part of the chromosome where the two sister-chromatids are attached to each other and where the microtubules bind during mitosis.

of the microtubules. Dynein molecules in the cortex of the cell orient the astral microtubules and help to pull the centrosomes apart. Dynein also holds the microtubules in the centrosome.

Once the nuclear membrane has broken down, the plus-end of kinetochore microtubules rapidly extend and shorten, until they have found a centromere to attach to. If a centromere hits the side of a microtubule, it can slide towards the plus-end, using kinesin-like motors. Once the kinetochore has caped the plus-end of the microtubule, the latter is stabilised against depolymerisation.

When microtubules are attached to kinetochores, they extend or shorten to align chromosomes in the equatorial plate. The exact mechanism is still a matter of research.

Once all chromosomes are bound and correctly aligned, the last mitotic check point is passed and cells enter anaphase. During **anaphase A** (or early anaphase) the kinetochore microtubules shorten by depolymerisation at the plus-end, pulling the chromosomes towards the centrosomes. This process is not energy dependent, so it is not powered by motor proteins.

During **anaphase B** (late anaphase) the centrosomes are pushed further apart by elongation of polar microtubules and pulled toward the cell cortex by astral microtubules. Motor proteins are involved in this process as it can be inhibited by ATP-removal. Sliding forces between polar microtubules are created by plus-directed kinesin, pulling of astral microtubules by minus-directed dynein. Antibodies raised against these molecules can inhibit anaphase B, but not anaphase A.

Cytostatica

Because microtubules are required for mitosis, substances interfering with microtubule formation and dissociation are important anti-cancer drugs. **Vinblastine** (from the Madagascan periwinkle *Vinca roseus* (L.)) leads to depolymerisation of microtubules and to arrest in metaphase. **Taxol** (now called **Paclitaxel** , a compound found in the bark of the pacific yew tree *Taxus brevifolia* NUTT) on the other hand leads to uncontrolled tubulin polymerisation, so that not enough monomers are available to form the mitotic spindle.

Cytogenetics

Colchicine (from the autumn crocus *Colchicium autumnale* L., an alpine flower) is used to treat acute gout, but can also arrest cells in metaphase by tubulin-depolymerisation. Metaphase-cells have their chromosomes fully condensed and aligned at the equatorial plate of the cell. Thus white blood cells or amniotic cells, stimulated into mitosis by **phytohaemagglutinin** and

Colchicine

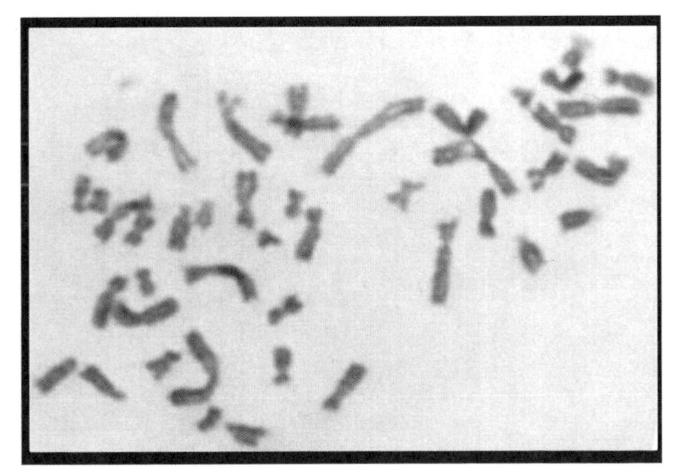

Figure 12.8. The autumn crocus (*Colchicium autumnale* L., top left) produces colchicine (top right), a substance that can depolymerise microtubules. This leads to cell cycle arrest in metaphase, when the chromosomes are condensed, an effect used in cytogenetics. Bottom: Metaphase spread prepared from human white blood cells, which were cultured *in vitro* after stimulation with phytohaemagglutinin.

treated with colchicine can be used to examine chromosomes under the microscope for anomalies (missing, additional or broken chromosomes, exchange of DNA between chromosomes). This is called **cytogenetics** and valuable for the diagnosis of some inherited diseases and some forms of cancer. An example is presented in fig. 12.8.

Scleroderma

Auto-antibodies against kinetochore proteins are frequently found in patients suffering from scleroderma (thickening of the skin due to collagen overproduction, accompanied by atrophy of sebaceous glands). The pathogenesis of scleroderma is unknown, however, there are familial cases.

Exercises

12.1. Each contraction of skeletal muscle requires:

1) increased phosphorylation of myosin.
2) Ca^{2+}-binding to myosin light chains.
3) polymerisation of G-actin to F-actin.
4) Ca^{2+}-binding to troponin C.

12.2. Which of the following statements is/are *correct* about skeletal muscle?

1) If muscle fibres lack ATP, myosin heads are released from the actin filaments.
2) The thick filament is a polymer containing myosin, tropomyosin and Cap Z.
3) During contraction the myosin heads of the thick filament bind covalently to the actin of the thin filament.
4) High sarcoplasmic $[Ca^{2+}]$ is maintained by active transport through a P-type ATPase.

12.3. Which of the following statements is *false*? Actin filaments

a) are produced by the spontaneous polymerisation of G-actin to F-actin.
b) contain bound GTP.
c) from the sperm acrosome pierce the jelly-layer of the egg during fertilisation.

d) are anchored to transmembrane proteins by filamin, ankyrin or spectrin.

e) are stabilised against depolymerisation by phalloidin, the poisson from the "angel of death" mushroom.

12.4. Which of the following statements is *false*? Microtubules

a) start with their minus ends from a central hub.

b) grow at the plus-, and shrink at the minus-end.

c) occur a in typical flagella in 9+2 arrangement.

d) form the "motor ways" for vesicular transport in cells, powered by kinesin and dynein.

e) pull chromosomes toward the cell poles during mitosis.

13

Cell-cell interactions

Metazoans consist of many different cell types, whose functions are closely interlinked. Only together can they perform all the functions that characterise a living organism: Attaining **shape, movement, perception, metabolism and reproduction**.

We will now see how all these cells are knitted into a single organism.

Before we start first a few definitions: If several cells of the same type come together, they form a **tissue**, of which there are 4 basic types: epithelial, connective, muscular and neural. Several different tissues together can form **organs**, which are functionally and anatomically distinct parts of our body.

13.1 Extracellular matrix

Cells secrete a network of proteins and carbohydrates, which is called **extracellular matrix (ECM)**. It has 4 major components:

Proteoglycan form a viscous matrix of carbohydrate chains with many negative charges bound to a protein core. The carbohydrate binds a lot of water, which gives the ECM elasticity and the ability to absorb pressure. **Hyaluronic acid**, a polysaccharide, also serves this function.

Collagen is a fibrous protein, which provides resilience against pulling forces. Thus it acts like glass fibers in a polyester composite material.

Elastin forms elastic fibers, which allow tissue to elastically give to pulling forces, and to regain their original shape once the force stops. Arteries are a typical example, they expand when the heart contracts and blood pressure rises, when the blood pressure drops, they shrink again.

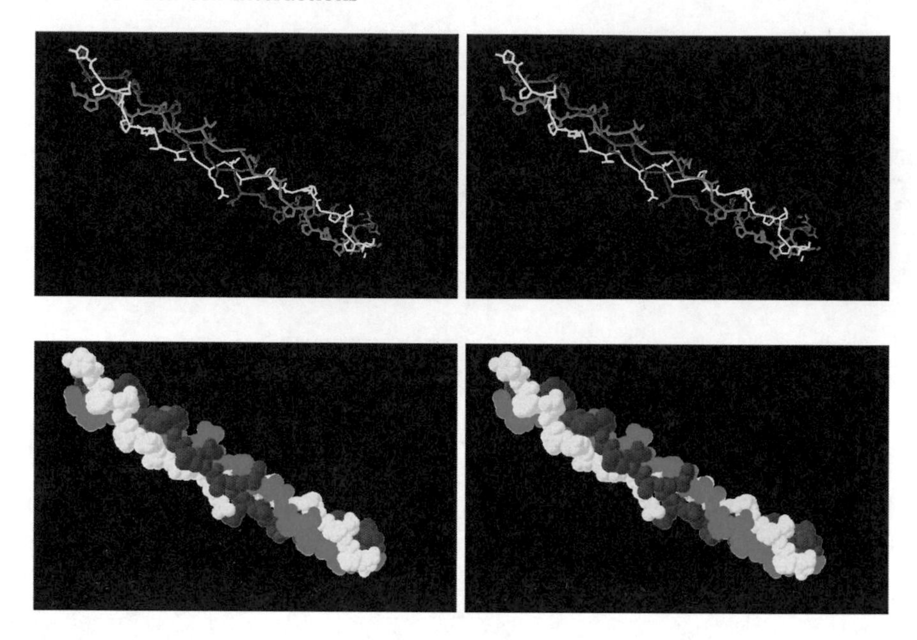

Figure 13.1. The collagen triple-helix (PDB-code 1BKV) as ball-and-stick and VAN DER WAALS model, both as stereo views.

Multiadhesive proteins link collagen, proteoglycans and cells together. The most prevalent of these proteins is laminin[1], a heterotrimeric protein of 820 kDa.

These components are mixed in different proportions, depending on what the purpose of a tissue is: **Tendons** have a high collagen content to transmit forces between muscle and bone. **Cartilages** on the other hand have a high proteoglycan content, thus they can cushion the shocks caused in movement.

Differentiation of cells is controlled in part by their ECM environment. Cell receptors interacting with ECM components stimulate the expression of appropriate genes. Additionally the ECM can bind and slowly release hormones.

13.1.1 Collagen

Collagen is the most abundant protein in animals. There are 16 known isoforms, but almost 90 % of all collagen in our bodies belongs to the isoforms I, II and III. These form fibrils. Type IV forms 2D-networks, for example in the basal lamina. The remaining isoforms cross-link collagen fibrils with each other or their environment.

[1] do not confuse laminin in the extracellular matrix with lamin in the nucleus.

Fibrous collagens have a distinct structure (see fig. 13.1). Collagen is a triple-helical coiled-coil of three protein chains, each exactly 1050 amino acids long. The core consists of repeating Gly-X-Pro/Hy-Pro motives. X can be any amino acid, except Trp. These amino acids have specific functions:

Glycine is the smallest amino acid, with just a hydrogen as side chain. Only Gly allows the required tight packing of the collagen triple helix. Mutation of just a single Gly residue in the protein leads to a non-functional collagen, resulting in serious inherited defects like *Osteogenesis imperfecta* (defect in bone formation because of mutated collagen I) or EHLERS-DANLOS-syndrome (weak joints caused by mutation in collagen-II). These diseases are potentially fatal.

Proline The fixed sharp angle of the C-N bond in the ring of proline stabilises the triple helix. Note that Pro destabilises most other ordered secondary structures, like β-strands. The tight wrapping of the chains around each other causes the high tensile strength of collagen, which is actually higher than that of steel (by weight and even by cross-section).

Collagen fibrils appear striped under the electron microscope, this pattern is caused by the gaps between adjacent triple helices (see fig. 13.2). Dark stripes appear where many gaps come together, these are filled with uranyl acetate during preparation. Parts with few gaps exclude uranyl acetate and appear translucent.

Medical aspects

Correct triple helix formation (see fig. 13.2) depends critically on the post-translational hydroxylation of proline to hydroxyproline by a hydroxylase in the ER-membrane. If this reaction is not carried out, triple helix dissociation temperature drops below 20 °C, i.e. the helices are unstable at body temperature. Pro-hydroxylase contains a Fe^{2+}-centre which can be oxidised to inactive Fe^{3+} by a side reaction. Hy-Pro formation therefore requires the presence of **Vitamin C (ascorbic acid)** to reduce Fe^{3+} to Fe^{2+} and regenerate the enzyme. Otherwise weakening of connective tissue by lack of properly formed collagen will lead to **scurvy**.

Collagen I, the major collagen found in bone, tendons, skin and interstitial tissue is formed from 2 $\alpha 1(\mathrm{I})$ and one $\alpha 2(\mathrm{I})$ chains. As mentioned, mutations in one of the genes for this protein lead to the inherited disease *Osteogenesis imperfecta*. Mutations in $\alpha 1(\mathrm{I})$ tend to be much more serious than those in $\alpha 2(\mathrm{I})$. This is easy to understand: In a heterozygous individual 50% of all collagen molecules will contain a correct copy of $\alpha 2(\mathrm{I})$, if the mutation is in this protein. If the mutation is in $\alpha 1(\mathrm{I})$, only 25% of all collagen molecules produced will be functional, 50 % will have one, and the remaining 25 % two non-functional copies. This form of inheritance is called **pseudo-dominant**.

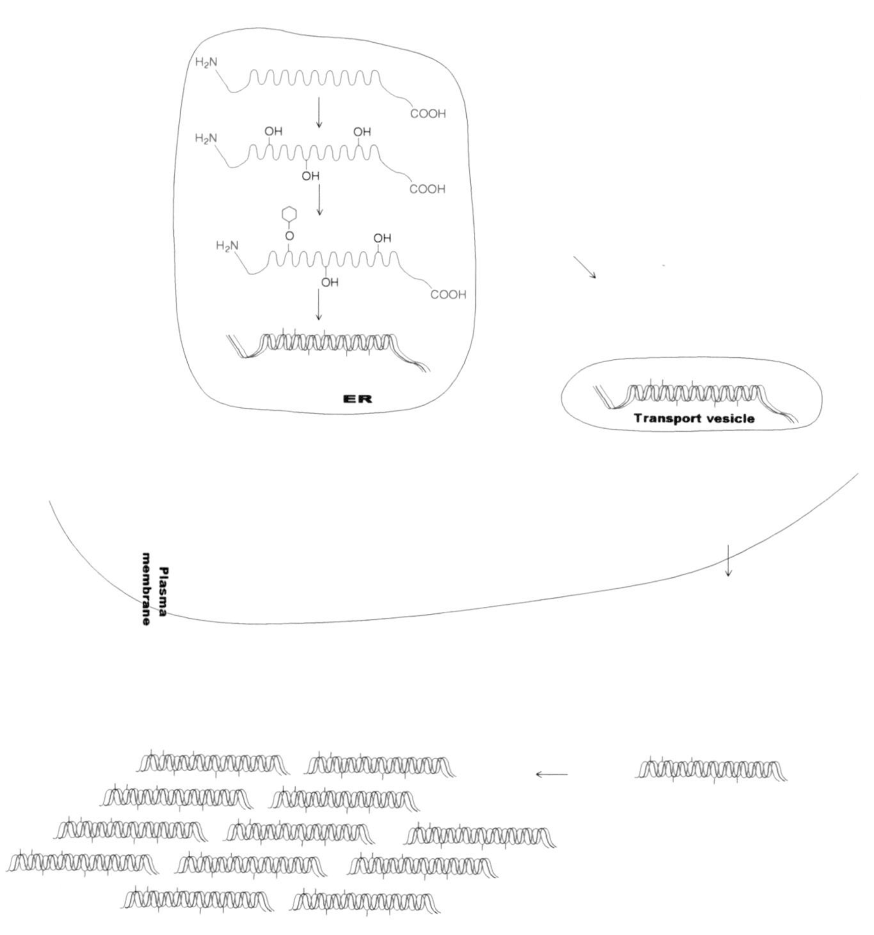

Figure 13.2. Biosynthesis of collagen. Pre-pro-collagen is transported into the ER, some Lys and Pro residues are hydroxylated. Some of the Hy-Lys residues are glycosylated, then the single collagen chains form triple helices. This is aided by disulphide bonds which form in the C-terminal pre-peptide, i.e. formation of triple helix starts at the C-terminus. The pre-pro-collagen is clipped to pro-collagen and exported across the plasma membrane. In the extracellular space the N- and C-terminal pro-sequences are cut off, allowing the tropo-collagen molecules to associate into fibrils, 67 nm thick cables. In a final step, cross-links are introduced between Lys and Hy-Lys residues (not shown), both between the three chains of a collagen molecule and between molecules. Note the stripes in the developing collagen molecule.

Cell-collagen interactions

Cells can change the morphology of the collagen they secrete by exerting pulling forces on the molecules, this leads to a tighter packing. If two explantates of embryonal fibroblasts are placed apart on a random collagen matrix, they will organise this matrix into a tight band between them. Along this band cells will move between the explantates. This is a model experiment demonstrating the origin of connective tissue during embryogenesis.

13.1.2 Elastin

Elastin is a Gly- and Pro-rich protein of 750 amino acids. It forms elastic fibers which are about 5 times as extensible as a rubber tape of equal cross-section. Unlike collagen, elastin is non-glycosylated, contains no Hy-Lys and very little Hy-Pro.

Elastin is secreted into folds in the cell surface, there elastic fibers are formed by cross-linking between Lys-residues (see fig. 13.3). The elastin molecules are normally curled, when a force is applied, the molecules straighten, gaining in length. The cross-links between molecules prevent them from sliding past each other, so they can regain their original shape when the force is released.

The elastin core in elastic fibers is surrounded by microfibriles made from glycoproteins like **fibrillin**. During embryogenesis microfibriles are formed first, elastin is introduced later. Apparently the microfibriles are required for correct organisation of elastin molecules.

MARFAN-syndrome is caused by a mutation in the fibrillin gene, which results in weak elastic fibers. In the worst cases patients die from a ruptured aorta. The most prominent sufferer from this disease was probably ABRAHAM LINCOLN.

13.2 Cell adhesion molecules

In order to function in an organism, cells need to specifically interact with each other. This is the purpose of several proteins, which are collectively known as **cell adhesion molecule (CAM)**. Some of these are found in electron-microscopically visible structures, the **cell junctions** (see fig. 13.4).

A number of important transmembrane proteins are involved in establishing cell-cell interactions:

Homophilic interactions between proteins of the same type:

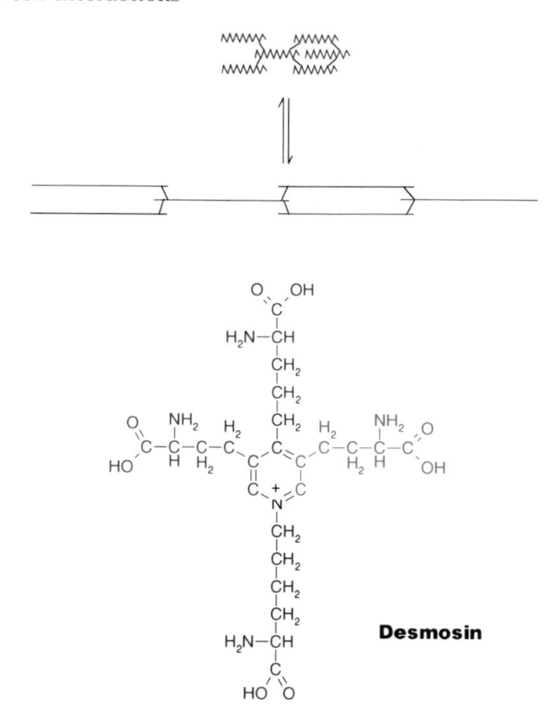

Figure 13.3. Elastin is a component of elastic fibres, which give tissue stretchability. The molecules are coiled up in the relaxed state, when stretched they become more linear. Cross-links between molecules prevent them from sliding past each other under force. These cross-links are made by the unusual amino acid **desmosin**, which is post-translationally synthesised from 4 lysine residues (marked in different colours).

Cadherins consist of 5 extracellular domains, with Ca-binding sites between them. The N-terminal domain dimerises with a second cadherin in the same cell (Ca-dependent), these dimers interact with cadherin-dimers on the other cell. More than 40 different types of cadherins are known, the best studied are:

E-cadherin is the most abundant cadherin and found in the pre-implantation embryo and in non-neuronal epithelia. It occurs in the lateral part of the cell membrane and holds the epithelia together. In cell culture experiments the addition of antibodies against E-cadherin or (cheaper!) removal of Ca causes cells to detach from each other.

P-cadherin is found in the trophoblast and required for nidation.

N-cadherin is found in neuronal tissue, lens, heart and skeletal muscle.

Cadherins are type 1 membrane proteins (see fig. 16.2 on page 233). Changes in cadherin expression often accompany metastasis and other pathological processes. Cadherins prefer to interact with other cadherins of the same type.

Ig-superfamily CAMs, also called **nerve-cell adhesion molecules (N-CAMs)**, have type III fibronectin domains followed by several Ig-domains (see fig. 10.2 on page 142). They cause **Ca-independent** cell-adhesion. They are all encoded by a single gene, but alternative splicing generates 3 isoforms. N-CAM-180 and -140 are type 1 transmembrane proteins, N-CAM-120 has lost the transmembrane domain and is anchored by glycosyl phosphatidylinositol (GPI) instead (type 6 membrane protein, see fig. 16.2 on page 233). N-CAMs in embryos contain up to 25% sialic acid, the many negative charges weaken N-CAM/N-CAM interactions, thus cell-cell interactions in embryos can be broken again during morphogenesis. Adult N-CAMs contain much less sialic acid. N-CAMs are required for nerve, glia and muscle cell differentiation.

Heterophilic interactions between different proteins:

Mucin-like CAMs are rich in carbohydrates and bind to **selectins**, which have a lectin (sugar-binding) domain. Selectins are important for **leucocyte**/tissue-interactions. When leucocytes are to leave the blood stream their **sialyl** LEWIS-**x-antigen** attaches to P-selectin on the blood vessel wall. The selectin is normally hidden in intracellular vesicles of the endothelial cells, but upon reception of **inflammatory mediators** they undergo rapid exocytosis. They then trap leucocytes passing by, but because the interaction is weak, leucocytes are slowed down rather than stopped. Thus leucocytes "roll" along the endothel.

Integrins are heterodimers with an α- (17 known isoforms) and a β-subunit (8 known isoforms). They bind mostly to extracellular matrix proteins like fibronectin or laminin. Integrin expression is also stimulated in leucocytes when **platelet activating factor (PAF)**, a paracrin hormone from blood vessel endothelial cells, is received. Integrin then binds to I-CAM-I and I-CAM-II on the endothel. This interaction is much tighter than that afforded by selectin, the leucocytes spread out on the endothel and start moving into the tissue. Deficiency in β-integrin expression prevents leucocytes from leaving the blood vessels, resulting in an inborn error, **leucocyte adhesion deficiency**.

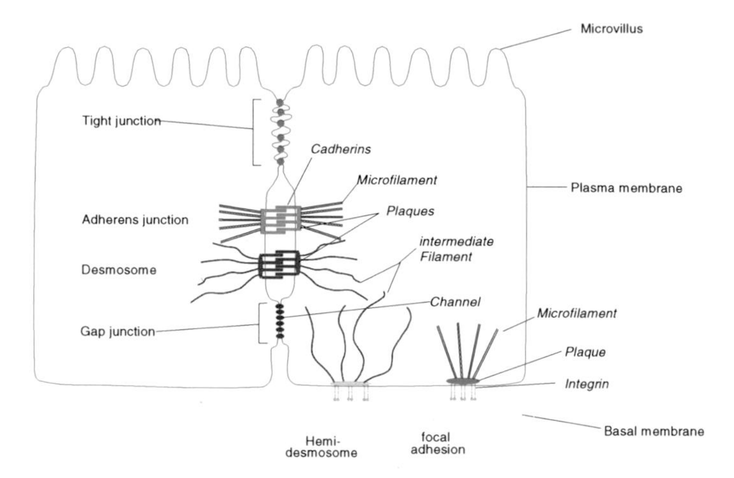

Figure 13.4. Cell-cell and cell-matrix junctions

13.3 Junctions

Tight junctions connect epithelial cells and prevent passage of fluid from one side of the epithelium to the other. For example, they separate the intestinal content from blood.

Gap-junction are distributed on the lateral surfaces of cells, they allow the exchange of small molecules (nucleotides, second messengers) between neighboring cells. Thus they help to integrate the metabolism of the cells in a tissue. In EM the gap junctions look like close juxtaposition of cell membranes, with the narrow gap between them filled by particles. These "particles" are actually water-filled channels, linking the cytosols. The pore size of these channels is about 12 Å, this is sufficient to allow free passage of molecules up to 1200 Da, and to completely prevent passage of molecules greater than 2000 Da.

High cytosolic Ca-concentrations cause these channels to close. Remember that in a living cell [Ca^{2+}] is very low (below 0.1 μM), whilst extracellular [Ca^{2+}] is much higher (1 – 2 mM). If a cell ruptures, cytosolic [Ca^{2+}] will increase and shut down the gap junction channels. Thus neighboring cells are protected from the leak.

Cell-cell junctions have a structural role, they hold cells together by forming a bridge between their cytoskeletons. In cell-cell junctions the plasma membranes of the two cells run parallel with a distance of only 15 – 20 nm. The close spacing is achieved by cadherins, adaptor proteins link them to the cell skeleton. There are 2 types of cell-cell junction:

Adherens junctions Cadherin E is located in a continuous band just below the tight junctions. These molecules are cross-linked to actin filaments by α- and β-catenins. This horizontal band of junctions links all cells within a layer of tissue and distributes forces between them.

Desmosomes Cadherins **desmocollin** and **desmoglein** are cross-linked to **keratin** filaments by adaptor proteins like **placoglobin**. These form a cytoplasmic plaque which is visible in EM. Because the keratin IF reaches through the cells, desmosomes can distribute forces between different layers of a tissue. *Pemphigus vulgaris* is an autoimmune disease caused by autoantibodies against desmoglein, one of the cadherins in desmosomes. This leads to disruption of desmosomes and thus to blistering of skin and mucus membranes.

Cell-matrix junctions Like cell-cell junctions, cell-matrix junction maintain the integrity of a tissue. The most important cell membrane proteins involved are **integrins**. The combination of α- and β-isoforms determines the binding specificity of integrins. Most integrin isoforms are expressed in several cell types and most cell types express several integrins. However, some isoforms are specific to certain cell types, the α_L, β_2-type in **leucocytes**, which makes contact with blood vessel endothelial cells (rather than extracellular matrix) has already been discussed.

Another important example is the α_{IIb}, β_3-type found in **platelets**. Normally it is inactive, but after stimulation of platelets it binds fibrinogen, fibronectin, VON WILLEBRAND-factor and other blood clot proteins. We do not yet know how this is regulated.

α_4, β_1-integrin keeps haematopoietic cells in the **bone marrow**, because it binds to fibronectin, which is secreted into the extracellular matrix by stromal cells. It also binds to the stromal cells directly, via V-CAM-I. At a late stage of their differentiation haematopoietic cells reduce integrin expression, so they may enter the blood stream.

Integrin/cell matrix interaction typically have low affinities (µM), but because of the large number of interactions involved, cells bind firmly into their matrix. The weak individual interactions however are important for moving cells. Affinity appears to be subject to regulation by the cell, the mechanism is unclear. We distinguish 2 types of cell-matrix interactions:

focal adhesions are used by moving cells to temporarily attach themselves to the matrix. They consist of a cluster of integrins, which are linked to actin stress fibers by a focal plaque.

hemidesmosomes are found on the basal surface of epithelial cells, they attach the cells to the basal lamina. They consist of α_6, β_4-integrin which links extracellular laminin to keratin IF inside the cell.

Disintegrins

In some cases normally immobile cells need to move, for example keratinocytes in wound healing. This is achieved by peptides, which compete with ECM proteins for binding to cellular integrins. These peptides are called **disintegrins.**

Some snake toxins also contain disintegrins to prevent blood clotting.

Disintegrins may be secreted at the same time as proteases, which dissolve either the extracellular matrix and/or the integrins on the cells surface. Examples are **fibrinogen** and **matrix-specific metalloproteases (MMPs)**.

Exercises

13.1. Posttranslational modifications of which of the following amino acids is important for cell matrix stabilisation: 1) Lysine, 2) Glycine, 3) Proline, 4) Tyrosine

13.2. Explain the clinical symptoms of scurvy from the molecular action of vitamin C.

13.3. Which of the following statements about collagen is *correct*?

1) For the formation of the collagen triple helix every third amino acid needs to be a glycine.
2) Mutations in the $\alpha 1(1)$ chain of collagen (which occurs in two copies) are inherited in a pseudo-dominant manner.
3) Mutations in the single $\alpha 2(1)$ chain of collagen will be inherited in a recessive manner.
4) Collagen triple helix formation requires a helicase activity.

13.4. Which of the following statements is *false* about junctions?

a) Tight junctions prevent the uncontrolled mixing of the fluid on both sides of epithelia.
b) Gap junctions allow the exchange of small molecules (< 1200 Da) between cells.
c) Adherens junctions link the cell membrane to the basal lamina.
d) Desmosomes are connected to the intermediary filament and distribute forces throughout the tissue.
e) Disintegrins break the interaction between hemidesmosomes and the extracellular matrix.

14

Aiding in folding: molecular chaperones and chaperonins

C. ANFINSEN could show that denatured (boiled or urea-treated) RNase could resume its correct folding once the temperature or urea concentration had been lowered [4]. He postulated that all proteins contain in their primary structure (amino acid sequence) the complete information which determines their secondary and tertiary structure (folding in 3-dimensional space). This hypothesis has proven fruitful (NOBEL Price 1972).

The correct conformation of a protein is usually the one with the **lowest free energy**, and given enough time, all proteins will eventually assume their correct fold. However, many proteins would require a very long time, because they can become trapped in miss-folded states. These states have a free energy which is higher than that of the native state, but lower than that of all neighboring conformations. Unfolding such miss-folded proteins therefore requires a high activation energy and hence is a slow process.

Additionally, inside a cell proteins are packed to very high densities, this increases the risk for inappropriate interactions between proteins, which would result in protein aggregation and precipitation. **Aggregation** is distinguished from **oligomer-formation** during protein assembly by the irregular, non-specific and disordered nature of protein-protein interactions, aggregates are devoid of biological function. Protein aggregation is involved in several diseases in man, like prion diseases (see page 131), ALZHEIMER, HUNTINGTON, PARKINSON and ALEXANDERS disease, sickle cell anaemia, cystic fibrosis, hepatic cirrhosis, desmin-related myopathy, cataract and diffuse LEWIS body disease.

In *E. coli* cytosol the protein concentration is 300–400 mg/ml. Assuming an average molecular weight for proteins of 50 kDa, this is a molar concentration of 6–8 µM. Since one mol contains $N_A = 6.022 \times 10^{23}$ molecules and 1000 l is equivalent to 1 m^3, this is equivalent to

361–482 × 10^{21} molecules/m^3, and each molecule has 2.1–2.8 × 10^{-24} m^3 space available. This is a cube of 12.8–14.1 nm. The unit cell volume of proteins is 1.68–3.53 $Å^3$/Da [49], and 1 $Å^3$ is equivalent to 1 × 10^{-30} m^3. Thus a protein of 50 kDa would have a volume of 84–177 × 10^{-27} m^3, equivalent to a radius of 2.72–3.48 nm. In other words, cytosol can be viewed as an assembly of protein spheres of 5.5–7.0 nm diameter with on average about 5.8–8.7 nm space between them. This space is filled with water, minerals and metabolites.

Nature has solved these problems by inventing two classes of proteins, called **molecular chaperones** and **molecular chaperonins** [24, 79].

Chaperones, "like their human counterparts, prevent unfavourable interactions between proteins during critical periods of their existence" (U. HARTL). Thus they prevent miss-folding and aggregation, but they can not unfold miss-folded proteins.

Chaperonins on the other hand will actively unfold miss-folded proteins, using energy from ATP-hydrolysis to break offending bonds.

It is important to remember that neither chaperones nor chaperonins can actively fold proteins, folding is spontaneous according to the ANFINSEN-hypothesis. Both chaperones and chaperonins require energy in the form of ATP to function. Chaperones and chaperonins may be constitutively expressed (always present) or stress-regulated (so called **heat shock proteins**, even though cold, starvation, poisoning and other forms of stress can also activate their synthesis). An interesting way to regulate the expression of some heat shock proteins has recently been described in *E. coli* and *Bradyrhizobium japonicum* [11]: the mRNA forms at low temperatures a secondary structure which masks its ribosome binding site (start codon and SHINE-DALGARNO-sequence). At elevated temperatures the secondary structure melts and translation becomes possible. The same mechanism is also used to regulate the expression of virulence factors in certain human-pathogenic bacteria like *Yersinia pestis* and *Listeria monocytogenes*. A change of environmental temperature to the relatively high 37 °C indicates successful infection of the host and is followed by the synthesis of proteins required for pathogen-host interaction. Below 30 °C these proteins are not expressed.

14.1 Hsp70 is an example for molecular chaperones

In the eukaryotes four isoforms of Hsp70 are found (see fig. 14.1 for an evolutionary tree of these proteins):

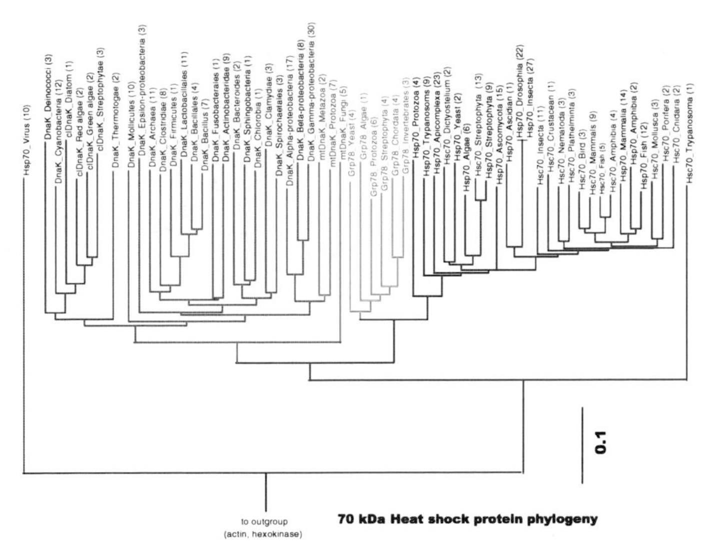

Figure 14.1. Evolutionary tree of the 70 kDa heat shock proteins (Hsp70). The tree has two major branches, a eukaryotic and a prokaryotic one. Prokaryotic Hsp70s are often called **DnaK**. Each of these major branches has several recognisable sub-branches. Mitochondrial and plastid versions of DnaK (magenta and green, respectively) can be distinguished from their bacterial homologues (orange). The eukaryotic sub-tree contains not only proteins expressed in stress-situations (**Hsp70** proper, blue), but also a cytosolic (red, Hsc70, 70 kDa heat shock cognate) and an ER-version (cyan, **Grp78^{BiP}**, 78 kDa glucose regulated protein/Ig heavy chain binding protein) which are constitutively expressed. Certain plant viruses have their own copy of Hsp70 (brown). The tree is rooted by inclusion of homologue proteins (actin, hexokinase) as outgroup. The number of proteins included in each group is given in brackets. Sequences were obtained from ExPASy (`http://au.expasy.org/`, clustering was performed with ClustalX (`http://www-igbmc.u-strasbg.fr/BioInfo/`).

Hsp70 proper which is expressed in the cytosol, but only in stress situations.

Hsc70 (70 kDa heat shock cognate protein) is also cytosolic, but constitutively (always) expressed.

Grp78^{BiP} (78 kDa glucose regulated protein/immunoglobulin heavy chain binding protein) is found in the ER (see chapter 16.1 on page 233) and has similar functions as Hsc70. It is always expressed, but its concentration rises in stress situations.

DnaK is the prokaryotic equivalent of Hsc70, found also in plastids and mitochondria.

Of these, Hsc70 is responsible for:

- binding to newly synthesised proteins as they come from the ribosomes to prevent them from miss-folding (see fig. 15.3 on page 228). About 15–20 %

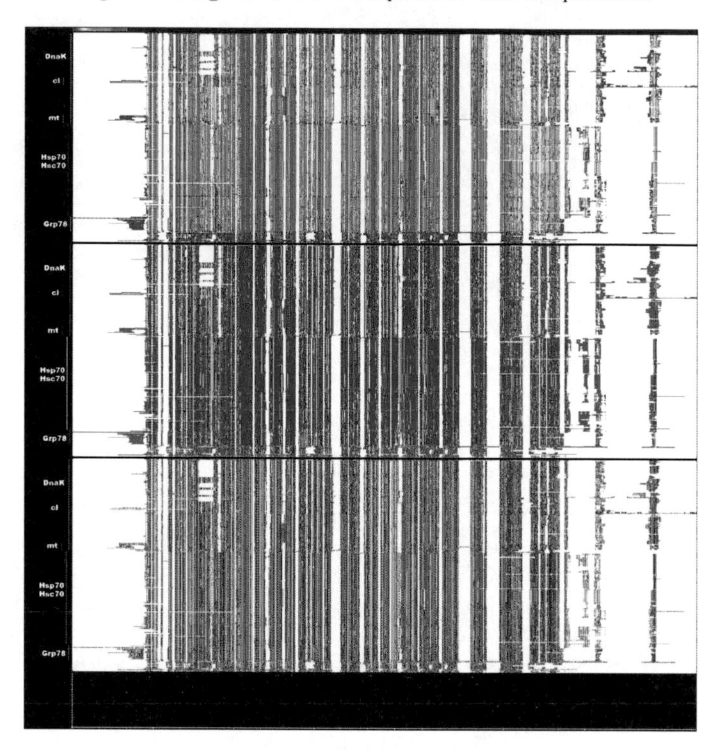

Figure 14.2. Homology plot of the 70 kDa heat shock proteins. Sequences were aligned into a matrix as described for fig. 14.1. Each row of the plot describes a protein sequence, each column a position of the consensus sequence. The colour depends on the amino acid present. In the top diagram pixels are coded by the frequency of the amino acid in that position (red rare, purple frequent), in the second by identity with the consensus sequence (red: identity, yellow: conservative substitution, blue: non-conservative substitution) and in the third by Shapely colours. In the bottom diagram the identity and similarity are plotted. White denotes gaps in all diagrams.Regions of high conservation are mainly found in the N-terminal ATPase domain, the protein-binding C-terminal domain is less conserved and presumably determines substrate specificity. Various clusters of protein sequences can be identified, in particular the separation between prokaryotic and eukaryotic sequences is apparent.

of all proteins produced by the cell transiently bind Hsc70, in pulse chase experiments it has been demonstrated that most protein/Hsc70 complexes dissociate within 10 min and essentially all within 30 min [24]. Interaction of the nascent protein with Hsc70 may be required in particular for apoproteins, which need to be maintained in a state that allows binding of cofactors like iron-sulphur centres.

- binding to actin and tubulin during their release from the minus end of microfilaments and microtubules and their delivery to the growing (plus) ends (see fig. 11.2 on page 176).
- removal of clathrin from coated vesicles. This allows them to fuse with endosomes during endocytosis (see section 16.4 on page 231).
- import of nascent proteins into organelles. Hsc70 and its prokaryotic counterpart, DnaK, work together as "molecular ratchet" (see page 227).
- cross-linking of glutamate decarboxylase (which makes the neurotransmitter GABA) to synaptic vesicles. This allows efficient loading of the vesicle with neurotransmitter.
- recognition of terminally miss-folded proteins. One of the co-chaperones, Chip, is a ubiquitin ligase which prepares such damaged proteins for degradation in the proteasome. Hsp70 and Hsc70 are also involved in the binding of ubiquitinated proteins to the proteasome, with the co-chaperone Bag-1 (see 159ff and [19]).
- shuttling of peptides from the proteasome to the ER (together with other chaperones), where they are taken up by the Tap1/Tap2 transporter and bind to the MHC-I for display at the cell surface (see fig. 10.14 on page 162).
- co-operation with other heat shock proteins as required for their function (see for example fig. 14.3). Thus Hsc70 forms a central part of the folding machine in our cells; accepting proteins for example from small heat-shock proteins and delivering them to the GroEL/GroES chaperonin.
- maintaining the apo-proteins of [Fe-S]-dependent cytosolic enzymes in a folding competent state, until the [Fe-S]-cluster (synthesised in the mitochondria) has been bound.

The reaction mechanism of Hsc70 is probably the same in all these diverse functions. When Hsc70 binds ATP, its protein binding site opens. Any proteins bound are released, new proteins can bind freely. Once the ATP is hydrolysed however, the protein binding site closes, any bound protein is trapped until the ADP has been exchanged for fresh ATP.

Exchange of ADP for ATP and the hydrolysis of ATP is probably controlled by **protein cofactors**. In the bacterial homologue of Hsc70, DnaK, these cofactors have been identified: **DnaJ** binding to Hsc70 stimulates the hydrolysis of ATP, **GrpE** on the other hand stimulates the exchange of ADP for ATP. Eukaryotic versions of these proteins have been isolated. DnaJ-like proteins (for example Hsp40 or auxillin) have in common the so called J-domain, a motive of 4 α-helices with the conserved sequence HPD in the loop between helix 2 and 3. The J-domain is often, but not always followed by a Gly/Phe-rich region and 4 Cys-repeats that together form 2 Zn-binding sites. This conserved stretch is followed by a non-conserved C-terminus which is responsible for substrate protein and Hsp70 recognition. ADP/ATP exchange factors (like Bag1 and HspBP1) have also been identified in eukaryotes, in addition proteins that seem to stabilise the Hsp70/ADP/substrate complex. However, we do not yet understand the regulation of Hsp70 function in eukaryotes.

Note that Hsc70 is involved in many functions that involve **directed transport** of proteins (clathrin from the uncoated vesicle to the growing pit, peptides from the proteasome to the ER, actin from the minus to the plus end of microfilaments and so on). If DnaJ-like cofactors were located at the source, and GrpE-like proteins at the destination, this vectorial action of Hsc70 could be explained.

Apart from the constitutively expressed Hsc70, cells under stress (heat, cold, lack of food, presence of certain poisons) can make also the stress-regulated **70 kDa heat-shock protein (Hsp70)**, to help maintain protein structure in the cell and to clear away any debris. For example Hsp70 binds to polyubiquitinated proteins (with some E_3-ubiquitin ligases (like Chip) acting as ATPase stimulating factors) and delivers them to the 26 S proteasome (see page 159 for a description). There the Bag1 protein allows recognition of the Hsp70/protein complex by the proteasome and acts as ATP/ADP exchange factor.

14.2 Other heat-shock proteins also have chaperone activity

There are several other families of heat-shock proteins, in many cases we know little about them. The following examples, Hsp90 and sHsp, were selected because some coarse picture of their function and reaction mechanism is emerging:

14.2.1 Hsp90

Hsp90 acts as a homodimer and is involved not so much in the *de-novo* folding of proteins but in several regulatory processes inside the cell by binding to key proteins like steroid hormone receptor (SHR, see fig. 14.3) or src-kinase. It usually acts together with Hsc70 and several other chaperone proteins. Still unclear is to what extend Hsp90 is involved in the cellular stress response.

Many details of its reaction cycle, and the function of several of its co-chaperones, are still unclear. These other proteins often have a characteristic binding site for Hsp90 and substrate proteins, called the **tetratricopeptide repeat**, a couple of helices which form a grove into which substrate peptides

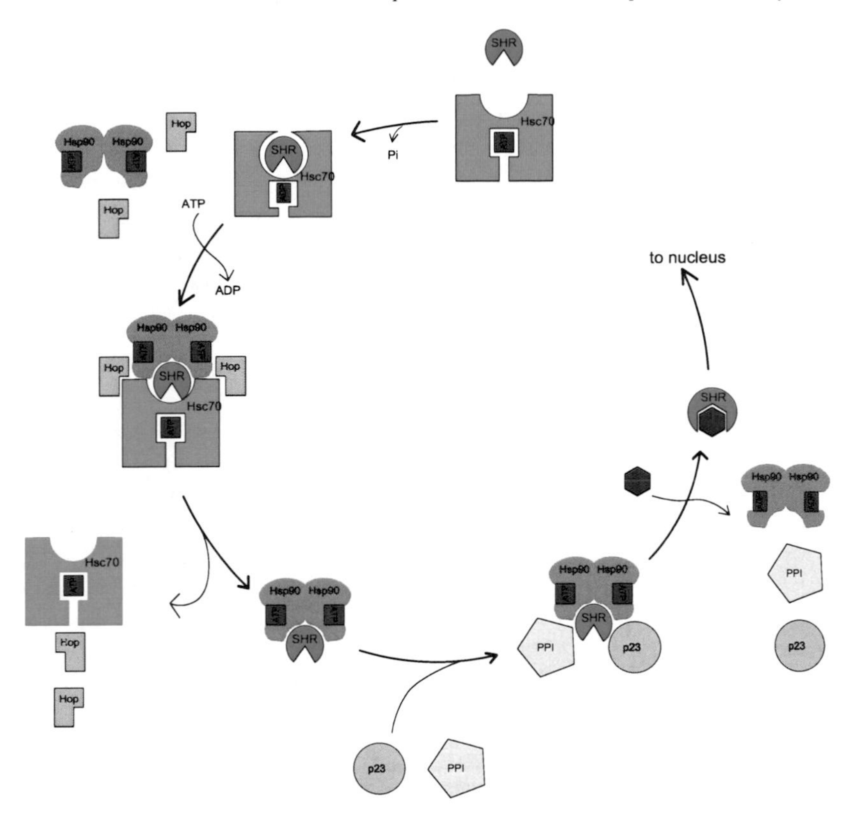

Figure 14.3. Cooperation between Hsc70 and Hsp90 is required to keep the **SHR** in a binding-competent state. Steroid hormones are very hydrophobic, they passively diffuse through the plasma membrane into the cytosol, where they are bound by SHR. The resulting complex is transported into the nucleus, where it regulates the transcription of certain genes. In order to bind the hydrophobic steroids, SHR needs exposed hydrophobic patches which, given the high protein concentration inside a cell, might lead to aggregation. Also, bound steroid is required to keep the free receptor in its native conformation. Hsc70-ATP binds empty SHR (either newly translated or after it has lost previously bound steroid). Hydrolysis of the ATP bound to Hsc70 leads to closing of its protein binding site, trapping the bound SHR. This complex binds Hsp90 and a co-chaperone, $\mathbf{p60^{Hop}}$, which stimulates ADP/ATP exchange in Hsc70. Thus its protein binding site opens and the SHR is transferred to Hsp90. Hsp90 not only keeps SHR in a binding competent state, but also acts as a lath, where other proteins can bind and perform repair work on SHR. Examples include **PPI**, which catalyses the conversion between the *cis-* and *trans-*conformation of proline residues in the protein chain. Also involved is a protein of unknown function, **p23**. Binding of steroid to SHR leads to its dissociation from Hsp90, during this process Hsp90-bound ATP is hydrolysed.

can bind in extended conformation. One example of such proteins is a family of Hsp90-dependent **peptidyl prolyl *cis/trans*-isomeras (PPI)**, which catalyse the conversion between the *cis*- and *trans*-conformation of proline residues in the protein chain (see fig. 2.1 on page 14). Thus Hsp90 appears somewhat like a lath, on which different tools (co-chaperones) can be used to work on substrate proteins.

In fig. 14.3 interaction of Hsp90 and the steroid hormone receptor is explained in detail. Like other chaperones, Hsp90 is an ATPase, mutations that prevent ATP hydrolysis destroy the chaperone function of Hsp90.

14.2.2 Small heat-shock proteins

In the presence of substrate proteins small heat-shock proteins (sHsp) form globular oligomers with 12–42 subunits, these oligomers have a central cavity lined with hydrophobic amino acids. Thus they can bind considerable amounts of substrate proteins, substrate binding and oligomer formation is a highly co-operative process. sHsp can not only prevent the precipitation of miss-folded proteins, but to some extend even solubilise aggregates. However, there seems no set pathway for releasing bound proteins, some sHsps do not even bind ATP. Instead, substrate proteins are transferred to Hsc70 or Hsp70. Little is known about the reaction cycle of small heat-shock proteins.

The most well known sHsp is the **α-crystallin** found in our eye-lenses [38]. Eye lenses are made from fibre cells, these cells are added to the outside of the lens throughout life. Once made, the fibre cells stop protein turnover and there is no exchange of proteins between cells. Thus the proteins in the core of the lens were made during embryogenesis. Protein concentration in fibre cells is very high (450 mg/ml) to achieve a high refractive index. Any protein unfolding would result in aggregation, the aggregate would scatter light, such a particle is called a **cataract**.

To protect the proteins if not from unfolding so at least from aggregation during a life span of several decades almost 90 % of the protein in fibre cells is α-crystallin, which comes in 2 isoforms, α_A (75 %) and α_B (25 %). α_B-crystallin (encoded on chromosomes 11) is found in most cells of the body, α_A-crystallin (encoded on chromosome 21) only in the lens.

After birth most of the α-crystallin in the lens is water soluble, with increasing age more and more is in the water-insoluble fraction (that is, bound to other proteins), beyond age 40 little soluble α-crystallin is found. Bound proteins include β- and γ-crystallins, house keeping enzymes, cytoskeleton and proteins of the plasma membrane.

α-crystallin subunits have a molecular weight of about 20 kDa, but they form polymers of 15–50 subunits, which in cryo-EM show a central cavity. Because of the large, irregular oligomers no X-ray structures are available. Phosphorylation of α-crystallin seems to reduce the size of the oligomers, functional significance unclear.

Although α-crystallin can bind ATP there is no ATPase activity, and the function of ATP-binding is not known.

Several mutations in α-crystallin have been linked to inherited cataracts, including R120G in α_B- and R116C in α_A-crystallin (desmin related myopathy, the sensitivity of crystallins to proteolysis is increased). In some cancers and some infections (both viral and bacterial) expression of α_B-crystallin is up-regulated, to ease cell division or pathogen multiplication, respectively. This also results in immune system stimulation, leading to hyperimmunity but, at the down-side, also to autoimmunity. Pharmaceuticals that increase sHsp-expression as adjuvant for immunisation, or that decrease sHsp-function to suppress autoimmunity are under development.

14.3 The GroEL/GroES-foldosome is an example for molecular chaperonins

In bacteria, the GroES/GroEL foldosome (the name indicates that mutations in these proteins inhibit the growth of certain bacteriophages) consists of 2 beakers of 45 Å inner diameter, each formed by 7 GroES (Hsc60) molecules. One of these beakers is covered by a lid, which is formed by 7 GroEL (Hsc10) molecules. This is called the *cis*-ring, all its GroES-subunits have ATP bound. The other beaker is called the *trans*-ring, it does not have a GroEL-lid and all its GroES-subunits contain bound ADP. In thermophilic archaea like *Thermococcus sp.* the chaperonin is a homo-octamere with build-in lid, apparently to increase stability against thermal dissociation. In eukaryotes a similar complex, called **T**ailless complex polypeptide 1 **Ri**ng **C**omplex (TRiC) exists, which is a hetero-oligomer of 8 different subunits. It is assumed that function and mechanism of the various chaperonins are similar.

Each GroES-subunit consists of 3 domains, called equatorial, intermediate and apical (see fig. 14.4 and 14.5). The equatorial domain contains the nucleotide binding site, the intermediate domain binds the substrate protein and the apical domain binds GroEL.

The two GroES beakers go through the same reaction sequence, but with a phase difference of 180 °. A miss-folded protein (indicated by W in fig. 14.4) is bound in the cavern of the *trans*-ring. Each of the 7 GroES-molecules forming the ring co-operatively exchanges an ATP-molecule for its bound ADP. This results in a change of conformation that leads to the unfolding of the substrate protein (the unfolded substrate is indicated by a C). By this conformational change what was the *trans*-ring now becomes the *cis*-ring.

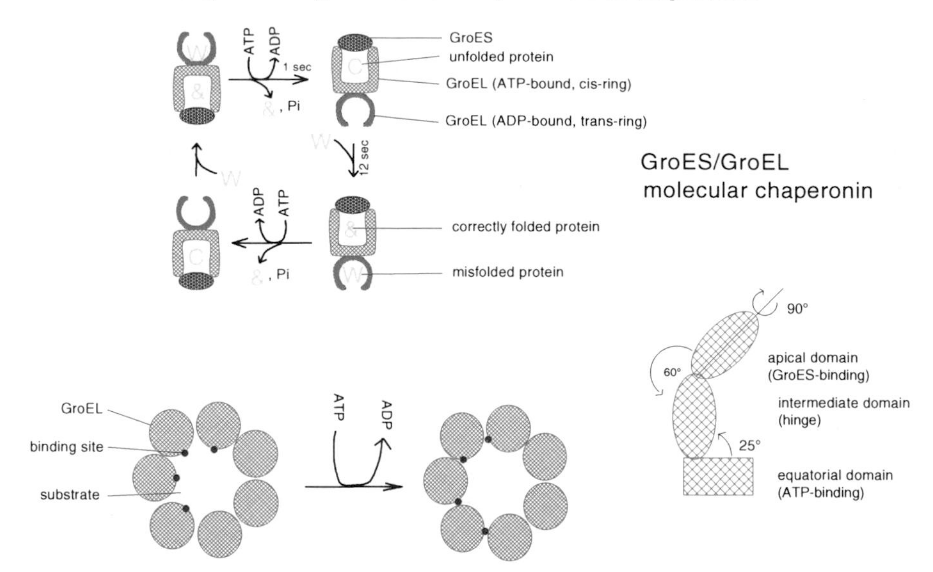

Figure 14.4. Top: The foldosome GroEL/GroES is a typical chaperonin. It can unfold miss-folded proteins (W) and provides a protected space for unfolded proteins (C) to achieve folding (&). **Bottom**: X-ray crystallography studies indicate that ATP binding results in considerable conformational changes which cause an increased distance and lower availability of hydrophobic protein binding patches. Thus mis-folded proteins can be actively unfolded and then released into the cavern of GroEL, where they can try to refold.

The *cis*-ring is covered by a GroEL-hat. Thus the substrate protein is enclosed in a protected space (ANFINSEN-cage), where it has time to fold correctly under „**infinite dilution**“. Note that the GroES/GroEL foldosome does not catalyse folding, it only prevents access of other proteins, which might aggregate with the substrate. Additionally, enclosure of the protein in a small space changes the thermodynamics of folding [75]. Correct folding is achieved by the substrate protein alone, in line with the ANFINSEN-hypothesis.

Once the substrate protein has achieved correct folding (& in the drawing), bound ATP is hydrolysed, the GroEL-lid is removed and the substrate released from what has become the *trans*-ring again. One ATPase-cycle takes about 15–30 s.

How can the foldosome decide when a protein is correctly folded? Miss-folded proteins usually have hydrophobic amino acid residues exposed at the surface, these are buried inside correctly folded proteins. Each GroES subunit has a hydrophobic patch, which can interact with hydrophobic residues in the substrate protein in the ATP-bound *cis*-state. In the ADP-bound (*trans*-) state these residues face the interface between the proteins of the beaker.

ATP/ADP-exchange in GroES results in a considerable conformational change, which twists the substrate binding sites (red patches in the lower left half of fig. 14.4) from the centre of the cavity to the side, thereby increasing the distance between these patches. This tears apart the inappropriate folding of the substrate. Once the conformational change in GroES is completed, the hydrophobic patches are buried between the interfaces of the GroES-subunits. In other words: no binding sites are available for the substrate protein, which lies now unfolded and unbound inside a protective cavity, which is lined exclusively with hydrophilic amino acids. In this state GroES also binds the GroEL lid, thus the protective cavity is totally closed. In this womb the substrate protein can try to fold correctly.

If the protein manages to achieve a correct conformation, it will no longer have exposed hydrophobic residues. Thus it can be released once the GroEL-lid is removed. If correct folding is not obtained, the substrate will rebind to the GroES in the next cycle. This can be repeated several times, until the substrate is correctly folded.

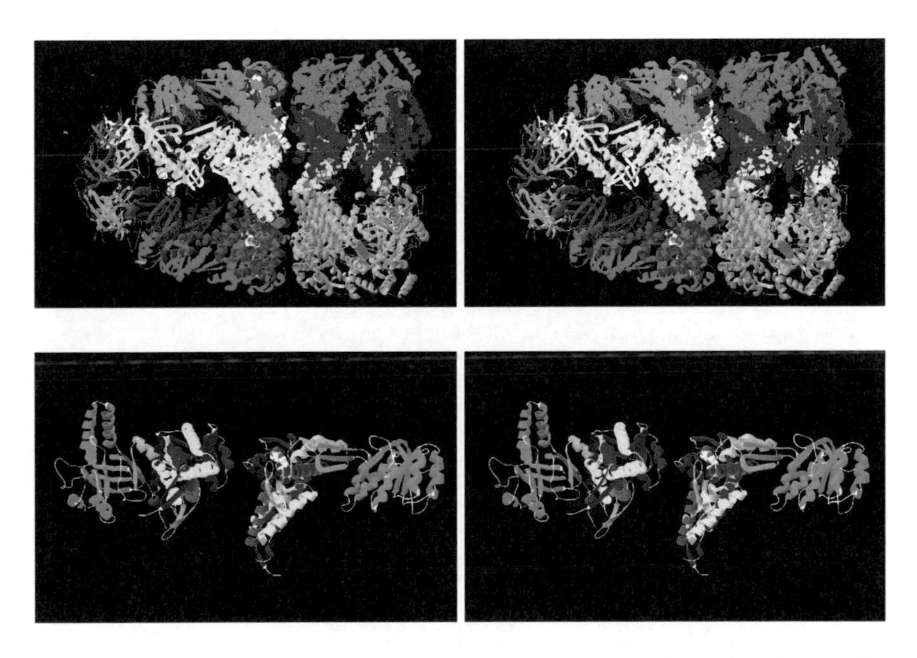

Figure 14.5. Top: Crystallographic structure of the GroEL/GroES-foldosome from *E. coli* at 2.9 Å resolution (PDB-code 1PF9). One ring of GroES has ADP bound in each subunit and is covered with the GroEL lid, the other ring is nucleotide free. **Bottom**: Comparison of the GroES-subunits with and without nucleotide (same molecule as above, but only one subunit from each ring shown). Note the considerable change in tertiary structure.

In *E. coli* about 10 % of all proteins produced transiently interact with GroES/GroEL, this figure triples under heat stress. The complex binds preferentially proteins between 10–55 kDa, about 60 kDa proteins would fit into the beaker. Larger protein can however also be bound if they consist of several independently folding domains. GroES/GroEL has a preference for proteins with α/β-structure (see chapter 2.3 on page 23), it may be required to produce the correct long-range hydrogen bonds required for β-sheet formation and for bringing the α-helices in a correct orientation with respect to the β-sheet.

Although we have looked at chaperones and chaperonins separately, in a living cell they work closely coordinated and sequentially as "folding machines". GroES/GroEL can accept substrate proteins from DnaK, which in turn can accept them from other heat shock proteins.

Exercises

14.1. Which of the following statements is *false* about molecular chaperones?

a) DnaK fulfills roughly the same function in prokaryotic cells as Hsp70 in eukaryotic.

b) DnaK requires two co-factors, the ADP/ATP-exchange stimulating GrpE and the ATPase-stimulating DnaJ.

c) DnaK unfolds miss-folded proteins using the energy of ATP-hydrolysis to to break hydrogen bonds.

d) The chaperone action of α-crystallin is required to prevent cataract formation in the eye lens.

e) Hsp90 keeps the steroid receptor in a binding-competent state.

14.2. Which of the following statements is *false* about the GroES/GroEL chaperonin?

a) GroES/GroEL can catalyse the correct folding of a protein.

b) The *cis*-ring of GroES/GroEL is covered by the GroEL lid and all its GroES molecules have ATP bound to them.

c) The *cis*-ring of GroES/GroEL acts as an ANFINSEN-cage.

d) All GroES molecules in the *trans*-ring have ADP bound.

e) In the *trans*-ring miss-folded proteins are actively unfolded.

Part IV

Membrane transport

15

Transport of proteins into mitochondria

15.1 The mitochondrium in the cell cycle

Mitochondria and plastids were derived from **endosymbiont** bacteria about 1–1.5 × 10^9 a ago, and contain their own DNA (see fig. 15.1). There may be several mtDNA molecules in each mitochondrium, the exact number is species-specific. The mtDNA is replicated and the mitochondria divide during the entire interphase. Division of mitochondria starts with an invagination of the inner membrane, the outer follows later. Mitochondrial division is not necessarily coupled to cell division: If for example a muscle is exercised, the number of mitochondria per muscle cell increases. The number of mitochondria per cell can be quite large, about 1000 in a rat liver cell.

Because of their large number, mitochondria are distributed randomly ("stochastically") between the daughter cells during mitosis, just like the vesicles derived from nucleus, ER and other internal membrane systems. Thus both daughter cells receive approximately, albeit not exactly, the same number of mitochondria.

Mitochondrial proteins and lipids

All proteins encoded by mtDNA are synthesised on mitochondrial ribosomes. However, most mitochondrial proteins are encoded by nuclear DNA, synthesised in the cytosol and post-translationally imported into the mitochondrium. No proteins encoded by mtDNA are exported from the mitochondrium.

Mitochondrial lipids are synthesised on the ER. Mitochondria only modify them later, cardiolipin is found only in mitochondria and bacteria (see fig. 15.2.)

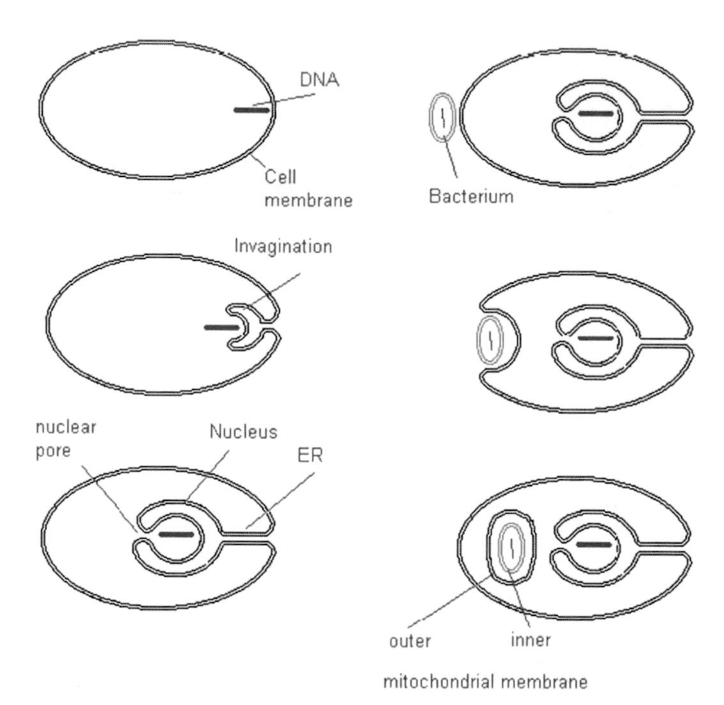

Figure 15.1. Origin of the modern eukaryotic cell: Organelles are derived either by invagination of the original prokaryotic cell membrane (left, ER, Golgi, nucleus) or by the uptake of endosymbionts (right, mitochondria, plastids). The inner mitochondrial and plastid membranes are derived from the cell membranes of the symbiotic organisms, the outer membranes are remnants of the phagosome. Note that the intermembrane space, as well as the inside of ER and Golgi apparatus, are topologically "outside". The environment in these compartments resembles that of the interstitial fluid (high Na^+ and Ca^{2+}, low K^+ and Mg^{2+}, oxidising).

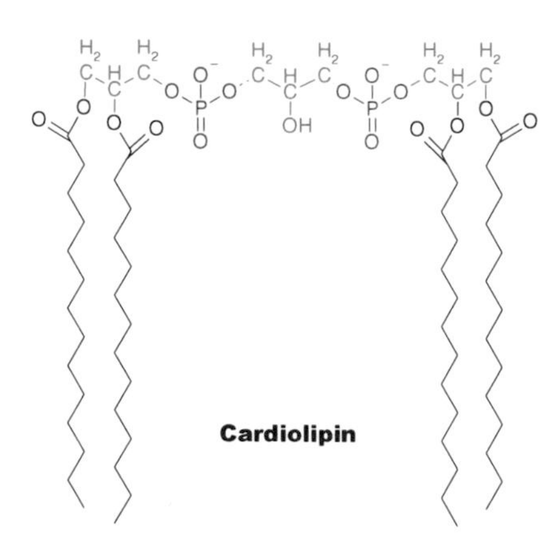

Figure 15.2. Cardiolipin makes up 20 % of the inner membrane of mitochondria. It consists of 2 phospholipids, linked by a glycerol bridge.

Mitochondria and antibiotics

Mitochondria have 70 S ribosomes, just like bacteria. Thus they can be affected by antibiotics which react with bacterial 70 S-, but not eukaryotic 80 S ribosomes (e.g. chloramphenicol). Such antibiotics may therefore damage mitochondria.

Inherited diseases of mitochondria

Because there are many mitochondria in a cell, which are independently transmitted to daughter cells, mutations of mitochondrial genes may affect only some of them. This is called **heteroplasmy**. As a result, the severity of mitochondrially inherited diseases can vary significantly not only from person to person, but also from tissue to tissue in the same person.

Such diseases will be most noticeable in tissues with high ATP-turnover like eye (LEBER's hereditary optic neuropathy, chronic progressive ophtalmoplegia), nervous system (KEARNS-SAYRE-syndrome) and muscle ("ragged red fibre" myopathies).

15.2 Synthesis and sorting of mitochondrial proteins in the cytosol

Many mitochondrial proteins are encoded by nuclear DNA, they are translated on ribosomes in the cytosol.

Proteins coming out of a ribosome are unfolded. In order to prevent fatal protein aggregation, and maintain them in an unfolded, transport competent state, they are complexed with Hsc70, a molecular chaperone (see fig. 15.3 and section 14.1 on page 212).

On the N-terminal side of mitochondrial proteins is a signal sequence, which is recognised by a receptor in the outer mitochondrial membrane. This sequence forms an amphipatic helix, with a positively charged hydrophilic, and an uncharged hydrophobic side (see fig. 15.4). This structural motive (rather than the amino acid sequence) is recognised by a specific receptor on the outer mitochondrial membrane. If this motive is transferred to a cytosolic protein by genetic engineering the chimeric proteins gets imported into the mitochondria. If the signal sequence is removed from mitochondrial proteins, they can no longer be imported.

Mitochondria have about 1000 TOM/TIM pairs each (transporter of the outer/inner) membrane, see fig. 15.5). The import of proteins from the cytosol into the mitochondria occurs in two distinct steps.

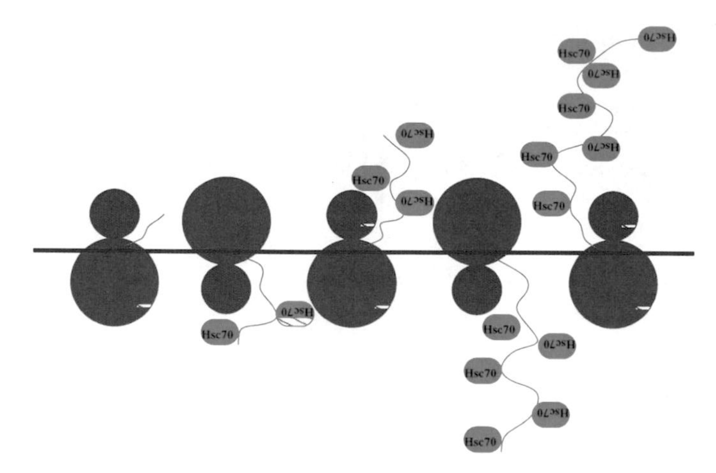

Figure 15.3. Proteins synthesised on the ribosome are immediately captured by the 70 kDa Heat shock cognate protein. This prevents miss-folding.

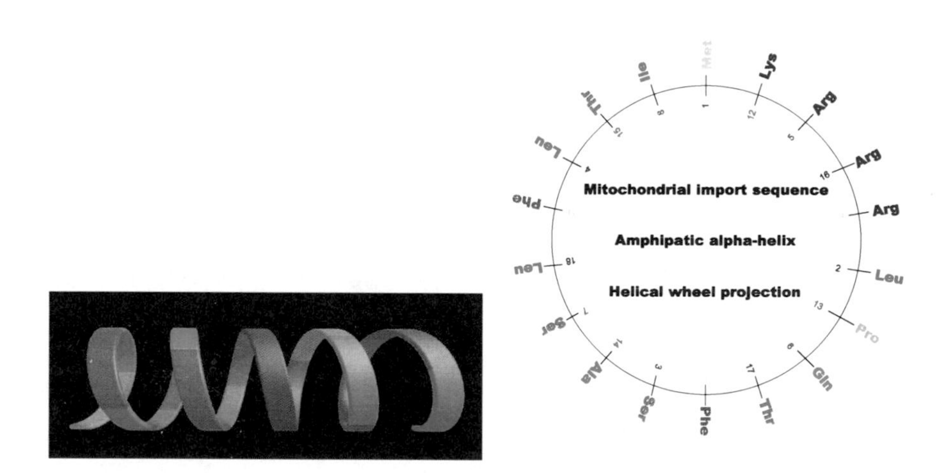

Figure 15.4. N-terminal signal sequence for import into mitochondria, here the subunit 4 of cytochrome c oxidase from yeast. The sequence MLSLRQSIRFFKPATRTL- forms an α-helix. Every third or fourth amino acid is positively charged (Lys or Arg), since there are 3.6 amino acids per turn in an α-helix all positively charged amino acids (blue) are in the same quadrant. The remaining quadrants are occupied by hydrophobic (green) or polar (orange) amino acids, also a few aromatic (dark grey) amino acids are present. This type of helix is called amphipatic. **Left**: Ribbon diagram of the alpha helix, the N-terminal Met is coloured yellow. Turning this helix by 90 degrees around the y-axis results in **right**: Helical wheel projection. The numbers inside the circle denote the position of the respective amino acid in the sequence. These numbers are coloured red for the first turn, orange for the second, yellow for the third, green for the fourth and blue for the fifth.

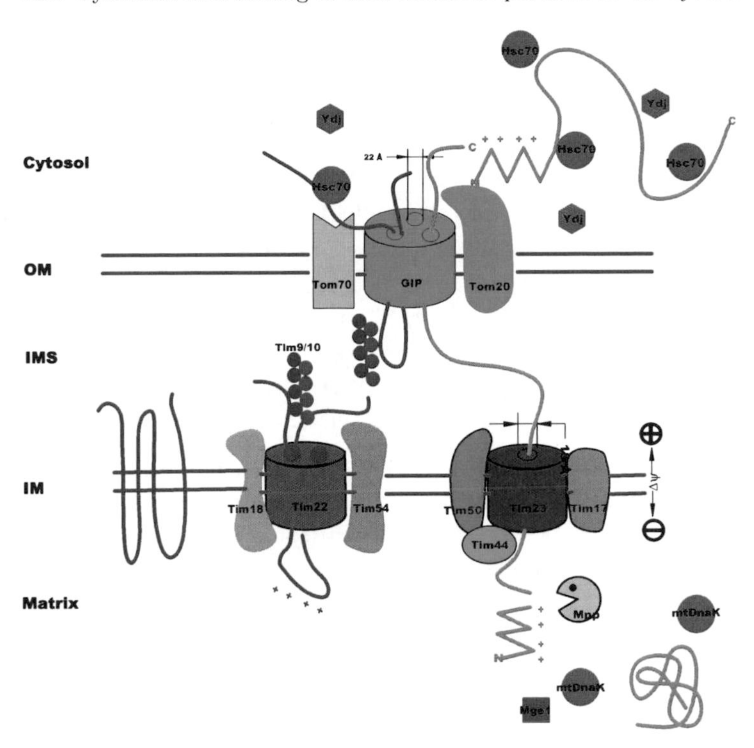

Figure 15.5. Mitochondrial proteins are transported from the cytosol into the mitochondria by two membrane transporters, called TIM and TOM respectively. Cytosolic Hsc70 and mitochondrial DnaK are also required. Import occurs only in places where the inner and outer membrane are close together, allowing concurrent transport of the protein across both membranes. The pore through the transporter has a diameter of approximately 2 nm.

First the signal sequence of the protein is bound to a receptor on the outer membrane of the mitochondrium. This signal sequence, as we have seen, has a number of positive charges. The matrix of the mitochondrium is negatively charged with respect to the cytosol (−200 mV, this is equivalent to a field strength of 400 000 V/cm), because endoxydation leads to the transport of protons from the matrix to the cytosol. As a result the signal sequence is electrophoretically pulled through the outer and inner membrane transporter. Experimentally, this step does not require ATP and can proceed at low temperatures. However, uncouplers like 2,4-dinitrophenol (DNP, see page 294), which dissipate the electrical potential difference across the inner membrane, prevent signal sequence transfer.

The rest of the protein is imported in the second step, which depends on ATP-hydrolysis and can occur only at 37, but not at 5 °C. The protein inside the transporters can move either into or out of the mitochondrium, and it will do so randomly (BROWNIAN motion). Thus, in the absence of other factors, no net transport of protein in either direction would occur. However, any stretch of protein that comes into the mitochondrium is immediately bound by mtDnaK. This prevents the protein from slipping back. On the other hand, on the cytosolic side the protein is released from Hsc70, which allows it to move into the mitochondrium. Thus a random movement is used to achieve vectorial transport by the binding and unbinding of 70 kDa heat shock proteins. This mechanism is called **molecular ratchet**.

Some researchers claim that binding of mtDnaK to the imported protein leads to a change in the way mtDnaK interacts with the inner membrane, allowing it to actively pull the protein through the transporter. Evidence for this additional mode of action of mtDnaK is currently limited, though.

Once the protein has been imported into the mitochondrial matrix, the signal sequence is cleaved of by **matrix processing protease (Mpp)**. Matrix proteins then fold, as mtDnaK dissociates from them. Folding may involve GroEL/GroES chaperonin action.

Proteins of the inner mitochondrial membrane or the intermembrane space are transported to their destination directed by a second signal sequence, which is unmasked once the first has been cleaved of. In other words, these proteins (for example ATP synthase subunits) are transported from the cytosol through both mitochondrial membranes into the matrix, and from there into the inner membrane or intermembrane space, respectively. This is a reminder of the origin of mitochondria as endosymbionts. Other proteins like cytochrome c are small enough that in their unfolded form (e.g. apocytochrome c, without the haem group) they can enter the intermembrane space directly through the outer membrane, presumably through a porin-like channel. Folding of these proteins either around a prosthetic group or by disulphide bond formation prevents them from returning to the cytosol.

Synthesis of proteins encoded by mitochondrial and nuclear DNA is synchronised, but the mechanism is still unknown.

16

Intracellular protein transport

In the last chapter we have seen how proteins are inserted post-translationally into mitochondria. Similar pathways exist for plastids, nucleus and peroxisome, although the signal sequences and transport mechanism vary. In most cases the signal sequence is cleaved off after the protein has reached its destination, the exception are nuclear proteins because the nucleus breaks down during each cell cycle, releasing its proteins into the cytoplasm. Hence after each cell division nuclear proteins need to be reimported.

Insertion of proteins into the ER however occurs co-translationally, from there it is then transported by specialised vesicles to GOLGI-apparatus, lysosomes, endosomes, secretory vesicles and plasma membrane.

16.1 Transfer of protein into the ER lumen

The signal for protein import into the ER is a positively charged amino acid near the N-terminus of the protein, followed by a box of 6-15 hydrophobic amino acids (mostly Leu, Ala, Val) followed in turn by polar amino acids plus a Gly or Ala marking the cleavage site.

Once this signal sequence emerges from the ribosome (since it is right at the N-terminus that happens early in synthesis) it is bound by the **signal recognition particle (SRP)**, which halts protein synthesis at the ribosome by also binding to its large subunit. This binding is GTP-dependent. SRP is a ribonucleoprotein, composed of six polypeptides and 1 RNA-molecule (see fig. 16.1).

The entire complex of SRP, ribosome, mRNA and nascent protein (about 70 amino acids long at this stage, of these about 40 protrude from the ribosome) is then bound by the **SRP receptor** which interacts with the **peptide translocation complex (translocon)** in the ER membrane. This leads to

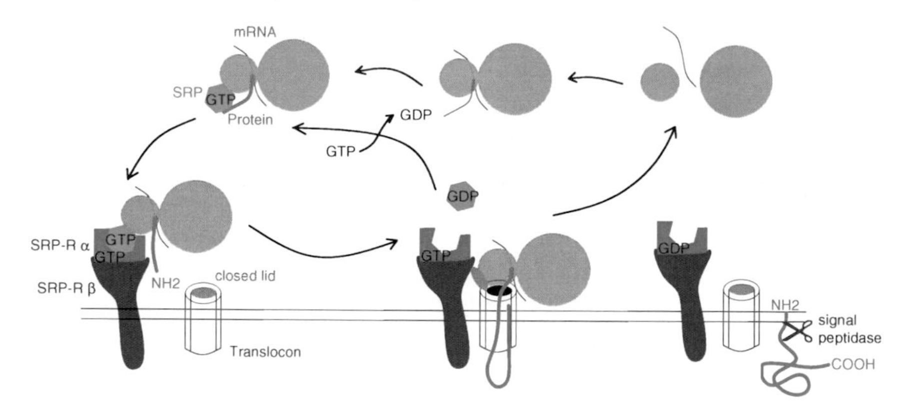

Figure 16.1. Model for the transfer of nascent proteins into the ER. For explanations see text.

hydrolysis of SRP-bound GTP, dissociation of SRP and resumption of protein synthesis. The hydrolysis of GTP provides the energy required for the formation of the ribosome/nascent chain/translocon complex.

Bound ribosomes give ER-membranes a characteristic appearance under the electron microscope, forming the **rough ER**. Thus rough ER is the site of membrane protein biosynthesis, **smooth ER** lacks ribosomes and fulfills other functions (e.g. lipid biosynthesis, biotransformation).

In the absence of SRP *in vitro* synthesis of proteins destined for the ER can occur normally in the absence of ER-membranes, if the membranes are added later the transport of the completed proteins into the ER is not possible (exception: proteins with less than 100 amino acids can be transported into the ER post-translationally).

If the *in vitro* synthesis is performed in the presence of isolated ER membranes (which form vesicles, so called **microsomes**), the newly synthesised protein is immediately resistant to proteases added to the outside medium, in other words they are during the process of synthesis sequestered inside the enclosed space of the microsomes, where the proteases can not reach them.

Interaction of the SRP/mRNA/ribosome/nascent protein complex with the translocation complex is via the hydrophobic box in the **signal sequence**, removal of this box from the gene sequence of an ER protein (or replacement of one of its amino acids with a charged one) prevents targeting to the ER, addition of that box to a cytosolic protein leads to translocation into the ER. Translocation is accompanied by removal of the signal sequence by the enzyme **signal peptidase**, which is located inside the ER.

The key component of the peptide translocation complex is a transmembrane protein, **Sec61p**, the product of the mammalian homologue to the yeast *sec61* gene. This protein forms a channel through the membrane. The growing polypeptide is threaded from the ribosome through this channel and reaches the lumen of the ER without exposure to the cytosol. Also required is the **translocating chain associated membrane protein (TRAM)**, which binds the signal sequence once it has been handed over by SRP. Two further proteins in this complex, **Sec61β** and **Sec61γ** bind the ribosome. These proteins are sufficient to reconstitute protein translocation into liposomes *in vitro*. Electron microscopy indicates that the translocon is 50 – 60 Å high and has a diameter of 85 Å, the channel is about 20 Å in diameter. In the absence of ribosomes, the cytosolic entrance of the channel is closed by a loop of Sec61p, which goes out of the way after ribosome binding.

In the ER lumen, the nascent protein is bound by an ER-specific homologue of the 70 kDa heat shock protein family, **78 kDa glucose regulated protein (Grp78)**. This maintains the nascent protein in a folding-competent state until translocation is complete.

16.2 Membrane proteins

There are 6 types of membrane proteins, as depicted in fig. 16.2:

Type 1 protein synthesis starts in the same way as that of proteins destined to the ER lumen. The ribosome/mRNA/protein complex attaches to the translokon, and the signal sequence binds to TRAM. The nascent protein is imported into the ER, until a **stop-transfer signal** is encountered.

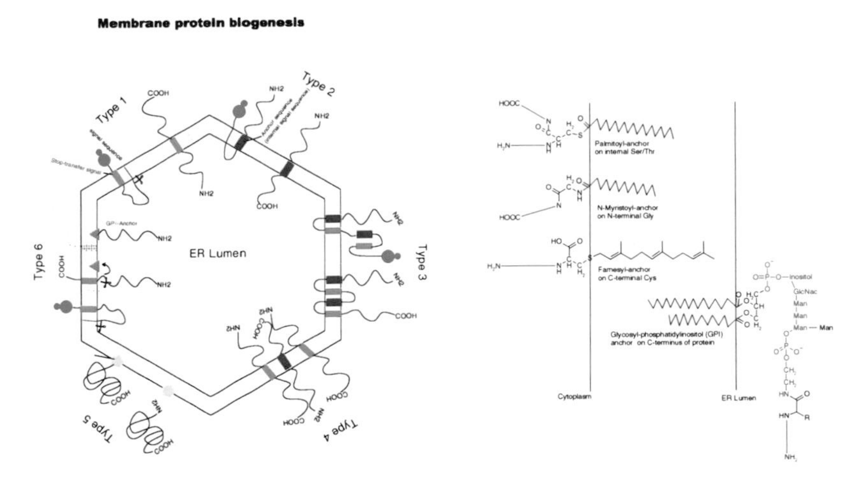

Figure 16.2. Left: There are 6 classes of membrane proteins. **Right**: Linkage of proteins to the membrane by hydrophobic anchors (for type 5 and 6).

Figure 16.3. Fluid-mosaic model of the biological membrane. **Phospholipids** form a bilayer with the hydrophobic tails of the lipids (yellow) buried inside, and the hydrophilic head groups (blue) exposed to the aqueous medium on both sides. **Cholesterol** (magenta) is also present. In this matrix **transmembrane proteins** are embedded (green), additionally proteins are bound to the membrane by **GPI-anchors** on the extracellular side (top) and **acyl-** or **prenyl-residues** (cyan) on the cytoplasmic side (bottom). **Sugar side-chains** (grey) may be attached to proteins and lipids on the extracellular side of the membrane.

This is a stretch of 20 – 25 hydrophobic amino acids, which form a transmembrane α-helix. Since this sequence, once inserted into the membrane, can not be pushed into the aqueous environment of the ER lumen, the amino acids C-terminal of it stay in the cytosol. Cleavage and digestion of the signal sequence results in a type 1 integral membrane protein, with the C-terminus on the cytosolic side and the N-terminus in the lumen of the ER. If the stop-transfer signal is deleted from the gene of a type 1 protein, or if one of the amino acids encoded by it is mutated into a charged one, the protein enters the ER lumen and is eventually secreted.

Type 2 proteins have an internal **signal-anchor sequence**, so the N-terminus stays on the cytosolic side of the membrane. All amino acids C-terminal of the signal-anchor sequence are translocated into the ER-lumen, the orientation of type 2 proteins is opposite to that of type 1. Signal-anchor sequences too form 20–25 amino acid transmembrane helices. As a general rule the cytosolic end of a membrane helix has positively charged amino acids, this situation is stabilised by the negative potential of the cytosol with respect to the ER-lumen. This "**positive inside**" rule determines wether a hydrophobic helix acts as a stop-transfer or signal-anchor sequence. Long hydrophobic stretches tend to function as stop-transfer signals, signal-anchor sequences tend to be shorter.

Type 3 proteins span the membrane several times. They have repeated signal- and stop-transfer sequences. Once the first of these pairs has left the translocon, anchoring the protein in the membrane, protein synthesis continues in the cytosol, further signal/stop transfer sequence pairs are inserted into the membrane using the translocon, but without SRP and SRP-receptor. Originally both N- and C-terminus face the cytosol, however, posttranslational proteolysis can change that.

Type 4 proteins are made by the combination of several polypeptides (either type 1 or -2) into a single quaternary structure.

Type 5 proteins are synthesised in the cytosol, they are chemically linked to hydrophobic molecules anchored in the cytosolic leaflet of the ER-membrane. 3 subtypes are possible:

- Internal Cys or Ser residues of the protein are chemically linked to palmityl-groups.
- N-terminal Gly-residues are linked to N-myristoyl-groups.
- C-terminal Cys is linked to farnesyl- or geranylgeranyl-groups.

Linking proteins to acyl- or prenyl-groups can convert an inactive cytosolic protein into a membrane-bound active one. This regulatory function of the enzymes responsible for prenylation or acylation has excited considerable pharmacological interest.

Type 6 proteins are linked to the luminal leaflet of the ER membrane by a **glycosyl phosphatidylinositol (GPI)** residue bound to their C-terminus. These proteins are synthesised like type 1 proteins, later the protein is cleaved near the transmembrane domain by an endoprotease, the new C-terminus is transferred to a GPI-anchor at the same time. This step removes the cytosolic bit of the original protein, which would interact with the cytoskeleton and prevent free diffusion of the protein in the membrane plane. The GPI-anchor in plasma membrane proteins also acts as a **sorting signal**, directing the protein to the **apical** domain of the cell.

16.3 Folding, posttranslational modification and quality control of membrane proteins

16.3.1 Disulphide bond formation

The cytosol is a **reducing** environment, and Cys-residues carry thiol (-SH) groups. The lumen of the ER on the other hand corresponds to the extracellular space, and is **oxidising**. Some of the thiol groups of proteins in the ER will be converted to **disulphide** bridges:

$$R - SH + HS - R + GSSG \rightleftharpoons R - S - S - R + 2GSH \tag{16.1}$$

Disulphide formation does not utilise oxygen directly for oxidation of thiols, but the tripeptide **glutathione** (γ-Glu-Cys-Gly, fig. 16.4). This redox-coupler can occur in both reduced (GSH) and oxidised (GSSG) form.

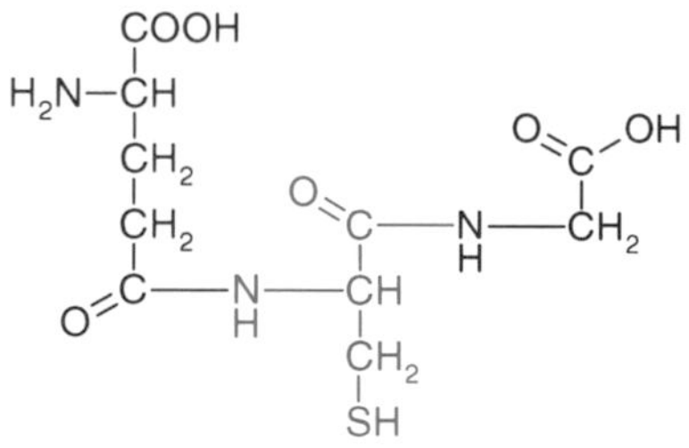

Glutathione (GSH)

(γ-Glu-Cys-Gly)

Figure 16.4. Glutathione is an important redox-coupler in cells.

This modification is important for correct **tertiary structure** of proteins. This means that neither are all thiols converted nor is it unimportant which -SH groups are involved in the formation of a disulphide bond.

One way of ensuring that correct thiols enter disulphide formation is to oxidise Cys-residues in the N-terminal part of the protein before more C-terminal Cys-residues have been incorporated. Immunoglobulins are an example for this method.

In other proteins disulphide bonds are formed across the molecule (for example insulin, Cys 1-4, 2-6 and 3-5). Wrongly formed disulphide bonds can be corrected by **protein disulphide isomerase (PDI)**, an abundant enzyme in the ER of secretory tissues.

Bacterial cells do not have an oxidising compartment like the eukaryotic ER. If one tries to express human proteins in bacteria for the production of **recombinant therapeutics** problems may occur. Proteins that require disulphide bond formation often do not fold correctly, instead they aggregate in the form of **inclusion bodies**. In some cases it is possible to isolate these inclusion bodies, completely unfold the proteins with urea or similar agents and refold *in vitro* under proper oxidising conditions by slow removal of the denaturant. However, in many cases such proteins will be produced in cultured animal cell lines to avoid this complication.

16.3.2 Glycosylation

Most secreted and transmembrane proteins contain sugar trees, which

- aid in protein folding. For this reason substances interfering with glycosylation (desoxinojirimycin, tunicamycin) may be used to **treat viral infections.**
- target proteins to the correct subcellular compartment.
- serve as receptors for cell-cell recognition and as **immunological determinants** (blood group antigens!).
- form the binding sites for bacteria and viruses.

Nuclear and cytosolic proteins on the other hand are usually not glycosylated.

The sugar trees can be added to the OH-groups of Ser, Thr and hydroxylysine (**O-linked oligosaccharides**) or to the amino-group of Asn (**N-linked oligosaccharides**). While O-linked oligosaccharides are generally short (1 – 4 sugar groups), N-linked oligosaccharides are quite complex. The same protein may have different N-linked oligosaccharides depending on the tissue where it was produced, and even the protein molecules from one cell type can be heterogeneous. Because some of the sugar groups (**sialic acids**) contain negative charges, the number of negative charges in a glycoprotein may vary, making their separation in the laboratory difficult (see Section 3.4 on page 38). The reasons for these differences are not understood, but subject to intensive study.

O-linked oligosaccharides

O-linked oligosaccharides are produced in the GOLGI-apparatus by adding one sugar at a time to the growing tree on the protein. The sugars are first activated by the addition of nucleotides (UDP for **galactose** (Gal) and **N-acetylgalactosamine (GalNAc)** , CMP for **N-acetylneuraminic acid** (NANA, sialic acid). Synthesis of these activated sugars occurs in the cytosol, they are transported into the GOLGI-apparatus by antiporters, which exchange them for the free nucleotides left after formation of oligosaccharides.

Transfer of the sugar residues is catalysed by membrane-bound glycosyltransferases, which are specific both for acceptor OH-group and donor sugar-nucleotide.

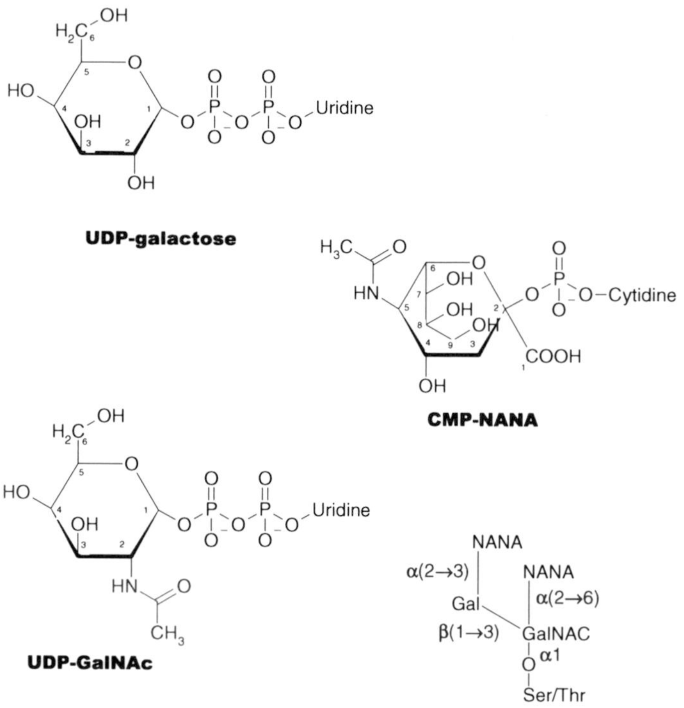

Figure 16.5. Formation of O-linked sugars in the ER and GOLGI-complex.

In most cases, synthesis of O-linked oligosaccharides starts with the transfer of **GalNAc** (from UDP-GalNAc) to Ser or Thr OH-groups by GalNAc-transferase bound to the luminal leaflet of the membrane of the **rough ER** or ***cis*-GOLGI network**. After transport of the nascent protein to the ***trans*-GOLGI network Gal** (from UDP-Gal) (in $\beta 1 \rightarrow 3$-linkage) and two **NANA**-residues (from CMP-NANA) are added in sequence.

Transfusion

Not only proteins, but also lipids are O-glycosylated. The **AB0-blood group antigens** are oligosaccharides linked to glycoproteins and glycolipids of erythrocytes (see fig. 16.6). The **0-antigen** has the sequence −Glc−Gal−GlcNAc−Gal−Fuc, which can be synthesised by all humans. People with blood **group A** produce a **GalNAc-transferase** that transfers a GalNAc-residue to the second Gal, people with blood **group B** produce a galactose-transferase instead. Those with blood **group AB** have both these enzymes and hence produce both the A- and the B-antigen, those with blood group 0 produce neither of these transferases and hence have the O-antigen only. The GalNAc- and Gal-transferases responsible for synthesis of A- and B-antigen respectively are very similar proteins which differ in only 3 amino acids.

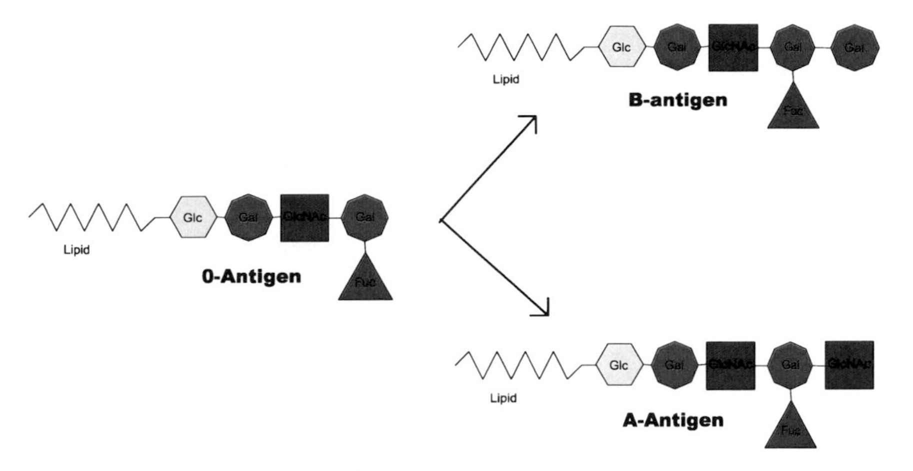

Figure 16.6. The AB0-blood group antigens are oligosaccharides bound to plasma membrane proteins and lipids. For details see text.

Intestinal bacteria (*E. coli*) also produce these antigens, hence each type of bacteria can grow only in humans with matching blood groups. This exposure to the common bacteria leads to immunisation to foreign blood group antigens even without exposure to foreign blood. Therefore a severe immune reaction is produced if a patient is transfused blood from a non-matching donor, even the first time.

Note that since donors with blood group O can produce neither the A- nor the B-antigens and since the O-antigen is present in all people, these donors can donate erythrocytes to patients of any blood group (**universal donor**).

On the other hand, patients with blood group AB do not produce antibodies against either O-, A- nor B-antigen and can receive erythrocytes from donors of any blood group (**universal recipient**).

This picture is made more complicated by the existence of several other antigens (most notably the rhesus-factor), which are caused by variations in the amino acid sequence of proteins in the erythrocyte membrane.

N-linked oligosaccharides

N-linked oligosaccharides are transferred to proteins as a preformed block and then modified (see fig. 16.7).

On the cytosolic side of the ER membranes nucleotide-activated sugar residues are added to the long-chain alcohol **dolichol** via a pyrophosphate linkage. Dolichol is a polyisoprene with 75 – 95 carbon atoms, which is enough to

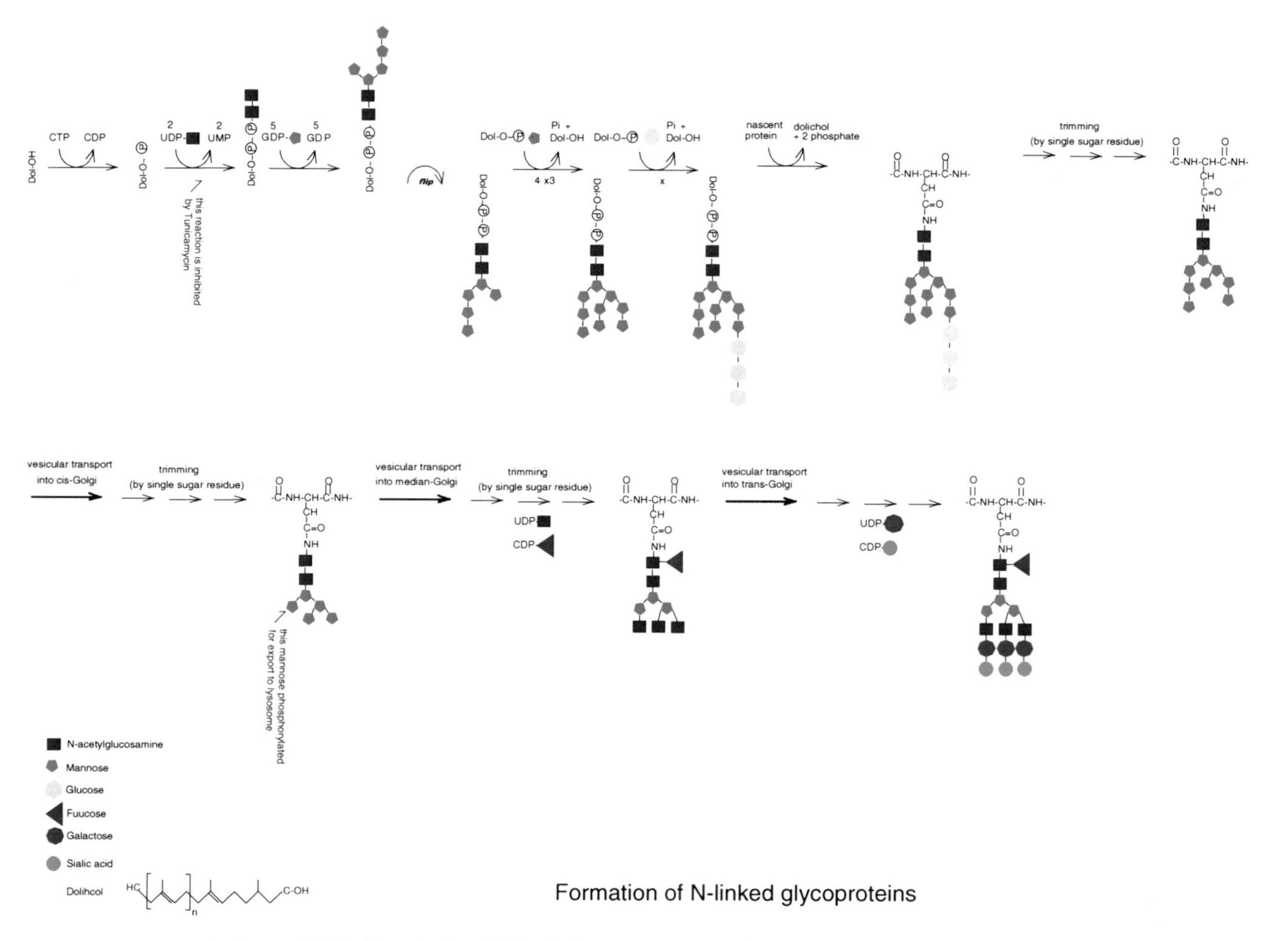

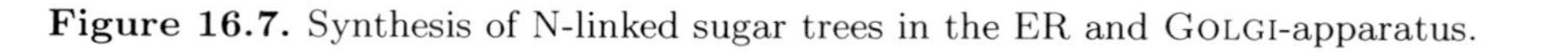

Figure 16.7. Synthesis of N-linked sugar trees in the ER and GOLGI-apparatus.

span the membrane 4 or 5 times. Hence dolichol is anchored very firmly in the lipid bilayer.

A special **flippase** transports this initial tree to the luminal side of the membrane, where further sugar residues (activated by linkage to dolichol) are added. These dolichol-intermediates too are produced on the cytosolic leaflet of the membrane, and flipped to the luminal side by a specialised transporter.

Transfer of the resulting **oligosaccharide** $(Glc)_3(Man)_9(GlcNAc)_2$- tree from dolichol pyrophosphate to Asn-residues of a nascent protein occurs during its entry into the ER. The receiving Asn must occur in a Asn-X-Ser or Asn-X-Thr sequence. The oligosaccharide-protein-transferase is an integral multi-subunit membrane protein which binds to the ribosome on the cytosolic side of the ER membrane and to its substrates on the luminal side. Thus the transferase is held near the nascent protein during its transfer into the ER.

The sugar tree is then trimmed back by glucosidases to $(Man)_8(GlcNAc)_2$- (**high mannose type**) by specific glucosidases. Apparently, the terminal glucose residues act as a signal to the transferase that the oligosaccharide is complete, after transfer to the protein they are no longer needed and removed. Removal of these sugar residues is however dependent on proper protein folding, in addition a specific glucose transferase can add back one glucose residue to unfolded or mis-folded proteins as part of the quality control system of the ER. This glucose residue serves as a binding site for the chaperones calreticulin and calnexin (see page 243).

The protein is then exported to the *cis*-GOLGI-apparatus, where further trimming to $(Man)_5(GlcNAc)_2$- (**core oligosaccharide**) occurs. The protein is then transported to *median*- and *trans*-GOLGI for further trimming and addition of different sugar residues to the final **complex** glycoprotein.

16.3.3 Protein folding: chaperones of the ER

We have seen how certain proteins, molecular chaperones and chaperonins, aid in the folding of cytosolic proteins and how they can in addition also help with import of proteins into compartments like mitochondria (see section 15.2 on page 229). Given the importance of the ER for protein synthesis it is hardly surprising that this compartment is particulary rich in chaperones.

One of them is the **78 kDa glucose regulated protein (Grp78)**, a homologue of the cytosolic Hsp70/Hsc70 proteins (see fig. 14.1). As mentioned above, Grp78 receives nascent proteins when they come out of the translocon and maintains them in a folding competent state.

Proline is an unusual amino acid because its α-amino group is part of a ring structure. In Pro the *cis*-conformation occurs more frequently than in other amino acids. Proper protein folding requires the conformation of the

Pro-residues to be correct, this is ensured by **peptidyl prolyl *cis/trans*-isomerase (PPI)**. This reaction is often the rate limiting step in protein folding in the ER. PPIs are often specific for a particular protein. Folding of the major histocompatibility complex (MHC) requires a specific PPI, which is one of the targets for the immunostatic drug **Cyclosporin A**, which is used to prevent organ rejection in transplant patients. Hsp70s also are target for this drug.

The sugar side chains of nascent proteins are binding sites for a couple of sugar-residue binding proteins (**lectins**), which act as chaperones. Such lectins can be membrane bound (like **calnexin**) or soluble (like **calreticulin**). This interaction is particularly important for **oligomeric proteins**, it allows the monomers to find their proper partner and prevents unspecific aggregation. Once the proteins are properly folded, outer glucose residues are clipped from the sugar tree by glucosidase II, which abolishes lectin binding.

16.3.4 Protein quality control in the ER

Only correctly folded proteins may leave the ER, those which have not yet obtained native conformation are retained for refolding or—if that proves impossible—for destruction.

Some mutations lead to the inability of a protein to attain its native conformation. **Cystic fibrosis** (see page 285) is an example for a disease caused by this mechanism. But lack of essential proteins is not the only possible consequence of such mutations: Accumulation of mis-folded proteins in the ER of secretory cells can cause serious damage. An example is α_1-antitrypsin, secreted mainly by the liver, but also by macrophages. Lack of this protein results in uncontrolled hydrolysis of elastin and is a cause of lung tissue destruction (emphysema). Accumulation of unprocessed antitrypsin results in the formation of a para-crystalline precipitate in the ER of the liver.

Unfolded proteins usually have chaperonins attached to them, this is one of the signals that prevent transport to the GOLGI-apparatus.

Destruction of mis-folded ER-proteins

Proteins that can not be folded correctly (and orphan subunits whose partner can not be folded) need to be destroyed, because their accumulation in the ER would eventually destroy the cell. This is called **ER associated protein degradation (ERAD)** [9].

Mis-folded membrane proteins are transported back into the cytosol through the **Sec61**-translocon. During this **retrograde transport** membrane-bound

ubiquitinating enzymes mark these proteins for destruction by the 26 S-**proteasome**. However, there appears to be also a proteasome-independent pathway for ERAD, about which little is known yet.

Recent research indicates that digestion of the **outer mannose** residue on the middle branch of the Asp-linked sugar tree (see fig. 16.7) by a highly conserved, ER-specific mannosidase I specific for unfolded proteins controls entry into the ERAD pathway for soluble proteins. As we have already mentioned, removal of glucose from the sugar tree by glucosidase II is prevented in unfolded proteins, this effect is amplified by removal of that mannose residue. Thus **calnexin** and **calreticulin** stay bound to the protein, targeting it to the ERAD pathway.

The unfolded protein response

Bound chaperones (Grp78 and calnexin) are one way for the quality control machine to recognise mis-folded proteins. Grp78, PPI and PDI expression is up-regulated in cells experiencing stress-conditions (like limited glucose supply, hence glucose regulated protein Grp). This is called **unfolded protein response** [74] and is caused by a lowered concentration of free Grp78, as more and more of this protein becomes bound to mis-folded substrates. This causes Grp78 to dissociate from its receptor **Atf6** which is then allowed to move into the GOLGI-apparatus, where it is split by two GOLGI-resident endoproteases, S1P and S2P. The proteolytic peptides increase the transcription of ER chaperones.

If this does not correct the situation, Grp78 dissociation from **Ire1**, a protein found in the inner nuclear membrane (which is continuous with the ER) leads to the splicing of HAC1 mRNA. **Hac1** activates the transcription of mRNA encoding not only for ER chaperones, but also for enzymes involved in the degradation of terminally mis-folded (glyco-)proteins.

The third step of unfolded protein response is dissociation of Grp78 from **Perk**, which then phosphorylates the **elongation factor 2α (elF2α)**, resulting in a stop of translation. This should give the cell time to deal with the problem, before it gets overwhelmed with unfolded protein. Continued activation of Perk finally results in **apoptosis** (cell suicide), to protect the rest of the organism.

16.4 Vesicular protein transport in eukaryotic cells

Each of the cellular compartments has a special function, and an appropriate set of enzymes to meet them. However, for the cell to survive, there has to

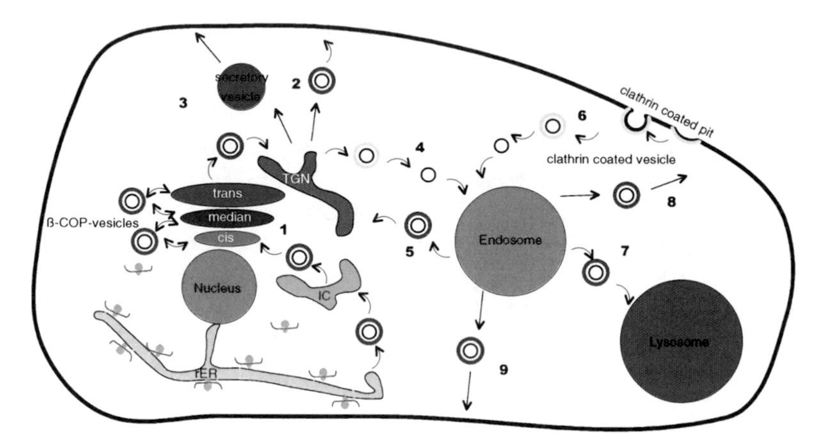

Figure 16.8. Vesicular protein transport in eukaryotic cells. The coats for the pathways depicted are:

1: rER - intermediate compartment - *cis-, median-* and *trans*-GOLGI-apparatus	β-COP
2: *trans*-GOLGI-apparatus - *trans*-GOLGI-network - cell membrane	unknown
3: *trans*-GOLGI-network - secretory vesicle, plasma membrane	unknown
4: *trans*-GOLGI-network - endosome	clathrin
5: endosome - *trans*-GOLGI-network	unknown
6: apical cell membrane - endosome	clathrin
7: endosome - apical cell membrane	unknown
8: endosome - basolateral cell membrane	unknown
9: endosome - lysosome	unknown

be constant transport between the compartments. For example, newly synthesised protein need to be transported from the ER to the Golgi apparatus for glycosylation, and from there to the target membranes. That needs to be carefully orchestrated, the proteins destined to the Golgi need to be included, but any proteins that belong into the ER need to be excluded from transport. Transport, once initiated, has to go specifically to the Golgi, and not to, say, the endosome.

Transport of proteins between cellular compartments can be achieved either by **transport vesicles** or by **maturation of compartments**. The maturation model is discussed in particular for the GOLGI-apparatus and the endosomal-lysosomal compartments. As we shall see in a moment, these models need not be mutually exclusive.

Maturation

For the GOLGI-apparatus the **cisternal maturation model** would state that as more and more lipid is produced on the smooth ER, new *cis*-GOLGI-cisternae are formed. As this happens, the old *cis*- becomes the new *median*-, and the old *median* the new *trans*-GOLGI-cisternae. In this model, transport of proteins between GOLGI-cisternae is required only for **retrograde transport** of resident enzymes back into the compartment where they belong, e.g. *cis*-GOLGI-resident glucosidases from the *median*- back to the *cis*-compartment. **Anterograde transport** of newly synthesised proteins however would occur by **bulk flow**.

In the case of the **endosome/lysosome** compartment similar ideas have been put forward. Endocytotic vesicles fuse to form new early endosomes, these mature into late endosomes and finally lysosomes (removal of proteins by transcytosis and recycling, acidification).

Vesicular transport

While in the maturation model compartments like the GOLGI-apparatus and the endosome are fairly dynamic structures, which are produced, age and are recycled, the vesicular transport model sees them as static entities, factories which receive raw materials, process them and release their final products, without changing much in the process.

Transport vesicles are produced by budding of membranes, caused by assembly of a cage of special proteins on the cytosolic side of the membrane. Two such proteins have been characterised so far, **clathrin** and **β-coat protein (COP)**.

Clathrin

Clathrin coated vesicles(CCV) are involved in endocytosis and in the transport of newly synthesised lysosomal proteins from the TGN to the endosome. **Clathrin triskelia** consist of three copies each of the heavy and light clathrin proteins. They are tetrahedral in shape, with an apex where all the C-termini of the protein chains meet. From there the proteins extend in three L-shaped arms downward and sideways. Free clathrin molecules can spontaneously assemble into sheets and baskets *in vitro*, no ATP is required at this step.

β-coat protein

Vesicles budding from the ER *en route* to the Golgi and vesicles budding from the Golgi for anterograde and retrograde transport are coated with **COP**.

We distinguish COP-I and COP-II. **COP-I** is responsible for **retrograde transport** between GOLGI-cisternae and between *cis*-GOLGI and rER. **COP-II** is responsible for **anterograde transport** from the rER to the *cis*-GOLGI-cisterna. There is a high degree of homology between the equivalent proteins from COP-I and COP-II.

Formation of COP-coated vesicles starts with **coatomers**, a complex of 7 different coat proteins, assembling on the cytoplasmic side of the membrane, a reaction that requires an additional protein, **ADP-ribosylating factor (ARF)** in COP-I (or **Sar1** in COP-II) in their GTP-bound form. Both proteins belong to the group of **small GTP-binding proteins**, which is involved in key regulatory steps in our cells. Hydrolysis of that GTP and release of the P_i is required for the uncoating of COP-coated vesicles, further rounds of vesicle formation require GTP/GDP exchange in ARF and Sar1, respectively.

Experimental evidence

With some very large material, like **pro-collagen fibers**, vesicular anterograde transport seems impossible because these proteins do not fit into vesicles. This would favour the cisternal maturation model.

With smaller proteins, like the **Vesicular Stomatitis Virus Glycoprotein (VSV-G-Protein)** observations seem to favour the vesicular transport model, transport through the compartments is faster than could be accounted for by cisternal maturation.

Note however that these models are not necessarily mutually exclusive: There may be a vesicular fast track for smaller proteins within the framework of a maturing cisternae.

16.4.1 The specificity of transport reactions

With all the vesicles budding of the various intracellular membrane systems and fusing with others, **specificity** of the fusion event must be maintained. In other words: Each vesicle must carry an **address label** that allows it to specifically fuse with its target membrane, and only with that one.

Exocytotic vesicle fusion with the plasma membrane is the case about which we know most, it is assumed that fusion events in other membrane systems are regulated in a similar way.

The first thing that became known about several types of fusion events in the eukaryotic cell was that a cytosolic factor is involved. Finally a small protein, **NEM-sensitive factor (NSF)**, was isolated. If this protein is inactivated with the SH-Reagent **N-ethylmaleimide (NEM)**, or if it is removed from cytosol by immunoabsorbtion, fusion no longer works.

It also became known that fusion of synaptic vesicles with the praesynaptic membrane could be prevented by incubation with either **tetanus** or **botulinus toxin**, because these act as specific proteases that destroy **synaptobrevin** and **syntaxin** respectively. Synaptobrevin is a protein in the plasma membrane, syntaxin occurs in synaptic vesicles.

Synaptobrevin and syntaxin are special cases of a group of proteins known as **SNAP receptor (SNARE)**. These always occur in pairs, one on the vesicle membrane (v-SNARE, syntaxin in the case of exocytosis) and the other on the target membrane (t-SNARE, synaptobrevin in the case of exocytosis). Complex formation between v-SNARE and its corresponding t-SNARE is believed to account for the specificity of membrane fusion, this complex also contains **soluble NSF-attachment protein (SNAP)** and NSF. NSF, SNAPs, vSNARE and tSNARE form a stable complex, complex formation is followed by ATP hydrolysis and membrane fusion, little is known about the exact mechanism. Then the vSNARE/tSNARE complex dissociates, vSNAREs are enriched into special vesicles and transported back to the donor membrane.

Vesicle fusion is controlled in a not fully understood manner by another set of small G-proteins, which are called **Rab**. It seems that for each vesicle/membrane fusion event, there is a specific set of vSNARE, tSNARE and Rab proteins.

SNAPs and NSF-tetrameres can be isolated after solubilisation of the membranes as 20 S complex. ATP hydrolysis in bound NSF is supposed to be involved in the actual fusion event, possibly by changing the conformation of the complex and bringing the membranes closer together.

How do proteins know where they belong?

There are several hypotheses for this question, which again need not be mutually exclusive.

ER-resident proteins are prevented from moving with this **bulk flow** by a special sorting signal (KDEL) which is also responsible for their transport from the GOLGI-apparatus back to the ER, should they accidentally move to the GOLGI-apparatus. Proteins not bearing this signal will automatically move with the bulk flow from the rER to the GOLGI-apparatus.

Proteins may form large oligomeric complexes with other proteins in the compartment where they belong. These complexes could be too big to be included into transport vesicles. Such aggregation can be demonstrated experimentally: If one of the sugar transferases from the *median*-GOLGI is fused with a protein bearing an ER retention signal, this enzyme will accumulate in the

ER instead of being transported to the GOLGI. But it will also retain other transferases which have not been manipulated, and in such a mutant cell the GOLGI apparatus will be very small or even non-existent. Formation of such complexes is made easier by the fact that most GOLGI-proteins have a very similar structure, an N-terminal cytoplasmic domain, a single transmembrane domain (where complex formation occurs) and a C-terminal catalytic domain facing the lumen of the cisterna.

Another hypothesis starts with the observation that the concentration of cholesterol in the membrane increases from ER over Golgi to plasma membrane. This leads to an increase in the thickness of the membranes in the same order. The transmembrane domains of Golgi proteins are slightly shorter than their counterparts in plasma membrane proteins. If such a protein were to be brought into the plasma membrane, hydrophilic sidechains from the cytoplasmic or external domains would be drawn into the hydrophobic membrane environment, which is energetically unfavourable.

16.4.2 Transport of newly synthesised proteins to their destination

Proteins destined to the cell surface or for excretion into the extracellular environment are packed into transport vesicles in the *trans*-GOLGI network (TGN). The coat protein on those vesicles has not been identified so far, initial evidence for the involvement of clathrin has turned out to be spurious.

It is important to note that some of the vesicles fuse with the cell membrane in a non-regulated, automatic fashion, while other fuse only after a specific stimulus has been received (hormones, digestive enzymes, neurotransmitter). The mechanism for protein sorting into these different classes of vesicles is not clear. However, proteins destined for regulated secretion appear to form aggregates with two proteins (chromogranin B and secretogranin II) under the low pH (6.5), high $[Ca^{2+}]$ conditions of the GOLGI-complex.

Protein secretion

Vesicles for regulated exocytosis shrink up to 200 fold after they have been formed, for this reason they show a dark core in the EM. This core is of irregular shape and can therefore be distinguished from the quasi-crystalline core of peroxisomes. During formation of the mature exocytotic vesicle cleavage of their cargo may occur to convert inactive pre-proteins to the final form. The signal for fusion between exocytotic vesicles and the cell membrane is usually an increase in (local) $[Ca^{2+}]$.

Vesicles are transported from the Golgi to the cell membrane along microtubuli. They may cover considerable distances, for example in nerve cells. This transport is inhibited by **Nocodazole**.

Plasma membrane proteins

Polar cells with distinct apical and basolateral domains sort their exocytosed proteins into different vesicles for the two domains. The signals responsible for sorting are not known. A particularly interesting case are Na/K-ATPase (normally in the basolateral domain of animal cells) and the homologous H/K-ATPase (in the apical membrane of stomach). Responsible for their sorting are the non-catalytic β-subunits. Expressing Na/K-ATPase α- (catalytic) subunit in the absence of the β-subunit leads to protein destruction by the ER-quality control system, as the protein can not fold. If the Na/K-ATPase α-subunit is expressed together with H/K-ATPase β-subunit, the chimaera is found at the apical side of the cell. Similarly, H/K-ATPase α-subunit can be directed to the basolateral side by expressing it together with Na/K-ATPase β-subunit.

Mis-directed proteins can be sorted between apical and basolateral domain by transcytosis. Only the basolateral membrane has an ankyrin/spectrin network attached to it, the cytosolic domain of proteins belonging there interact with this network, locking them to the cytoskeleton. Tight junctions (see fig. 13.4 on page 208) prevent direct exchange of proteins between apical and basolateral domain.

Protein transport to the lysosome

Lysosomal proteins are initially produced and glycosylated like those destined for export. However, in the *cis*-GOLGI one or two outer mannose residues of the high-mannose glycoproteins are phosphorylated by first adding GlcNAc-phosphate (from UDP-GlcNAc) to them and then removing the GlcNAc. The enzyme responsible for this reaction, GlcNAc-phosphotransferase, recognises lysosomal proteins by specific internal signal sequences.

The mannose-6-phosphate residues thus created tightly bind to the luminal domain of M6P-receptor in the *trans*-GOLGI-network. This receptor gets enriched in clathrin coated transport vesicles which bind to the **late endosome**.

In the slightly acidic environment of the endosome, the M6P-residues dissociate from their receptor, which is recycled back to the GOLGI-apparatus by transport (recycling) vesicles, via the cell membrane. The M6P-residues are cleaved of the lysosomal proteins by a specific phosphatase, thus preventing them from rebinding to the M6P-receptor.

From the late endosome the lysosomal enzymes move to the lysosome. It is not quite clear yet whether this happens by vesicular transport between late endosome and lysosome, or whether lysosomes are formed by maturation of the late endosome.

In any case, once the lysosomal proteins arrive in the lysosome, they undergo conformational changes in the acidic environment (pH 5.0) of the lysosome and by proteolytic removal of pre-sequences. This is required for them to become enzymatically active. The reason is obvious: The hydrolytic activity of lysosomal enzymes, turned loose in any other place, could do considerable damage to the cell.

Lysosomal storage diseases

Inherited inability to produce GlcNAc-phosphotransferase leads to **I-cell disease**, because the lysosome can not be supplied with the hydrolytic enzymes it requires for function. Instead, the proteins normally destined for transport into the lysosome are secreted. As a result glycolipids and other materials accumulate in the lysosome as inclusions.

Interestingly, the lysosomes in the hepatocytes of people with I-cell disease have a normal complement of lysosomal enzymes. Apparently in hepatocytes there must be a second pathway for sorting proteins into the lysosome, the mechanism is not known however.

There are other "lysosomal storage diseases" caused by the inability to produce one of the hydrolytic enzymes of the lysosome. The corresponding substrate will accumulate inside the lysosome in large amounts.

For example some glycogen is continuously turned over in lysosomes, the function of this pathway is unknown. In POMPE's disease (glycogen storage disease type II) the required enzyme, α-1,4-glucosidase, is defective, leading to the accumulation of glycogen in vacuoles inside the cells, for example in liver, heart and muscle. The resulting cardiomegaly leads to early death.

The degradation of proteoglycans also occurs in the lysosome and requires a chain of enzymes. If any of these enzymes is defective, **mucopolysaccharidoses** result.

The same enzymes may also be involved in the breakdown of glycoproteins and glycolipids, so that **glycoprotein storage diseases** and **sphingolipidoses** also result.

Depending on which of the enzymes is affected, the symptoms can range from bone and connective tissue deformation over corneal clouding and blindness to severe mental deficiency. Many of these disease lead to an early death.

16.4.3 Endocytosis, membrane protein recycling and transcytosis

Endocytosis is a process by which cells collect segments of plasma membrane, transmembrane proteins and ligands bound to these proteins and moves them to the endosome and lysosome. There are several reasons why a cell might want to do this:

- The plasma membrane grows by fusion with exocytotic vesicles. This excess lipid needs to be collected and recycled for vesicle formation. Such recycling is particularly important for **nerve** and **gland** cells.
- Cell surface proteins may have become damaged. They are collected and brought into the lysosomes for degradation to amino acids and sugars, which are reused.
- Cell surface receptors may have ligands bound which need to be brought into the cell. Examples would be iron or cholesterol.
- Certain plasma membrane proteins are stored in pools of intracellular vesicles in a resting cell, after hormonal activation of the cell these proteins are brought back to the plasma membrane. This avoids the delays involved in the re-synthesis of proteins, if they were destroyed after each use. Na/K-ATPase in the kidney would be an example for this type of regulation.

Formation of the clathrin coated pit

Proteins destined for endocytosis accumulate in special structures of the plasma membrane, the clathrin coated pits, where they may occur in a concentration about 1000 times higher than in the rest of the cell membrane. They bind **adaptor proteins (adaptins)** on the cytosolic side of the membrane, and clathrin assembles on these adaptors.

Three different types of adaptor proteins are known: AP1 is required for TGN to lysosome transport, AP2 for endocytosis and AP180 for recycling vesicles in brain.

AP1 and -2 consist of 2 large and 2 small subunits (see figure 16.9). One of the large subunits is required for interaction with clathrin while one of the small subunits interacts with the internalisation signal (Tyr-polar-Arg-hydrophobic) on the membrane proteins. The second large subunit becomes phosphorylated during the uncoating reaction, dephosphorylation is required before the adaptor becomes available again for coated vesicle formation. For this reason only adaptors isolated from CCV, but not from cytosol, can re-assemble into coated pits in *in vitro* assays. Nothing is known about the function of the remaining small subunit.

The clathrin molecules assemble in the clathrin coated pits initially as curved sheets, a process that can occur at 4 °C and in the absence of nucleotides. In these sheets clathrin molecules are arranged in a hexagonal lattice. As the sheet grows, this hexagonal lattice is converted into a pentagonal one, increasing curvature to form a globe, which is connected to the plasma membrane

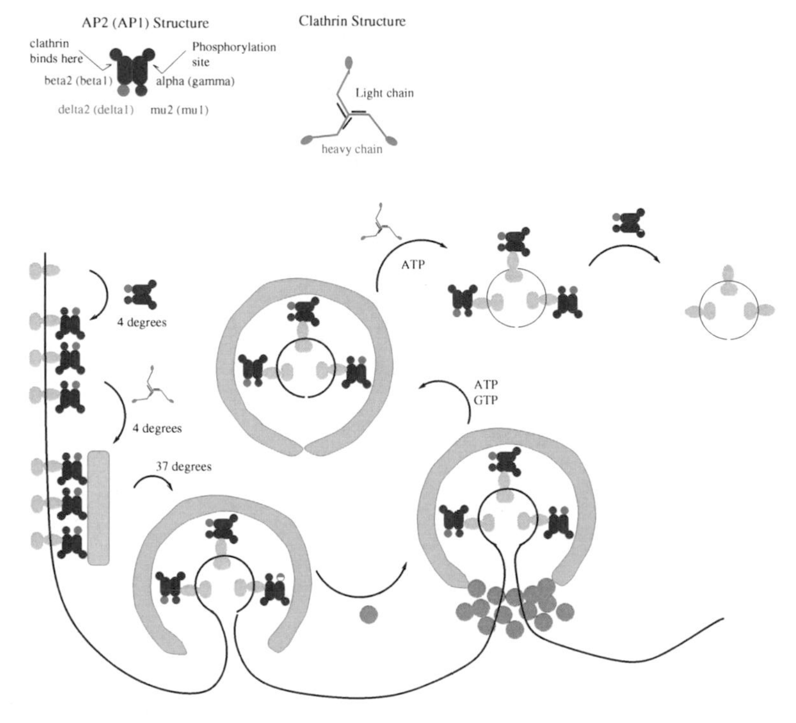

Figure 16.9. Clathrin coated vesicles in endocytosis.

by a stalk. This process is blocked by **anti-Hsc70 antibodies** and requires warming to 37 °C, but no nucleotides.

Formation of the clathrin coated vesicle

The stalk is surrounded by the protein **dynamin**, which is responsible for cutting the stalk (GTP dependent), so that free clathrin coated vesicles are formed. Dynamin binds to clathrin in its GDP form, but upon GDP/GTP exchange moves to the stalk where it self-assembles into helical stacks of rings, containing about 20 molecules of dynamin. Apparently then coordinated hydrolysis of GTP to GDP occurs to cut the stalk (thus freeing the CCV) and to release the dynamin-GDP molecules. In *Drosophila melanogaster* a temperature sensitive mutant of dynamin (*shibire^{ts}*) (jap. for paralysed) is known. At the non-permissive temperature these flies can no longer recycle lipids in their nerve cells, which leads to immotility. This makes it easy to collect affected animals for investigation. In mammals two isoforms of dynamin are known, dynamin-1 occurs only in neurons while dynamin-2 occurs in all cells.

Uncoating of clathrin coated vesicles

Because of their clathrin coats, clathrin coated vesicles can not fuse directly with the endosome. And because clathrin coats form spontaneously, energy is required to disassemble them again. This reaction is catalysed by a molecular chaperone, **Hsc70** (see page 212), also known as the uncoating ATPase. There is an equilibrium between bound and free clathrin arms. Hsc70·ATP binds to exposed clathrin arms, preventing them from rejoining the coat. This shifts the equilibrium from coat assembly to coat dissolution. Hydrolysis of bound ATP makes this process irreversible, until the ADP produced is exchanged against ATP. It is easy to see that specific ATP/ADP exchange factors at the plasma membrane would stimulate the formation of new clathrin coated pits, while ATP-hydrolysis stimulating factors at the CCV would stimulate the uncoating of CCV. Effective uncoating of CCV requires the presence of **auxilin** in the coat, and auxilin has a DnaJ-like domain. It is presently unclear however whether it actually functions as ATP-hydrolysis stimulating factor.

It has been shown that the formation of CCV requires a **phosphatidyl-inositol-3-phosphate kinase activity** in the plasma membrane to increase the stability of clathrin coats. A corresponding phosphatase (synaptojanin 1) is included in the CCV [51], its action provides a kind of timed fuse, decreasing the stability of the clathrin coats some time after vesicle formation. This starts the uncoating process.

One of the proteins endocytosed from the plasma membrane is **Na/K-ATPase**, which is therefore present in early endosomal membrane. It is recycled to the cell membrane and does not enter late endosomes and lysosomes. This has an important consequence:

The Na/K-ATPase is electrogenic, pumping 3 Na^+ from the cytoplasm into the exterior of the cell (or the endosomal lumen) for every 2 K^+ pumped into the cytoplasm. Na/K-ATPase activity therefore leads to a positive charge inside the early endosome, which limits the accumulation of protons —which also bear a positive charge— by a V-type H-ATPase. For this reason the pH inside early endosomes is 5.5 – 6.0. In late endosomes and lysosomes (which do not contain Na/K-ATPase) pH can be decreased to 5.0 by the H-ATPase.

Endocytosis and exocytosis are fast, high-throughput processes, about 2% of the cell membrane are covered with coated pits, each such pit turns into a CCV within about 1 min. About 2500 CCV are formed in a cell every minute, the whole plasma membrane surface is endocytosed (and exocytosed) statistically every 30 min.

The transferrin cycle

Transferrin is a soluble protein found in serum. Its function is to bind iron and transport it to the cells in which this nutrient is needed, for example in the bone marrow. These cells express a receptor on their surface, which, under neutral conditions, has a high affinity for the iron-carrying (holo-), but a low affinity for the iron free (apo-) transferrin. This is important, because in healthy people only about 1/3 of the transferrin molecules in serum carry iron. Once the receptor has bound holo-transferrin, it is collected in clathrin coated pits and is transported to the endosome via clathrin coated vesicles.

Inside the endosome the pH is acidic, around ≈ 5.5. Under these conditions, Fe is released from the receptor and becomes available to the cell. The acidic conditions have a second effect however: The affinity of the transferrin receptor for the apo-transferrin increases. Thus receptor and transferrin stay together while they are brought back to the cell surface by transport vesicles.

On the cell surface the receptor/apo-transferrin complex is exposed to a neutral pH again. As mentioned before, the transferrin receptor has a low affinity

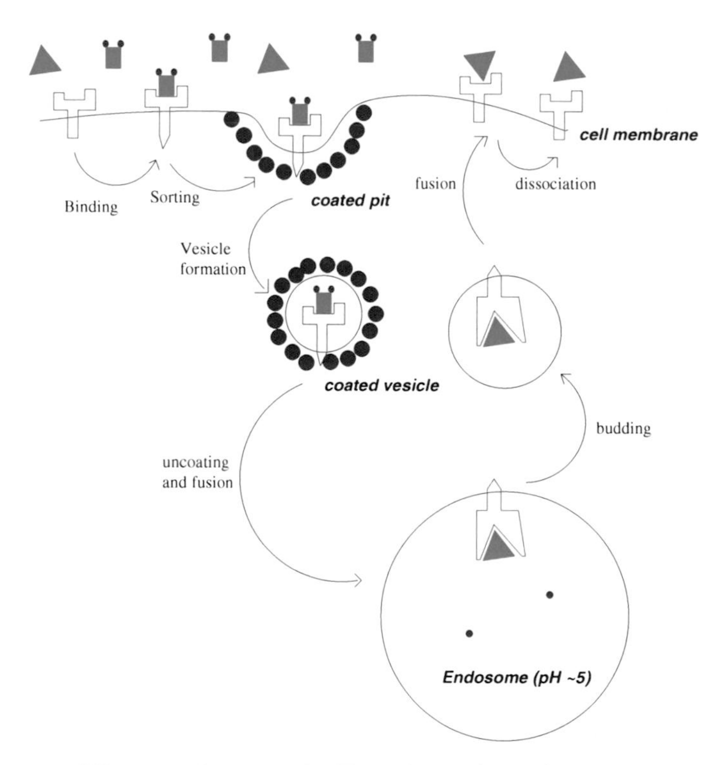

Figure 16.10. The transferrin-cycle: Transferrin (green) with bound iron (red) binds to the transferrin receptor (pink), which is then sorted into clathrin (blue) coated pits. The complex moves into endosomes via clathrin coated vesicles, the low pH inside this organelle releases the Fe and changes the conformation of the transferrin receptor. After return to the cell surface the complex is exposed to a neutral pH and the receptor assumes its normal conformation, leading to dissociation of the apo-transferrin

for apo-transferrin at neutral pH, thus the complex dissociates. The receptor is now available again for binding of holo-transferrin, while the apo-transferrin returns to the liver to bind new iron.

Cholesterol uptake

Cholesterol is an essential component of the cell membrane in eukaryotes. It is either taken up from food and transported to the liver, or synthesised there. Cholesterol in blood is transported as **very low-density lipoprotein (VLDL)**, which contain cholesterol, phospholipid and proteins. They are bound on the surface of cells by a specific low-density lipoprotein (LDL)-receptor. Receptors with bound VLDL accumulate in clathrin coated pits and are endocytosed. The lipid portion goes the endosome-lysosome route and is made available to the cell. The receptors are recycled to the cell surface, proteins and some remaining lipids are released back into the blood stream. Removal of the (low density) lipids results in an increased concentration of (relatively high density) proteins, thus **high density lipoprotein (HDL)** can be distinguished from the VLDL by centrifugation. HDL returns to the liver for reloading with cholesterol. If the cell can not use all the cholesterol supplied, some of it is returned to the surface too and lipoproteins of intermediate density are formed: **low-density lipoprotein (LDL)** ("bad cholesterol"). LDL is involved in the formation of arteriosclerotic plaques. HDL can bind free cholesterol and thus prevents arteriosclerosis (hence "good cholesterol").

Patients with mutated LDL-receptors can not efficiently take up cholesterol by this route. This leads to the accumulation of LDL in the blood, known as **familial hypercholesterolaemia**. Cause can be the inability to produce LDL-receptor, the inability of the receptor to bind LDL or failure to sort the receptor into clathrin coated pits.

Downregulation of receptors

Some cell types require **epidermal growth factor (EGF)** to initiate cell division. Binding of EGF to its receptor leads to activation of its **tyrosine kinase** activity, signaling the cell to divide (see fig. 8.1 on page 122). The receptor-hormone complex is internalised by endocytosis, unlike the transferrin or LDL receptors it is not recycled to the cell membrane again, but both hormone and receptor are digested in the lysosomes. Thus an increase in EGF concentration in blood will lead to an increase in receptor internalisation and a decrease in the receptor number in the cell membrane. This "**downregulation**" of receptor density prevents overstimulation of the cells. G-protein coupled receptors also are endocytosed in a manner that appears similar to tyrosine kinase receptors like EGF-R.

Internalisation of receptors occurs by clathrin coated vesicles, but the mechanism appears to be somewhat different than that of the normal endocytotic pathway. Cytosols from different cell types have similar ability to support transferrin receptor endocytosis, but differ in their ability to support the endocytosis of EGF-R. EGF-R endocytosis is not limited by the availability of AP-2.

Recycling vesicles in brain

Release of neurotransmitter in the synapses of nerve cells occurs by fusion of transmitter-carrying vesicles with the synaptosomal membrane. This obviously results in an increased lipid content of the praesynaptic membrane. That lipid needs to be recycled, and clathrin coated vesicles are used to achieve this.

The synaptic vesicles are then formed by budding from the endosome, and filled with neurotransmitter *en route* to the plasma membrane. Thus for exocytosis in nerve cells the GOLGI-apparatus is not directly involved (although of course the proteins required ultimately come from there).

If one looks carefully, recycling vesicles isolated from brain have some properties which are quite different from those of "normal" clathrin coated vesicles (which occur in all tissues including brain).

Analytical ultracentrifugation of CCV from most tissues (placenta is a convenient source) shows a fairly homogeneous population of about 157 S. If the same experiment is performed with brain CCV, two population with sedimentation constants of 145 and 124 S are seen. While the larger vesicles presumably correspond to the "normal" CCV, the smaller ones are the recycling vesicles.

The recycling vesicles are less stable than normal CCV against *p*H changes, and they are a better substrate for Hsc70. Uncoating by Hsc70 of recycling vesicles can occur even under conditions where ATP hydrolysis is blocked, this is not the case for CCV from placenta or liver.

If isolated clathrin coats from placenta or liver are incubated with purified Hsc70, the uncoating occurs in a stoichiometric manner, while coats from brain are uncoated catalytically by Hsc70.

It would thus appear as if the clathrin coated recycling vesicles from brain are unusual and adapted to rapid turnover. This makes evolutionary sense, as it reduces the refractory period in nerve cells.

Transcytosis

In the examples discussed above receptors were internalised from the apical surface of the cell and returned to there after unloading their ligand in the endosome for recycling. However, in some cases the proteins are send to the basolateral surface instead. For example, antibodies in milk bind to specific receptors in the apical (luminal) surface in the intestine of a suckling, are internalised with their receptor in CCV and transported to early endosomes. There the receptor-antibody complex is packed into transport vesicles which fuse with the basolateral surface of the cell. The neutral *p*H of the extracellular fluid leads to receptor-ligand dissociation and the antibody is released into the sucklings blood stream.

In the mother the transport occurs by a similar mechanism, but in the opposite direction: from blood antibodies are bound to the basolateral surface of the epithelium in the milk gland and transported to the apical surface and into the milk.

Epithelial cells have different early endosomes for their apical and basolateral surfaces, these are linked by a common late endosome.

Phagocytosis

Endocytosis as described above (also known as **pinocytosis**, gr. cell drinking) works only for small particles ($\leq 150\,\mathrm{nm}$), bigger particles are taken up by phagocytosis (gr. cell eating). While most cells in our bodies are capable of pinocytosis, phagocytosis is carried out by specialised cells, in particular macrophages. Phagosomes are not coated, and their size is determined by the size of the prey.

A/B-toxins

The retrograde pathway of protein transport was highjacked by a group of bacterial and plant toxins, known as **A/B-toxins**. Examples include the **cholera-toxin** (from *Vibrio cholerae*) and **ricin** from the castor bean, *Ricinus communis*. These toxins consist of two protein chains, linked by a disulphide bond. The non-toxic β-subunit binds to the plasma membrane, the entire toxin is endocytosed and transported to the early endosome. From there it is transported back to the ER, since the β-subunit contains the K/HDEL-sequence required for this route. In the ER α- and β-subunit separate, the α-subunit is ubiquitinated and transported into the cytosol using the ERAD-pathway. By an unknown mechanism the α-subunit evades hydrolysis in the proteasome, instead it causes havoc in the cell.

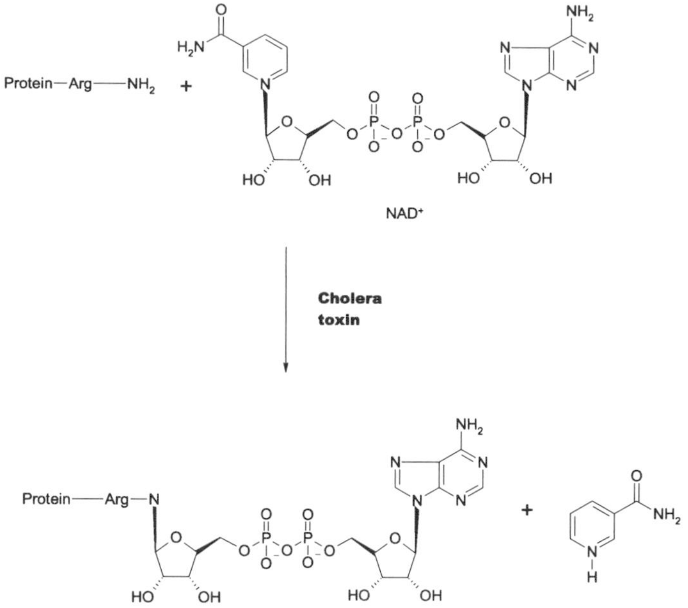

Figure 16.11. ADP-ribosylation of the α-subunit of G_s by cholera toxin blocks its GTPase function, leading to constant activation.

Cholera toxin acts by transferring ADP from NAD^+ to an Arg-residue in the α-subunit of G_s (see fig. 16.11), blocking its GTPase activity and rendering G_s permanently active. The chronic activation of cAMP-synthesis leads to the secretion of bicarbonate and chloride ions into the intestinal lumen, the resulting passive water efflux into the lumen leads to diarrhoea, which aids in the spread of the pathogen (but can kill the patient from dehydration and mineral loss).

A single molecule of ricin can prevent protein synthesis in a cell by enzymatically depurinating a specific adenosine in the 23 S-rRNA, making it one of the most potent poisons known to man. This high toxicity has kindled the interest of researchers working on anti-cancer drugs. If the α-subunit could be brought specifically into a cancer cell (for example using antibodies against tumor markers as vehicle), it would act as a "magic bullet". Unfortunately despite intense efforts of many groups, no success has been achieved yet.

Exercises

16.1. Which of the following statements is *false*: Disulphide bonds

a) are formed by oxidation of thiol groups.

b) are formed between cysteine residues.

c) hold the heavy and light chains of immunoglobulins together.

d) formation in the ER is controlled by specific enzymes.

e) stabilise the T-form of haemoglobin.

16.2. Which amino acid side chains bear O-linked carbohydrate trees? 1) Threonine, 2) Tyrosine, 3) Serine 4) Asparagine

16.3. The following peptide is a reducing agent
because
a disulphide bond can form between two such molecules.

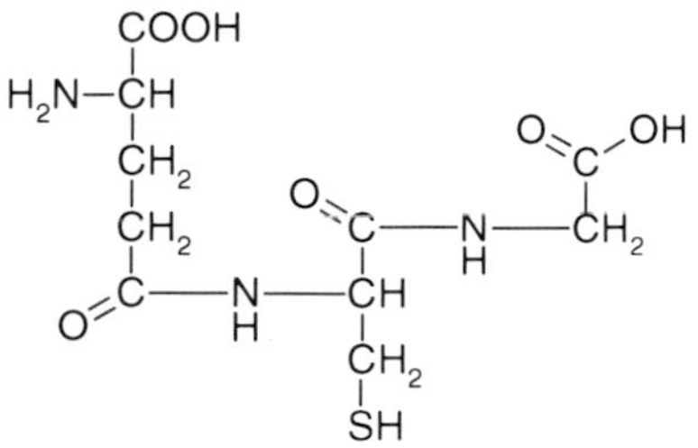

COOH
H_2N–CH
CH_2
CH_2
O=C–N(H)–CH–C(=O)–N(H)–CH_2–C(=O)OH
CH_2
SH

16.4. Which of the following statements is/are *true* about protein transport in cells?

1) Nuclear encoded mitochondrial proteins carry a signal peptide which forms an amphipatic α-helix.

2) Proteins to be secreted are synthesised on ribosomes which are attached to the endoplasmic reticulum.

3) Plasma membrane and secreted proteins usually are glycosylated.

4) Transport from the *trans*-GOLGI-network to the endosome occurs by clathrin coated vesicles.

16.5. Which of the following statements is *false* about membrane proteins?

a) Proteins with a single stop-transfer sequence have a cytosolic C-terminus.

b) The cytosolic end of transmembrane segments contains positively charged amino acids.

c) Glycosyl-phosphatitylinositol- (GPI-) residues can link proteins to the luminal leaflet of the ER membrane

d) Fatty acid residues (palmitoyl- or myristoyl-groups) or isoprenoids (farnesyl- or geranylgeranyl-groups) can link proteins to the cytosolic leaflet of the ER membrane.

e) Proteins with several membrane spanning domains (type 3 membrane proteins) interact with the signal recognition particle once for each transmembrane domain.

16.6. Q: Cardiolipin is characteristic for the membrane of

a) erythrocytes.

b) mitochondria.

c) the endoplasmic reticulum.

d) myelin sheets.

e) lysosomes.

16.7. Which of the following lipids is a major component of the outer leaflet of the plasma membrane?

a) phosphatidyl inositol

b) phosphatidyl ethanolamine

c) cardiolipin

d) cholesterol

e) triacylglycerol

16.8. Which of the following statements is *false*? Proteins insert into or attach to bilayer membranes by

a) hydrophobic α-helices.

b) non-covalent bonds.

c) sugar chains in N-glycosidic bonds.

d) lipid anchors.

e) β-barrels.

16.9. Which of the following statements is *false* about the plasma membrane?

a) Outer and inner leaflet of the plasma membrane have different composition.

b) The cytosolic part of transmembrane proteins may have tyrosine kinase activity.

c) The sugar chains of glycoproteins may face the cytosol.

d) The plasma membrane contains particularly high concentrations of cholesterol.

e) The plasma membrane of animal cells separates the extracellular medium with high $[Na^+]$ and low $[K^+]$ from the cytosol with low $[Na^+]$ and high $[K^+]$.

17

Transport of solutes across membranes

Cells are surrounded by a cell membrane, which consists of a double layer of lipid molecules. Only small, lipophilic (fat-loving) substances (gases, ethanol) can pass this membrane unaided, most biologically relevant molecules are hydrophilic (water-loving), they can not pass the cell membrane freely. Water too can penetrate the membrane only slowly. This allows the cell to actively maintain an internal composition that is different from the outside environment, one of the characteristics of life.

Even in equilibrium ion concentrations on both side of the plasma membrane are not equal, because proteins (most of which are charged at physiological pH) can not diffuse across. Since the number of positive and negative ion species must be equal on both sides of the membrane, this also influences the movement of small ions, resulting in an electrochemical potential named DONNAN-potential after its discoverer[1].

17.1 Passive diffusion

Hydrophobic molecules can enter a cell by first partitioning into the extracellular leaflet of the plasma membrane, flipping to the cytosolic leaflet and then partitioning into the cytosol. Of these steps, flipping across the membrane is the slowest, because of the high viscosity of phospholipids (about 1000 times the viscosity of water) .

Partitioning is determined by the relative equilibrium concentration of a substance in water $[S]_a$ and in phospholipid $[S]_l$ (in the lab, octanol is often used as a model for phospholipids). This ratio is called **partitioning coefficient** $K_p = \frac{[S]_a}{[S]_l}$.

[1] See `http://entochem.tamu.edu/Gibbs-Donnan/index.html` for a nice video explaining the DONNAN-potential.

The rate ($\frac{dn}{dt}$, in $\frac{\text{mol}}{\text{s}}$) of transport of a substance across a membrane depends on the area of membrane (A), the partitioning coefficient (K_p), the **diffusion coefficient** inside the membrane (D, depends on the viscosity of the lipid and the size and shape of the substrate), the thickness of the membrane (d, 2.5 – 3.0 nm for a phospholipid bilayer) and the concentration difference across the membrane according to FICK's law:

$$\frac{dn}{dt} = \frac{A}{d} * K_p * D * ([S]_i - [S]_o) \tag{17.1}$$

Most biologically relevant substances have similar diffusion, but quite different partitioning coefficients. Thus in first approximation passive transport of a substance depends on its partitioning coefficient.

As apparent from eqn. 17.1, the rate of passive diffusion through the membrane increases linearly with the concentration difference. If on the other hand the transport is mediated by transport proteins, the rate increases hyperbolically with concentration difference. Only a limited number of such proteins are present in the membrane, and each transport process needs a finite amount of time. Thus transport proteins follow HENRI-MICHAELIS-MENTEN-kinetics and passive diffusion and transport can be distinguished simply from their kinetic behaviour.

17.2 Transporters

In general, cell content is higher in K^+ and Mg^{2+} and lower in Na^+, Ca^{2+} and Cl^- than the interstitial fluid (or the substrate in case of single cell organism). Also, the cytosol is reducing.

However, passive leakage through the membrane is not zero, and over time those gradients would collapse if they were not actively maintained. At the same time, nutrients need to enter the cell and waste products need to leave.

For these purposes, cell membranes contain proteins which allow the selective passage of molecules and ions through "channels". Polarised cells (for example in epithelia) need to have different sets of transporters on the apical and basolateral surface in order to achieve substrate transport across the cell.

Transporters come in 3 basic flavours:

primary active transporters ("pumps") use the energy obtained from the hydrolysis of molecules like ATP, GTP, pyrophosphate or phospho*enol*pyruvate to actively pump molecules across the membrane against the concentration gradient. In animals, the most important pump is the Na/K-ATPase, which hydrolyses ATP to ADP and P_i to transport Na^+ out of, and K^+ into the cell. In plants, bacteria and fungi a similar function is performed by a H^+-ATPase, which pumps hydrogen ions out of the cell. Some bacteria can use Na-ions instead of protons. Under the right conditions, ATP-driven pumps can run in reverse and use the energy of an ion gradient to convert ADP and P_i into ATP (ATP-synthase).

secondary active transporters (co-transporters) use the energy of the gradients established by the pumps to drive transport of nutrients, hormones and waste products into or out of the cell. We distinguish

antiporters where the transport of the substrate occurs in the opposite direction from the coupling ion

symporters where both substrate and coupling ions flow in the same direction.

Secondary active transport can work against concentration gradients just like primary active transport.

passive transporters speed up the equilibration of concentration gradients across the membrane by facilitated diffusion. They can work either

as gated channels where the substrate flows through a pore in the protein which controls the kind of substrate that can pass, or

as shuttles which bind the substrate on one side of the membrane, then flip around to release the substrate on the other side

17.2.1 Primary active transporters (pumps)

Most pumps use the hydrolysis of ATP to ADP and P_i as energy source. However, there are some important exceptions. In the plant tonoplast membrane, a H^+-pyrophosphatase splits pyrophosphate (which is a by-product of metabolism) and uses the energy to pump protons into the vacuole. Since the pyrophosphate needs to be split into 2 phosphate ions anyway, the plant can conserve some energy by using it as fuel.

Phospho*enol*pyruvate is a super-energy-rich compound produced during the breakdown of sugar. It can be used to produce ATP from ADP, but some of the energy stored in PEP is lost as heat. Using PEP directly to drive transport can reduce this loss. Transporters able to do that are found in some bacteria.

Transport ATPases can be grouped into 5 families, based on subunit composition, protein sequence, reaction mechanism and inhibitor profiles.

E_1E_2-ATPases are found in the plasma membrane and in the endoplasmic reticulum. These enzymes occur in two principal conformations, called E_1 (with high affinity for ATP) and E_2 with low affinity. The phosphate group of ATP is transferred onto an aspartic acid residue of the enzyme,

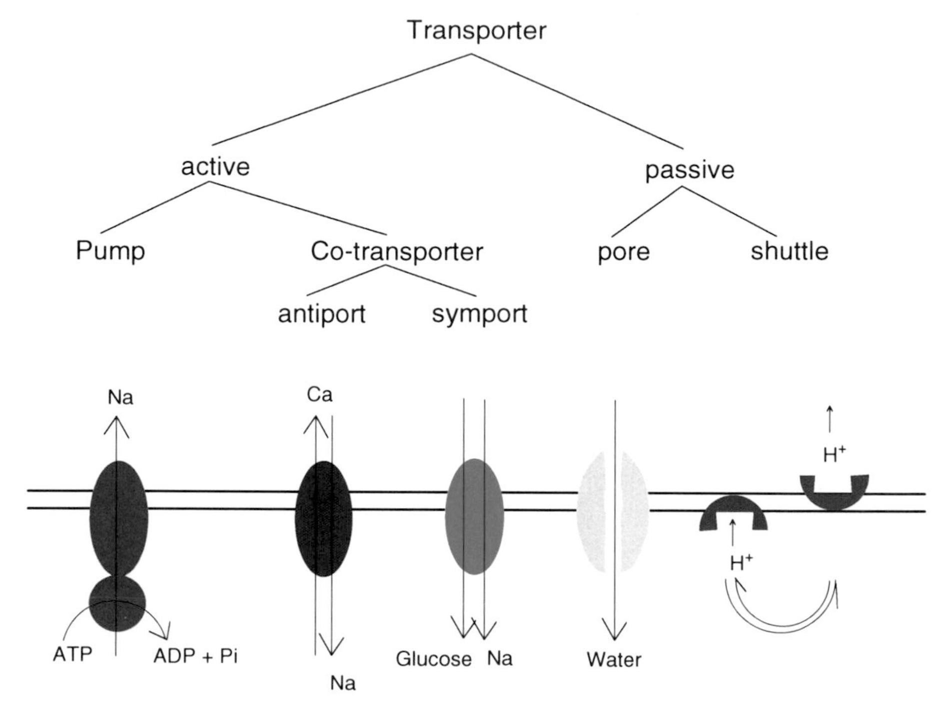

Figure 17.1. The basic mechanisms of transmembrane transport. Primary and secondary active transport can work against a concentration gradient across the membrane, passive transporters merely speed up the equilibration of gradients.

forming a phospho-intermediate. Hence this class of ATPases is some times also called **P-type**. **Vanadate** is able to replace the phosphate in the E-P·ADP state, inhibiting the enzyme with very high affinity.

F_1F_o-ATPases are found in the membranes of bacteria, mitochondria and plastids. This enzyme uses a proton (in some bacteria also Na^+) gradient across the membrane to drive the formation of ATP from its precursors ADP and P_i, i.e. it acts as an ATP-synthase rather than an ATP-hydrolase. The enzyme consists of an transmembrane part (called F_o) which is driven by the ion gradient like a turbine by water and drives ATP-synthesis in the F_1-head. Each of these parts consists of several protein subunits. The transmembrane part binds the inhibitor **oligomycin** with high affinity, hence the name.

A_1A_o-ATPases are very much like F_1F_o-ATPases, but are found in archaebacteria. The protein-sequences of their subunits are different enough from F_1F_o-ATPases that A_1A_o-ATPases are singled out.

V-type ATPases pump protons into the organelles of eukaryotes, like vacuoles (name!), endosomes, lysosomes or exocytotic vesicles, but not the ER.

They also occur in the plasma membrane of some acid secreting cells like osteoclasts (cells that degrade bone). V-type ATPases are inhibited by **bafilomycin A** and by **nitrate**. Structurally, V-type ATPases are similar to F-type, but a considerable evolutionary distance of the sequence results in different behaviour towards their respective inhibitors.

ABC-type ATPases (ATP-binding cassette) form the largest family of active transporters, with some 500 identified members. Originally, these transporters consisted of 2 transmembrane subunits, 2 cytosolic subunits which bind and hydrolyse ATP, and a regulatory subunit. Nutrient importers have an additional extracellular binding protein, which binds the substrate with high affinity and makes it available for transport. No specific inhibitors of this family is known, but they are inhibited by vanadate, even though apparently they do not form a phospho-intermediate. ABC-transporters pump nutrients into the cells, waste products and toxins out and metabolites from cytosol to organelles or *vice versa.* Mutations in some ABC-ATPases cause some important inherited diseases like cystic fibrosis.

E_1E_2-ATPases

P-ATPases occur in cell membranes and in the endoplasmic reticulum.

In the sarcoplasmic reticulum (ER of muscle cells) a **Ca-ATPase** is responsible for the accumulation of Ca inside the ER, upon activation of the muscle this Ca is released into the cytosol and causes contraction. Two different forms of this sarcoplasmic Ca-ATPase exist, found in slow-twitch and fast-twitch fibres, respectively.

One of the **aminophospholipid transporter**, responsible for maintaining the difference in composition between inner and outer leaflets of the cell membrane also is a P-type ATPase.

Copper is taken up in the intestine and excreted into bile by two different P-type ATPases. Deficiency in these transporters causes MENKES- and WILSON-disease, respectively.

Some bacteria have P-type transporters that export **toxic heavy metals** like Cd or Hg from their cytosol. Others take up nutrients like the **KDP**-system in *E. coli*, which transports potassium.

In the acinar cells of the stomach a H/K-ATPase pumps **protons** into stomach lumen, causing the low *p*H of stomach juice.

The stomach H/K-ATPase used to incite considerable pharmacological interest, as over-acidification of the stomach causes ulcer, a potentially life-threatening disease. Blockers of H/K-ATPase were used to treat this condition, until it turned out that the primary cause of over-acidification is an

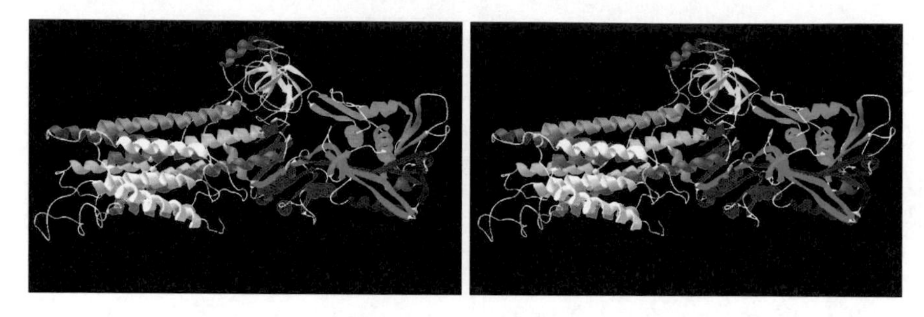

Figure 17.2. Stereo view of the crystal structure of sarcoplasmic Ca-ATPase from rabbit muscle at 2.6 Å resolution (PDB-code 1EUL). 2 Ca-ions (purple) are bound in the transmembrane part, which is formed by α-helices. On the right is the cytoplasmic section with binding sites for ATP.

infection of the stomach with the bacterium *Helicobacter pylori*, which can be treated with antibiotics (NOBEL-Price for Medicine 2005 to B. MARSHALL and R. WARREN). This discovery was originally heavily disputed by the pharmaceutical industry, because H/K-ATPase blockers needed to be taken for the rest of the patients life, while antibiotics treatment takes only about two weeks, a considerable loss of revenue.

Today H/K-ATPase blockers are still used to treat psychogenic (stress-related) overacidification of the stomach, of course a change in patients life style would be the more appropriate response.

The plasma membrane of all eukaryotes is energised by P-type ATPases, Na/K-ATPase in animals, H-ATPase in plants and fungi. Na/K-ATPase pumps 3 Na out of, and 2 K ions into the cell for each molecule of ATP hydrolysed. Because more positive charges are pumped out than in, Na/K-ATPase is **electrogenic** and partly responsible for the 70 mV potential difference across the cell membrane. Most secondary active transporters in eukaryotes are powered by the Na-gradient. Additionally, the electrochemical gradient across the cell membrane can be quickly dissipated by passive channels, their opening is the molecular basis for the **excitability** of nerve, gland and muscle cells.

We have some knowledge about the reaction mechanism of P-type ATPases, most of our understanding was gained from work on Na/K-ATPase (see fig. 17.3).

Na/K-ATPase in its E_1-conformation binds 3 cytosolic Na ions and ATP with high affinity. ATP is hydrolysed, the terminal phosphate group is transferred onto an aspartic acid residue of the enzyme.

When substrates are transported, they need to pass a **channel** through the enzyme. This channel must never be open at both ends at the same time,

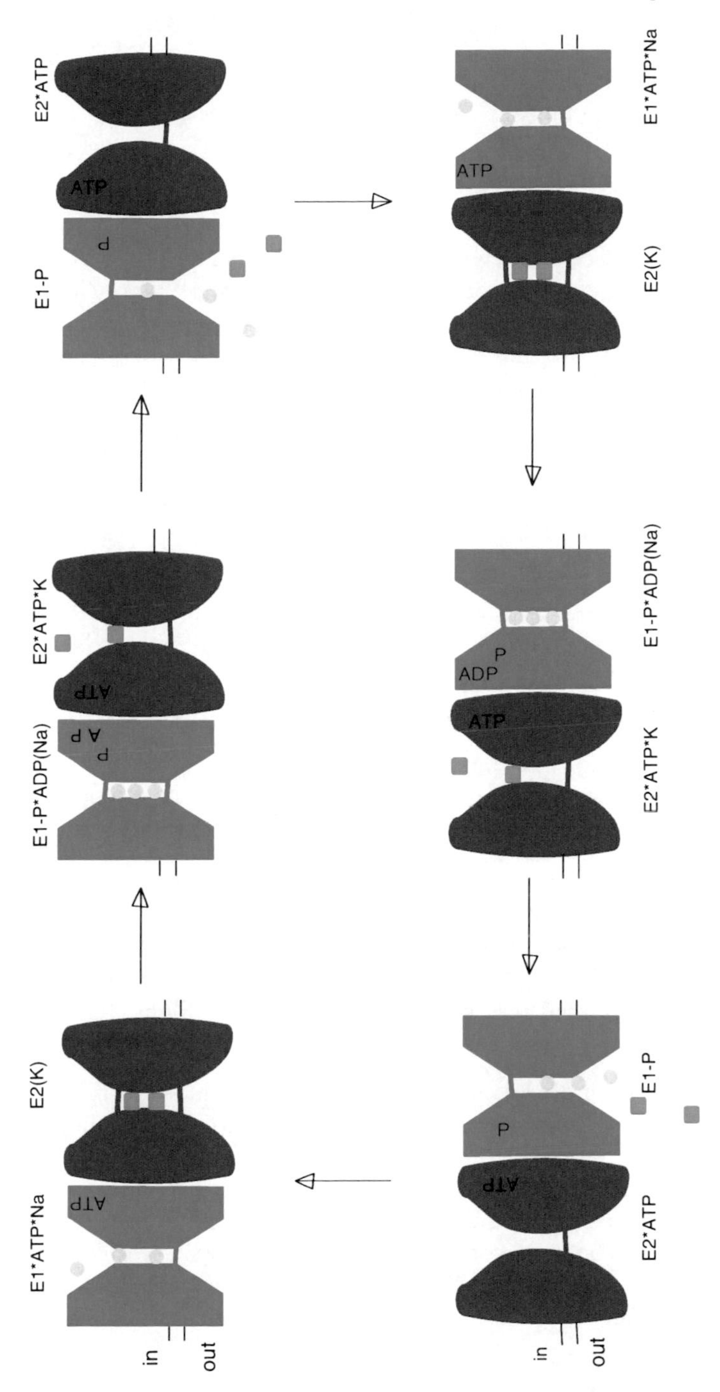

Figure 17.3. Reaction cycle of Na/K-ATPase as an example of P-type ATPases. Note the collaboration of 2 protomers, one in E_1, the other in E_2-conformation. Some investigators claim that P-type ATPases work as even larger oligoprotomers.

as this would result in free movement of substrate and the break-down of any concentration gradient. Thus there need to be gates at both ends of the channel, and at least one of them must always be closed (similar to an airlock). This implies that there must be a state were the substrate is inside the channel and both gates are closed, so that the substrate is not accessible from the outside. This is called **occlusion**.

The enzyme releases ADP while delivering the Na ions to the extracellular space. It converts to the E_2-form, which binds 2 extracellular K ions. These are occluded while the phosphate group is transferred to water and released. Low-affinity ATP-binding causes release of the K-ions to the cytosol and conversion of the enzyme to the E_1-form, which closes the cycle.

In actual fact there are at least 2 molecules of Na/K-ATPase co-operating in this cycle, while one is in the E_1-conformation, the other is in E_2 and *vice versa*. Thus the energy of ATP hydrolysis from the E_1-protomer can be used to release K ions from the occluded state. Some investigators even claim that 4 or more subunits have to cooperate, this is a topic of ongoing research.

Na/K-ATPase is the receptor for the hormone ouabain, which is produced in the adrenal glands and involved in the regulation of blood pressure. The Na-gradient build by Na/K-ATPase across the cell membrane is used to drive a Na/Ca-exchanger, which reduces the cytosolic Ca-concentration. If Na/K-ATPase is inhibited by **ouabain**, Ca transport out of the cell is reduced and the cytosolic Ca-concentration will increase. This will activate muscle contraction in the heart and in arterial walls, thus increasing pumped blood volume (**inotropic effect**) and blood pressure. Also, the hearts beating frequency is reduced.

Na/K-ATPase occurs in our bodies in several **isoforms**, which are expressed in different tissues and have different affinities for ouabain. Kidney and erythrocytes for example express the α_1-isoform with low affinity for ouabain, while heart and arterial wall muscle cells express the highly ouabain-sensitive α_3-isoform.

Ouabain itself is not absorbed in the intestine, and thus of little pharmacological use. However, some plants (most notably *Digitalis ssp.* the foxglove [see fig. 17.4]) produce related compounds which can be used to increase pumped blood volume in elderly patients with congestive heart failure. Foxglove has been in use for this purpose for some 400 years and has prolonged the life of many a person. Today of course preparations of purified digitalis-glycosides are used, which allow more precise dosing. This is important as the therapeutic bandwidth (difference between effective and lethal dose) of digitalis-glycoside is very small.

On the other hand, excessive secretion of ouabain into the plasma leads to primary high blood pressure, a serious condition that can result in heart failure. This can be reproduced in animal experiments by injection of ouabain.

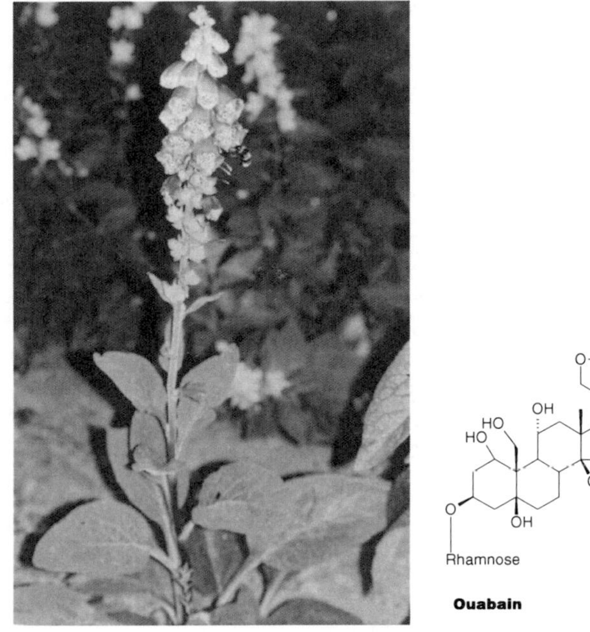

Figure 17.4. The foxglove (here red foxglove, *Digitalis purpurea*) has been used for centuries to increase heart output in elderly patients. It contains digoxin, which acts (at least in part, see text for further discussion) like the hormone ouabain which regulates the activity of our heart. Both are glycosides of modified steroids, the sugar moiety increases solubility, but has little effect on interaction with Na/K-ATPase.

Interestingly, therapy with digoxin does not lead to high blood pressure, the reason for that difference is not understood. Some effects of ouabain in animal experiments occur at concentrations that are lower than those required for measurable inhibition of Na/K-ATPase, possibly there are other receptors as well.

Most P-type ATPases consist of a single subunit (α-subunit) of about 112 kDa, but Na/K-ATPase in the animal cell membrane and the H/K-ATPase in stomach have a second, non-catalytic β-subunit of 47 kDa, the KDP-system in *E. coli* has three subunits.

The β-subunit acts as a scaffold during folding of the catalytic α-subunits. If either Na/K-ATPase or H/K-ATPase are expressed in genetically engineered cells without a β-subunit, the proteins are made on the ribosomes, but can not fold correctly in the ER and are destroyed by the quality control system of the cell. The β-subunit also acts as a sorting signal. If Na/K-ATPase α-subunit is expressed in cells together with the H/K-ATPase β-subunit, functional Na/K-ATPase enzyme is found on the apical, rather than the basolateral side of the cell. Similarly, if H/K-ATPase α-subunit is expressed together with

Na/K-ATPase β-, we get functional H/K-ATPase, but at the basolateral rather than the apical side of the cell.

F_1F_o-ATPases

The F_1F_o-ATPase is a splendid molecular machine. (P. BOYER)

F_1F_o-ATPases (or F-type ATPases for short) are found in the plasma membrane of bacteria and in the organelles derived from them (mitochondria and plastids). They act as **ATP-synthases**, using the chemosmotic energy of a proton gradient across the membrane to drive ATP-synthesis (MITCHELL-hypothesis, NOBEL-Price 1978). Remember that reactions are always written in the direction of negative ΔG^0, thus ATP is the substrate and ADP + P_i are products, even if the reaction occurs in the opposite direction under physiological conditions.

In the mitochondria food is oxidised to carbon dioxide and activated hydrogen (NADH + H^+ and $FADH_2$, respectively). The activated hydrogen is then oxidised to water, the resulting energy is used to transport protons across the inner mitochondrial membrane into the intermembrane space:

$$NADH + \frac{1}{2}O_2 + 11H_i^+ \rightarrow NAD^+ + H_2O + 10H_o^+ \tag{17.2}$$

In plastids and photosynthetic bacteria the energy of light is used to split water:

$$H_2O \rightarrow \frac{1}{2}O_2 + 2H_i^+ + 2e^- \tag{17.3}$$

The electrons are used to reduce NAD^+ to NADH + H^+ for sugar synthesis, during this process further protons are pumped from the stroma into the thylakoid lumen. So 9 – 12 protons assemble inside the thylakoid lumen for every molecule of water split, and can be used to drive ATP-synthesis.

The difference between mitochondria and plastids is that in mitochondria proton flow is from the intermembrane space into the matrix (outside to inside), while in plastids the flow is from thylakoid lumen out into the stroma. Accordingly, orientation of the F_1F_o-ATPases is reversed.

F-ATPases consist of a water soluble head (the F_1-part) and a transmembrane part (called F_o, because it binds the inhibitor **oligomycin**). F_1 can be isolated and has ATPase-, but no ATP-synthase activity. F_o alone is a unregulated proton channel through the membrane, it can not bind nucleotides. Only when both parts are brought back together can ATP-synthesis (or ATP-driven

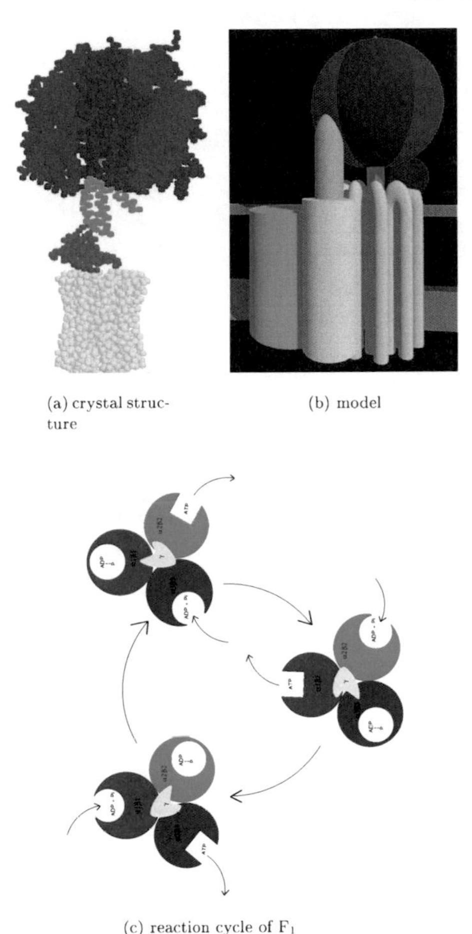

(a) crystal structure

(b) model

(c) reaction cycle of F_1

Figure 17.5. A: F_1F_o-ATPase from yeast mitochondria, crystallised with polyethylene glycol in the presence of ADP and the non-hydrolysable ATP-analogue AMP-PNP (PDB-code 1QO1). Resolution is 3.5 Å, allowing only the carbon backbone to be traced, and that only in some of the subunits. **B**: Schematic diagram. α- and β-subunits (3 each, red and magenta) form the bulk of the F_1-head, the γ and ϵ subunits (green and blue) connect this head to the hollow transmembrane cylinder formed by 10 molecules of subunit c (yellow, in some other species up to 14 c-subunits are present). During turnover the γ-subunit rotates around its long axis. The proton channel is formed by subunits a (cyan) and c (yellow), the subunits b (white) and δ (kaki) prevent rotation of the F_1-head. **C**:Reaction cycle of the F_1-part of ATP-synthase. Rotation of the γ-subunit (driven by proton flow through the F_o-part) makes each of the $\alpha\beta$-protomers assume 3 different conformations in sequence: An open, substrate binding conformation, a closed conformation where ATP-synthesis occurs and an open, product-releasing conformation.

proton transport against a concentration gradient under certain experimental conditions) occur (see fig. 17.5).

The human body produces and uses about 40 kg of ATP each day, each molecule of ATP is converted to ADP and back about 1000 times a day (about once every 9 s). Transmembrane movement of 3–5 protons is required for the production of one molecule of ATP, depending on the number of c-subunits. One additional proton is required for phosphate import into the mitochondrial matrix.

ATP synthesis occurs in the nucleotide binding sites of the β-subunits. The α-subunits have nucleotide binding sites as well, but these have regulatory function and do not directly participate in ATP synthesis.

For many years people have worked at solving the reaction mechanism of F-ATPases, by now we have some understanding how this molecular machine works, and it is probably the most unusual and exciting scientific discovery of recent years (NOBEL-price 1978 to P. MITCHELL and 1998 to J.E. WALKER and P. BOYER, together with J. SKOU for the discovery of Na/K-ATPase).

The key finding was that during ATP synthesis the substrates are hidden into a solvent-inaccessible pocket of the F_1-head, inside this pocket formation of ATP from ADP and P_i is almost spontaneous (K_{eq} for ATP hydrolysis is 2.4, rather than 1×10^5 as in solution!). This is caused by the tight binding of ATP to the enzyme, K_d for ATP is 1 nM in the catalytic site, for ADP only 10 µM. The difference in binding energy of 40 kJ/mol drives the formation of ATP. The energy of the proton flow through the membrane is required not for ATP-synthesis, but for opening the pocket and releasing the bound ATP. This is achieved by reducing the dissociation constant for ATP from 1 nM in the catalytic to 1 µM and 30 µM respectively in the two other sites.

We have X-ray crystal structures of the F_1-head, in the presence of ATP-analogues and ADP, which give us a good idea how ATP-synthesis and -hydrolysis happen. The conversion between the open and closed conformations of the β-subunit nucleotide binding site is driven by rotation of the γ-subunit around its long axis, which brings it into contact with each of the β-subunits in turn. This results in conformational changes, which affect the affinity for ATP (see fig. 17.5c).

Rotation of the γ-subunit (together with ϵ) during ATP-turnover has been demonstrated directly by attaching a fluorescently labelled actin-filament to its lower end. Each molecule of ATP hydrolysed in F_1 resulted in a turn of the γ-subunit by 120°, and the torque generated (40 pN/nm) is consistent with the energy released by ATP-hydrolysis. By the same method it can be shown that the ring formed by the c-subunits also rotates in the membrane, driving the γ-subunit. It is currently not clear however how proton flow through the a-subunit (or possible between a- and c-subunits) drives the turbine of c- and γ-subunits. With gold particles (which are smaller and experience less drag during rotation) a rotation speed of 134 min^{-1} was observed (at 23 °C, 2 mM ATP), which agrees with the rate of ATP hydrolysis. It appeared as if the 120 ° rotation for each step could be separated into two steps of 90 and 30 degrees, respectively, with a short dwell time in between. These steps may correspond to substrate binding and product release.

The number of subunits in the transmembrane rotor of the F_o-part (see fig. 17.5) varies from species to species between 9 and 15, each subunit binds one proton per turn, each turn produces 3 molecules of ATP. Thus the number of protons required to produce one molecule of ATP depends on the number of subunits in the rotor, a simple way for evolution to adjust the ATP-synthase to different electrochemical potentials across the membrane. Note however that the number of subunits in the rotor is constant for a given species and does not depend on environmental conditions, even in organisms like bacteria, where the environmental conditions can vary significantly.

How does proton flow drive the rotor (subunits c)? It appears that these subunits have a normally protonated Asp-residue, which gets deprotonated in that subunit which faces a positively charged Arg-residue of the stator (subunit b). The released proton moves downhill of the proton gradient, i.e. into the mitochondiral matrix (H^+_m). Salt bridge formation between the Arg and Asp residues keeps the c-subunit which faces the stator in a fixed position, while the rotor continues to move. Thus a considerable mechanical tension is build up, similar to the spring in a mechanical clock (see fig. 17.6). When a proton from the intermembrane space (H^+_i) binds to the Asp-residue, making it uncharged and breaking the salt bridge to the stators Arg, this tension is released by driving the rotor forward. Thus the rotor/stator assembly of the ATP-synthase resembles the turbine of a water-driven power plant, with the F_1-part being the generator.

Na-powered F_1F_o-ATPases

As discussed, F_1F_o-ATPases are usually powered by protons. In some organisms however, Na^+ is used instead, either facultatively or even obligatory. This occurs for example in *Vibrio alginolyticus*, a bacterium that lives in algal mats. As these algae remove carbon dioxide from the water for photosynthesis, the water becomes alkaline. If the bacterium would export protons during endoxydation, it would effectively try (and fail) to neutralise the medium instead of establishing a useful pH-gradient across its cell membrane. Under these circumstances, *Vibrio* will use Na^+ instead, and the resulting Na-gradient is used to power ATP synthesis. This is a nice adaption to a specialised environment.

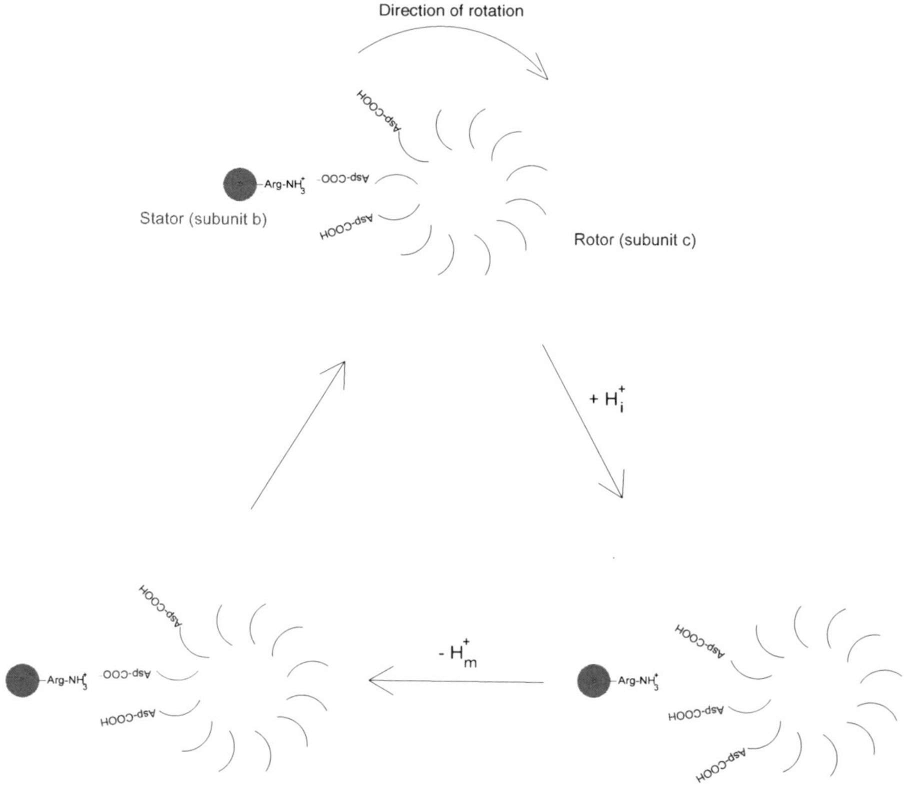

Figure 17.6. The oldest wheel in the world: How the proton gradient across the mitochondrial membrane drives the rotor of the F_o-subunit of the ATP-synthase. For details see text.

ABC-ATPases

Physiologically ABC-transporters work always as ATPase, but they can work either as importers for nutrients (for example trace elements, sugars, vitamins or amino acids), or as exporters for waste products (e.g. bile transporters in the liver), xenobiotics (multi-drug resistance transporters) and hormones (for example the transporter for the yeast pheromone α-factor). 15 ABC-transporters are encoded in intracellular parasites like *Mycoplasma*, 70 ABC-transporters (about 5 % of the genes) are encoded by the genome of the free-living *E. coli*, in the more complex human genome 50 out of 30 000 genes (0.2 %) encode ABC-transporters. None of the mammalian ABC-transporters are involved in the uptake of nutrients into the cytosol, one of their major roles in bacteria.

Some members of the ABC-transporter family have lost their transport capacity and act as receptors, which regulate the activity of other membrane

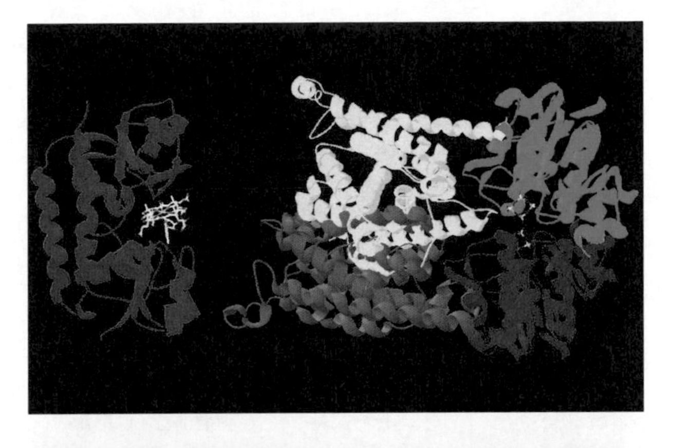

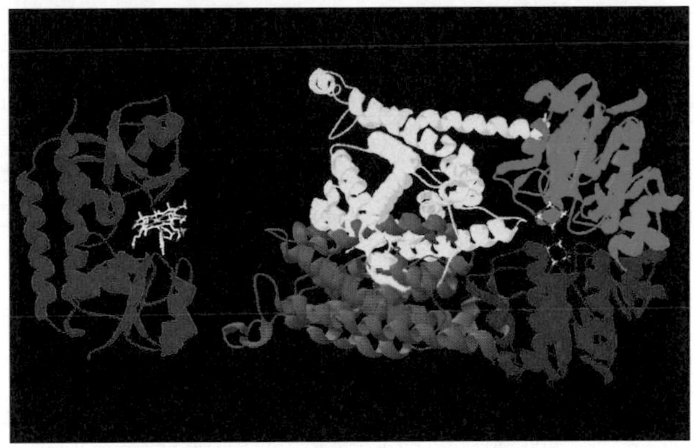

Figure 17.7. Stereo view of the bacterial vitamin B_{12} importer (Btu for B Twelf Uptake, E.C. 3.6.3.33, PDB-code 1l7v). Btu is an ABC-type ATPase whose crystal structure has been solved to a resolution of 3.2 Å. Note the 2 permeases (BtuC, blue and yellow) with their transmembrane α-helices and the 2 cytosolic ATPases (BtuD, red and green). The enzyme was crystallised as complex with **tetravanadate ions**, which mark the ATP-binding site. The **periplasmic binding protein** (BtuF, magenta) binds the substrate with high affinity and makes it available for transport. Unfortunately, only BtuC and BtuD have been crystallised together, BtuF was crystallised separately. Thus we do not know how binding of BtuF allows transport.

proteins, like CFTR and SUR. Others have lost their transmembrane domains and act as soluble enzymes.

Archetypical ABC-transporters consist of 2 transmembrane proteins (TMD), 2 cytosolic proteins with ATPase-activity (nucleotide binding domain, NBD), sometimes a cytosolic regulatory protein and, in case of bacterial importers, a high-affinity periplasmic binding protein (see fig. 17.7). During evolution, some of these proteins may have fused however, in the most extreme cases (like the multiple drug resistance ATPase Mdr1) all subunits have fused into a single multi-domain protein.

As far as X-ray crystallographic studies have revealed so far, the NBD-domain is a L-shaped molecule with the WALKER-B motive (a sequence characteristic for ATP-utilising enzymes) in the hinge-region. ATP-binding occurs across both arms of the L and results in conformational changes in line with the induced-fit hypothesis. The adenine ring stacks with a Tyr-residue of the proteins, this contributes significantly to the binding energy of ATP. ATP-hydrolysis leads to a 15 deg change in the angle between the two arms of the L, bringing the phosphate-binding Q-loop out of reach of the β-phosphate of the ADP. Absence of any nucleotide leads to destabilisation of the NBD-structure. Very little is known about how these conformational changes during the ATPase-cycle are transmitted to the transmembrane domain for coupling with transport.

Our knowledge about the reaction mechanism of ABC-transporters is still very sketchy (see fig. 17.8). In particular, it is not yet clear whether the oligomer described above is indeed the catalytically active unit, or whether several such oligomers have to work together.

ABC-transporter in humans

In humans the genes for ABC-transporters are designated with the letters ABC followed by a further letter and a number, like ABCA1. Several of them are involved in inherited diseases. The following table gives an overview of the human ABC-transporters:

Gene	Alias	Type	Gene locus	Tissue	Subcellular	Substrate	Related disease
ABCA1	ABC1, TDG, HDLDT1, CERP	full	9q31.1	ubiquitous	apical plasma membrane	cholesterol efflux onto HDL, lipid	Tangier-Disease
ABCA2	ABC2	full	9q34.4	brain, kidney, lung, heart	lysosomal membrane	Cholesterol?	drug resistance
ABCA3	ABC3	full	16p13.3	lung		surfactant	
ABCA4	ABCR, RmP, ABC10, STGD1	full	1p21.3	retina	outer rod segment rim	N-retinylidien-PE-efflux	Stargart-disease, Age related macula degeneration
ABCA5		full	17q24.3	muscle, heart, testes			
ABCA6		full	17q24.3	liver			
ABCA7		full	19p13.3	spleen, thymus			
ABCA8		full	17q24.3	ovary			
ABCA9		full	17q24.3	heart			
ABCA10		full	17q24.3	muscle, heart			
ABCA12		full	2q34	stomach			
ABCA13		full	7p12.3	ubiquitous, but low			
ABCB1	MDR1, PGP, GP170	full	7q21.12	adrenal, kidney, brain, liver	apical plasma membrane	xenobiotics, steroids, platelet activating factor, glucosylceramide	multidrug resistance

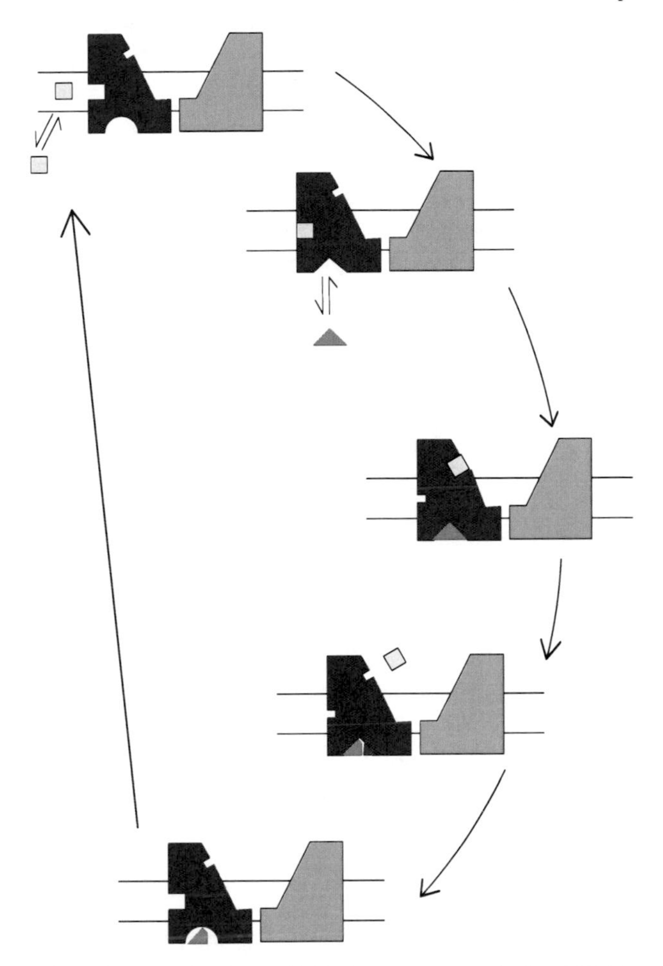

Figure 17.8. Reaction mechanism of Mdr1 as an example for ABC-type exporters. The substrate of Mdr1 are hydrophobic xenobiotics (yellow square), which partition into the cytoplasmic leaflet of the plasma membrane. From there they are bound to the substrate binding site of Mdr1. This changes the conformation of the protein and allows binding of ATP (orange triangle). The free energy of ATP-binding is used to translocate the substrate across the membrane (Mdr1 with the non-hydrolysable ATP-analogue AMP-PNP has an outward facing, low affinity substrate binding site). Dissociation of the substrate leads to splitting of ATP, the phosphate (red small triangle) leaves the nucleotide binding site (the resulting Enzyme-ADP complex can bind vanadate, which is a phosphate analogue). ADP-release then closes the cycle. Note however that the Mdr1 molecule consists of 2 almost identical halfs (blue and cyan), it is currently not known how they co-operate in the transport cycle. Indeed, co-operation of 4 ATP-binding sites can be demonstrated by enzyme kinetics, and the distance of fluorescent antibodies against Mdr1 changes during its catalytic cycle. This would indicate that two Mdr1 molecules, with 4 ATP-binding sites, co-operate during turnover.

ABCB2	Tap1	half (ABCB3)	6p21	ubiquitous	ER	antigenic peptides	
ABCB3	Tap2	half (ABCB2)	6p21	ubiquitous	ER	antigenic peptides	
ABCB4	MDR3, PGY3, PFIC3	full	7q21.12	ubiquitous, especially liver	apical plasma membrane	phosphatidyl choline flipping	PFIC-3, cholestasis of pregnancy
ABCB5		half (?)	7p21.1	ubiquitous			
ABCB6	MTABC3	half (?)	2q35		mitochondria	iron import	
ABCB7	ABC7	half (ABCB8)	Xq21-q22		mitochondria	Fe/S-cluster	
ABCB8	MABC1	half (ABCB7)	7q36.1		mitochondria		
ABCB9		half (dimer)	12q24.3	heart, brain	lysosome		
ABCB10	MTABC2	half (?)	1q42.3		mitochondria		
ABCB11	sPGP, BSEP, PFIC-2, PGY4	full	2q24.3	liver	apical plasma membrane	bile salts	PFIC-2
ABCC1	MRP1	full	16p13.12	lung, testes, PBMC, ubiquitous	basolateral, endosomes	conjugated anionic drugs	multi-drug resistance
ABCC2	MRP2, cMOAT	full	10q24.2	liver, intestine, kidney	apical PM	bile salts, bilirubin-glucuronide	Dubin-Johnson-syndrome
ABCC3	MRP3	full	17q21.33	lung, intestine, liver, adrenal	basolateral PM	bile salts, anionic drug conjugates	
ABCC4	MRP4	full	13q32.1	prostate		nucleosides	
ABCC5	MRP5	full	3q27.1	ubiquitous		nucleosides	
ABCC6	MRP6	full	16p13.12	kidney, liver			
ABCC7	CFTR	full	7q31.31	exocrine glands		regulated chloride channel	
ABCC8	SUR	full	11p15.1	pancreas		regulation of insulin secretion	
ABCC9	SUR2	full	12p12.1	heart, muscle			
ABCC10	MRP7	full	6p21.1	ubiquitous but low			
ABCC11	MRP8	full	16q12.1	ubiquitous but low			
ABCC12	MRP9	full	16q12.1	ubiquitous but low			
ABCD1	ALD, ALDP	half (dimer?)	Xq28	ubiquitous	peroxisomes	very long chain fatty acids	adrenoleukodystrophy
ABCD2	ALDL1, ALDR	half (?)	12q11	ubiquitous	peroxisomes		
ABCD3	PXMP1, PMP70	half (?)	1p22.1	ubiquitous	peroxisomes		
ABCD4	PMP69, P70R, PXMPIL	half (?)	14q24.3	ubiquitous	peroxisomes		
ABCE1	OABP, RNS4I		4q31.31	ovary, testes, spleen		oligoadenylat-binding	
ABCF1	ABC50		6p21.1	ubiquitous			
ABCF2			7q36.1	ubiquitous			
ABCF3			3q27.1	ubiquitous			
ABCG1	White	half (?)	21q22.3	ubiquitous		cholesterol? amphipatic drugs	
ABCG2	ABCP, MRX1, BCRP1	half (dimer)	4q22	placenta, intestine, liver, endothel	apical PM		
ABCG4	White2	half (?)	11q23	liver			
ABCG5	White3	half (ABCG8)	2p21	liver, intestine	apical PM	plant sterols	Sistosterolaemia
ABCG8		half (ABCG5)	2p21	liver, intestine	apical PM	plant sterols	Sistosterolaemia

The following list contains the medically important ABC-transporters, their physiological role and the inherited diseases caused by their defect:

ABCA All members of the **ABCA**-family are full transporters (both TMDs and NBDs fused into a single polypeptide). They are involved in the transport of lipophilic substances.

ABCA1 Cells synthesise phospholipids, especially phosphatidyl choline (PC), and take up cholesterol from the blood by receptor-mediated endocytosis. Both are incorporated into the plasma membrane, cholesterol mostly in special regions of the membrane, the **caveolae**. The external leaflet of the membrane contains preferentially PC and sphingomyelin, the inner leaflet phosphatidyl serine (PS) and phosphatidyl ethanolamine (PE). Some of the lipids in the outer membrane are recycled to the liver as high density lipoprotein (HDL). The ABCA1-encoded protein serves as a binding site for apolipoprotein A-I (apoA-I) and catalyses the ATP-hydrolysis dependent transfer of PC from the membrane to the apoA-I by an unknown mechanism, followed by transfer of cholesterol from the caveolae to the apoA-I/PC-complex.

Apart from its role in lipid metabolism ABCA1 is also required for the uptake of apoptotic cell fragments by macrophages, possibly by catalysing the outward-flip of PS in the plasma membrane, which serves as an "eat me"-signal to macrophages.

Deficiency in ABCA1-expression leads to Tangier-Disease (analphalipoproteinaemia), characterised by the deposition of cholesterol esters in peripheral cells (**foam cells**) and absence of HDL in the blood. Lymph nodes, spleen and tonsils are enlarged, also the liver. Corneas become cloudy, peripheral neuropathy is also observed. The disease is named after an island in the Chesapeake Bay, where the first family suffering from it was described.

ABCA4 transports N-retinylidien-PE (conjugated visual yellow) from retinal rod outer segments into the cytosol, acting as a lipid floppase (flippases catalyse transport from the cytosolic to the extracellular leaflet (flip), transport in the reverse direction is called a flop). There the compound is converted back to visual purple for re-uptake by the rods. If the visual yellow can not be regenerated in this fashion, accumulation of this compound and its degradation products leads to **retinitis pigmentosa** (homozygotes), STARGARDT-disease (heterozygotes) or **age-related macular degeneration** beyond the age of 60 (partially functional transporter).

ABCB is the only family of human ABC-proteins which contains both full and half-transporters. Half-transporters can form either homo- or heterodimers.

ABCB1 encodes for the **multiple drug resistance transporter Mdr1**, also known as **permeability glycoprotein (Pgp)**. Mdr1 is involved in the protection of our body against xenobiotics, which may be present in our food. It is expressed in intestine, kidney, liver and the cells of the blood-brain and blood-testes barriers.

Unlike most other enzymes, Mdr1 is not specific for a single (or at least a small group of) substrate, but transports a wide variety of compounds. They all seem to be amphiphilic molecules, which partition between the cytosol and the cytoplasmic leaflet of the membrane. The binding site of Mdr1 seems to open to the membrane, rather than to the cytosol. Thus cellular metabolites (which are usually hydrophilic and do not partition much into the membrane) do not act as Mdr1 substrates. Molecules bound at the cytoplasmic leaflet are then transported across the membrane and released into the extracellular medium (**not** the extracellular leaflet of the membrane!) under expenditure of ATP. This is called the "hydrophobic vacuum cleaner model of Mdr1" (see fig. 17.8).

Unfortunately, during chemotherapy of cancer patients, some cancer cells which by chance express a higher concentration of Mdr1 in their membrane are selected for (note that cancer cells are aneuploidic). Thus the cancer will initially vanish, but reappear after a few years, when those few cells that survived chemotherapy have multiplied. Such cells can express impressive amounts of Mdr1, which may account for 10 % of all the membrane proteins.

Because of the broad spectrum of drugs that is transported by Mdr1, these cells are resistant not only to the drug originally used to treat the tumour, but against many alternative drugs as well. This so called multi-drug resistance can be fatal for the patient.

Polymorphisms in the ABCB1-locus have been linked to variations in the pharmakokinetics of drugs like digoxin.

ABCB2/ABCB3 encode for the Tap1/Tap2 transporter, which is responsible for the transport of immunogenic peptides from the cytosol into the ER, where they bind to MHC-I (see fig. 10.14 on page 162). The peptides transported are preferentially 8–16 amino acids long (but up to 40 is possible) and have the sequence [K/N/R]R[W/Y]...[F/L/R/Y]. Experimental evidence indicates that ATP-hydrolysis in one and ADP-release in the other NBD is associated with peptide transport, and the reverse with regeneration of the cytosolic peptide binding site. Thus two molecules of ATP would be hydrolysed for each molecule of peptide transported. Even if this were true for Tap1/Tap2 we do not know whether it might be true for all ABC-proteins.

Congenital defects in either Tap1 or Tap2 results in **bare lymphocyte syndrome**, that is the inability to make MHC-I. In virus-infected and in tumor cells the expression of Tap1/Tap2 is sometimes down-regulated to prevent detection by the immune system. Thus Tap1/Tap2 expression in tumor biopsies can be used for grading and corresponds to the prognosis.

ABCB4 encodes Mdr3, which is highly homologous with the Mdr1 multiple drug resistance transporter. However, its main function appears to be the flipping of phosphatidyl choline (PC) from the inner to the outer leaflet of the canalicular membrane. From there PC is extracted by cholate micelles. If Mdr3 is not present, the PC-free micelles can attack the membranes of the cells lining the canaliculi and the bile duct, resulting in **progressive familial intrahepatic cholestasis type 3 (PFIC-3)**. The resulting non-suppurative inflammatory cholangitis leads to liver cirrhosis and possibly liver tumors. Females with +/- genotype may develop the disease during pregnancy. Some mutations in ABCB4 have been linked to **gall stones**.

ABCB6/ABCB7 encode for a FeS-cluster transporter in the inner mitochondrial membrane. The proteins apparently form homo-dimers with differential tissue expression.

The mitochondrial matrix contains an **iron-sulfur cluster assembly machine (ISU)** of about 10 proteins, which have not been fully characterised yet. Fe^{2+} is taken up into the matrix by a membrane-potential dependent secondary transporter, S^0 is synthesised inside the mitochondrium from cysteine. On ISU first [2Fe-2S] and then [4Fe-4S]-clusters are assembled in a NAD(P)H-dependent reaction. They can be used directly for transfer to mitochondrial apo-proteins, but also must be exported to the cytosol for union with (Hsc70/Hsp40-stabilised) apo-proteins there. Export possibly requires chelation (to glutathione?) and is mediated by the ABCB6/ABCB7-encoded transporter (see fig. 17.9).

Mutation in ABCB7 results in **X-linked sideroblastic anaemia/-cerebrellar ataxia (XLSA/A)**, where the iron-laden mitochondria of bone marrow cells form sideroblasts (high iron bodies that can be stained by the prussian blue reaction) located around the nucleus. Mutations in ABCB6 result in **lethal neonatal syndrome**.

ABCB8/ABCB10 encoded transporter is responsible for the export of peptides from the mitochondrial matrix to the inter-membrane space.

ABCB9 is expressed in the lysosomal membrane, its function is unknown.

ABCB11 encodes for the **sister of Pgp (Spgp)** protein, a bile salt exporter in the apical (canalicular) membrane of hepatocytes. The protein is a full transporter with considerable sequence homology to the ABCB1-encoded Mdr1/Pgp. Mutations result in **progressive familial intrahepatic cholestasis type 2 (PFIC-2)**.

ABCC genes encode for full transporters of hydrophobic substances. Because many drugs are hydrophobic, ABCC-encoded transporters may be involved in multiple drug resistance, hence they are called **Multidrug re-**

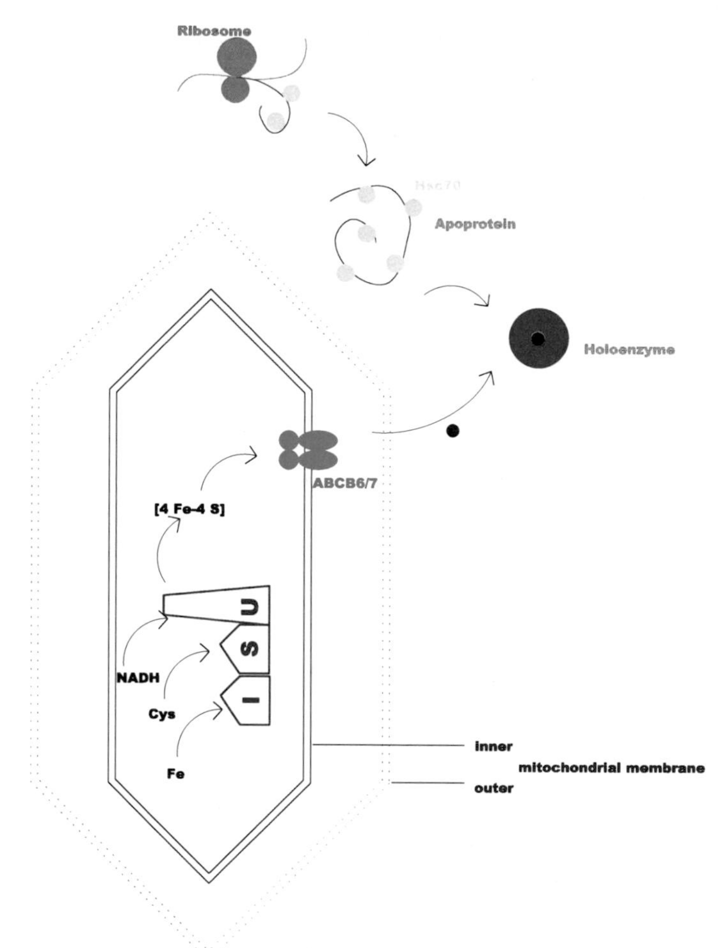

Figure 17.9. Assembly of FeS-cluster-dependent proteins. The clusters are synthesised in the mitochondrial matrix. The clusters not required inside the mitochondrium are exported through the ABCB6/ABCB7-transmembrane ATPase into the intermembrane space, from there they move to the cytosol by passive diffusion (the outer mitochondrial membrane is leaky enough to allow that). In the cytosol the clusters are united with apo-enzymes which were kept in a folding competent state by Hsc70 and its co-chaperone Hsp40.

sistance related protein (Mrp). In addition, the ABCC-family of genes also encodes for 3 proteins which are not pumps, but ion channels (CFTR, ABCC7) or regulators of ion channels (Sur1 and -2, ABCC8 and ABCC9).

ABCC1 encodes for Mrp1, a transporter for xenobiotics conjugated to glutathione. In some cases transported substrate and glutathione need not be actually linked by a chemical bond, but only present at the same time (aflatoxin B_1, nitrosamines from tobacco smoke). Arsen and antimon can be transported as complex with glutathione. In addition, Mrp1 is involved in the IgE-mediated efflux of leukotriene C_4 (LTC_4) from mast cells, important for the regulation of inflammation.

ABCC2 encodes for **canalicular multiple organic anion transporter (cMOAT)**, also known as Mrp2, which is located in the apical membrane of hepatocytes as well as in kidney, intestine, gall bladder and lung. Substrates for cMOAT are amphiphilic substances conjugated to glutathione or glucuronic acid. One of the most important functions of cMOAT in the liver is the excretion of bilirubin into the gall. Deficiency in cMOAT causes DUBIN-JOHNSON-syndrome (chronic idiopathic jaundice, high bilirubin-levels in serum, excretion of bilirubin-derivates in urine, hepatomegaly, melanin deposition in liver), a relatively mild disease that does not require treatment.

ABCC3 encodes for Mrp3, a protein expressed in the apical membranes in liver, intestine, adrenal gland and gall bladder. Substrates are glucuronlylated or sulfated bile salts and acidic drugs. This transporter is involved in the cholehepatic and enterohepatic cycling of bile salts, it also exports drugs modified in the liver into the blood stream for excretion with urine. High expression in the adrenal gland may indicate involvement in steroid hormone transport.

ABCC4, ABCC5 encode Mrp4 and Mrp5, which transport nucleoside monophosphates (both pyrimidine and purine), cAMP and cGMP. The exact physiological role is unclear. Interestingly these transporters are inhibited by phosphodiesterase blockers like Sildenafil (Viagra®). Overexpression of Mrp4/5 causes resistance against some nucleoside analogues used in anti-viral and anti-cancer therapy.

ABCC6 encodes a transporter in the basolateral membrane, the physiological substrate is unknown, possibly a peptide. Deficiency causes a rare autosomal recessive disorder, **pseudoxanthoma pigmentosum**. Although protein expression is low in most tissues except liver and kidney, patients suffer from calcification of the elastic fibres in multiple organs, resulting in loss of skin elasticity, arterial insufficiency and retinal haemorrhage. Thus the substrate of ABCC6p may be an unknown hormone.

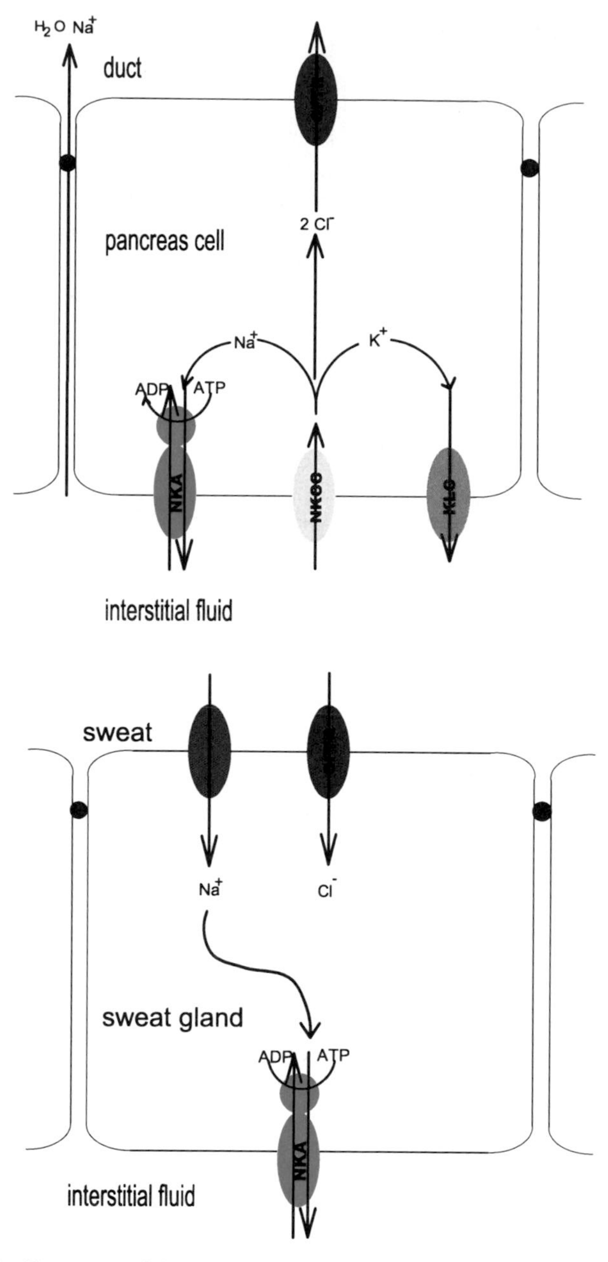

Figure 17.10. Function of CFTR in the pancreas and lung (top) and in the sweat gland (bottom). For details see text.

ABCC7 encodes for the cystic fibrosis transconductance regulator (CFTR), a non-rectifying anion channel of low conductivity (7–10 pS). CFTR is unspecific, but because Cl^- is the most common anion inside our bodies CFTR is essentially a chloride-channel (bicarbonate may also play a role). In exocrine glands (see Fig. 17.9) Na^+, K^+ and 2 Cl^- pass together from the interstitial fluid into the cytosol through the NKCC-transporter in the basolateral membrane. The sodium is removed from the cytosol by Na/K-ATPase (NKA), the potassium by the potassium leakage channel (KLC), both also in the basolateral membrane. As a result high $[Cl^-]$ are build up in the cytosol, which leave the cell passively through CFTR at the apical membrane. Water and sodium follow passively by paracellular transport. Cl^--transport is the rate limiting step in the secretion of glandular products.

In sweat glands Na-ions are reabsorbed by an apical Na-channel and the basolateral Na/K-ATPase. Passive flow of Cl^- through CFTR into the cytosol is required for electroneutrality.

CFTR has 9 sites for regulatory phosphorylation by protein kinase A (PKA), and 9 for phosphorylation by PKC. Their phosphorylation changes the secondary structure of CFTR, increasing the probability of the open (conducting) state. Rapid down-regulation of chloride flow is possible by a PP2C-like phosphatase. Additionally, P_{open} can also be increased by binding of ATP to CFTR. ATP binding to the first NBD is fairly stable, in the second NBD ATP is hydrolysed quickly, ADP and P_i are released. The reason for this difference is not understood.

In cystic fibrosis (CLARKE-HADFIELD-syndrome, mucoviscidosis), an autosomal recessive inherited disease, CFTR is mutated. There are several possible mutations, the most frequent (70%) is the deletion of phenylalanine 508 (ΔF^{508}-CFTR). This slows down protein folding in the ER so much that most CFTR molecules are degraded and do not reach the cell surface. Those molecules that do are, interestingly, fully functional despite the mutation, but the CFTR concentration in the cell membrane is too low.

Other mutations are found enriched in specific populations like W1282X in Ashkenazi Jews and 1677delTA in Georgia, Bulgaria and Turkey.

Patients with cystic fibrosis make a very though, viscid mucus, which easily becomes infected by bacteria. These recurrent lung infections are often fatal. Pancreas juice and bile secretion are also affected. The re-absorbtion of NaCl from sweat is prevented, resulting in salty sweat, which is a diagnostic marker for the disease.

Since heterocygotes for the CFTR-mutations loose less water in their intestinal secretions, they may be protected from the diarrhoea following infection with intestinal parasites like *Vibrio cholerae*, *Shigella ssp* or *Entamoeba histolytica*. This selective advantage may account for the relatively high frequency of mutated CFTR in the general population.

ABCC8 encodes the sulphonylurea receptor Sur1. 4 molecules of Sur1 and 4 molecules of **Kir6.**x (tissue specific) form an ATP-dependent potassium channel (K_{ATP}) in the plasma membrane. The complex is assembled in the ER, only complete molecules can reach the plasma membrane. **Kir6.**x can bind ATP and ADP, but not AMP, without Mg (very unusual!) and is closed by it. However, binding of MgATP to Sur1 opens the channel, thus the regulation of K_{ATP} by ATP is complex. Sur1 hydrolyses ATP.

In the β-cells of the pancreas high blood glucose levels lead to glucose uptake by the GluT2 transporter and to increased glucose metabolism. This results in a change in the ATP/ADP-ratio (in normal cytoplasm [ATP] $\approx$ 1–5 mM, [ADP] $\approx$ 100 µM), a decrease in [ADP] leads to dissociation of ADP from Sur1 and allows ATP-rebinding. That in turn closes the K_{ATP}-channel, depolarising the membrane and opening voltage dependent Ca-channels. The resulting influx of Ca^{2+} causes release of insulin from the β-cell.

In α-cells glucagon release is inhibited and in δ-cells somatostatin release stimulated by a similar mechanism. Sur1 is also found in GABAergic cells of the *substantia nigra*.

A defective Sur1 protein results in **persistent hyperinsulinaemic hypoglycaemia of infancy (PHHI)**. If the low blood glucose levels are not treated, severe neurological damage results, possibly leading to death. Treatment is either with K_{ATP}-channel openers like diazoxide or by subtotal pancreatectomy. PHHI is rare in normal populations, however, in inbred populations its frequency can rise to 1:2500. Usually inheritance is recessive, but dominant mutations have been described. PHHI may also result from mutations in **Kir6.**2, glucokinase or glutamate dehydrogenase.

A mutation in **Kir6.**2 (E23K) has been implicated in **type 2 diabetes mellitus**. However, at least in mice, this mutation can cause diabetes only in association with **obesity**. The increased blood lipid levels in obesity change the ATP-sensitivity of the K_{ATP}-channel, resulting in hyperinsulinaemia followed by apoptosis of the β-cells. Inhibitors of the K_{ATP}-channel (imidazolines, sulphonylureas and benzamidoderivates) are used to treat diabetes, as they can be given orally. K_{ATP}-openers like chromakalim have little clinical use at present.

ABCC9 encodes for the Sur2 protein, which has similar function as Sur1. It is found together with **Kir6.**2 in dopaminergig SN neurons and seems to protect against seizures in hypoxia. Sur2 is apparently also involved in the regulation of blood glucose levels by the ventromedial hypothalamus.

ABCC10, ABCC11, ABCC12 encode for Mrp7–9, proteins of unknown function and low expression in all tested tissues. No associated disorders have been described.

ABCD -genes encoded proteins are closely related half-transporters, which form homo- or heterodimers. The functional consequences of partner choice have not been investigated yet. Also the reasons for tissue specific expression of ABCD-isoforms is unknown.

All ABCD-transporters transport long chain (14–24 carbon atoms) and very long chain ($>$ 24 carbon atoms) branched or straight fatty acids (or their CoA-derivatives) into the peroxisomes for β-oxidation. Short- and medium-chain fatty acids are catabolised in mitochondria.

The peroxisome can be formed from other peroxisomes by enlargement and fission, or by *de novo* from the ER. Peroxisomal proteins are expressed in a coordinated manner under the control of the **peroxisomal proliferator activation receptor α (PPARα)**, they have C-terminal - SKL and/or N-terminal RKX_5[Q/H]L- import sequences. Diseases related to the peroxisome can be caused by general defects of peroxisome formation (like ZELLWEGER-syndrome) or by specific defects of a single enzyme, for which adrenoleukodystrophy is a typical example.

The VLCFA-CoA-synthase is located at the matrix leaflet of the peroxisome and associates with ABCD1.

ABCD1 is mutated in **adrenoleukodystrophy**, an X-linked recessive inherited disease. Activity of the synthase is also reduced. Failure to metabolise LCFA and VLCFA results in their accumulation in plasma and tissue (brain, adrenals). Myelin degeneration in the white matter of the brain leads to mental deficiency. Until the cause of the disease was understood, patients died in adolescence from neuronal degeneration, which today can be controlled by a special diet. The film "*Lorenzos oil*" is about adrenoleukodystrophy, and probably the only film Hollywood has ever made on lipid metabolism. In this context it is worth mentioning that the mixture of erucic ($C_{22:1}\omega9$) and oleic ($C_{18:1}\omega9$) acid reported in the film does normalise VLCFA in plasma, but fails to significantly slow the progression of the disease.

In yeast PXA-1 and -2 form heterodimers which are required for growth on oleic ($C_{18:1}$) but not octanoic ($C_{8:0}$) acid as sole carbon source. β-oxidation of oleic acid is possible in the peroxisomes of such cells if their membrane is permeabilised with detergent.

ABCE proteins have no transmembrane domain at all, the two nucleotide binding domains are fused and act as a soluble enzyme.

ABCF These proteins too lack TMDs.

ABCG -encoded proteins are half-transporters, which form heterodimers. In the fused proteins the NBD is N-terminal, the TMD C-terminal.

ABCG2 is expressed in the trophoblast. Its function is unknown.

ABCG5/ABCG8 encode a transporter for plant cholesterol-derivatives (phytosterols, in particular sistosterol, 24-ethyl cholesterol) in the apical membrane of liver and intestinal cells. In the intestine, transport of phytosterols out of the cell reduces their uptake into the body, in the liver the transporter increases the rate of excretion into bile. Thus the concentration of phytosterols in our blood is kept low. **Sistosterolaemia** results from a genetic defect in either half of the transporter. The elevated blood level of phytosterols leads to arteriosclerosis, haemolytic episodes, arthritis and tendinous xanthomas (yellowish nodules of lipid-laden histiocytes forming in tendons and fascia).

17.2.2 Secondary active transporter (co-transporter)

Secondary active transport uses the energy stored in an electrochemical gradient across the cell membrane for one substrate to transport another substrate against a concentration gradient. As mentioned in the previous section, the plasma membrane of animals is energised by a sodium, that of bacteria, fungi and plants usually by a proton gradient. Internal membranes are also energised by proton gradients. As these ions "fall downhill", they can "lift uphill" other substrates, for example nutrients that have to be transported into, or waste products and secreted compounds, that have to be moved out of the cell.

The electrochemical gradient of an ion contains free energy according to the NERNST equation:

$$\Delta G = R * T * \ln\left(\frac{[Ion]_i}{[Ion]_o}\right) + z * F * \Delta\mathfrak{E} \qquad (17.4)$$

with R = universal gas constant (8.28 J/(mol * K)), T = absolute temperature (310 K for mammals), z = number of charges per ion, F = FARADAY-constant (96 102 J/(mol * V)) and $\Delta\mathfrak{E}$ = electrical potential across the membrane (≈ -70 mV for most animal cells). In mammals, $[Na]_i \approx 12$ mM, $[Na]_o \approx 145$ mM. Thus moving Na downhill the concentration gradient across a mammalian cell membrane provides ≈ -6.4 kJ/mol $+ -6.7 kJ/mol = -13.1$ kJ/mol in free energy, with the chemical and the electrical gradient contributing about half each. For comparison, the free energy of ATP-hydrolysis under the conditions of the cell is about -52 kJ/mol (standard free energy only -32 kJ/mol).

The concentration gradient for an uncharged substrate like glucose, which can be created using Na-coupled flow is

$$\frac{[Glc]_i}{[Glc]_o} = \exp\left(\frac{n * \Delta G}{R * T}\right) \tag{17.5}$$

$$= \exp\left(\frac{2 * 13\,100\,\mathrm{J{\cdot}mol{\cdot}K}}{8.28 * 310\,\mathrm{J{\cdot}mol{\cdot}K}}\right) \tag{17.6}$$

$$\approx 27\,600 \tag{17.7}$$

Many co-transporters belong into the **major facilitator superfamily (MFS)**. These consist of a single polypeptide chain of around 50 kDa with 12 transmembrane helices. They are connected by long cytoplasmic and short extracellular loops (see fig. 17.11). They usually bind their substrates with µM dissociation constant, transport is reversible under suitable experimental conditions.

Crystal structure and reaction mechanism of LacY

Of the MFS-proteins currently only one example structure at molecular resolution (3.6 Å) is available [1], that of the *E. coli* lactose permease (LacY, see fig. 17.11). The molecule consists of two domains (helices 1–6 and 7–12), which are folding homologues although sequence homology is low. A mutant protein (C154G) was used for crystallisation, as this mutant is locked in the inward-facing conformation. This reduction in conformational freedom makes crystallisation possible, at the resolution of 3.6 Å major structural features can be recognised but finer details (for example the position of bound water molecules involved in indirect hydrogen bonds) are not visible.

From previous biochemical studies it was known that Glu^{126} and Arg^{144} are involved in substrate binding, Arg^{302} and Glu^{325} in proton translocation and Glu^{269} and His^{322} in coupling between proton and sugar flow.

In the crystal hydrophobic interaction between Trp^{151} and the pyranosyl ring is apparent, which orients the sugar molecule in the binding site. Met^{23} forms a van der Waals-bond with C6, Arg^{144} a hydrogen bond to O3 of the pyranosyl-group. Glu^{126} may indirectly form a hydrogen bond via a water molecule, however, at the resolution of 3.6 Å this can not be verified. Lys^{358} forms a hydrogen bond with O4', Asp^{237} forms a hydrogen bond with Lys^{358} and possibly an indirect hydrogen bond with

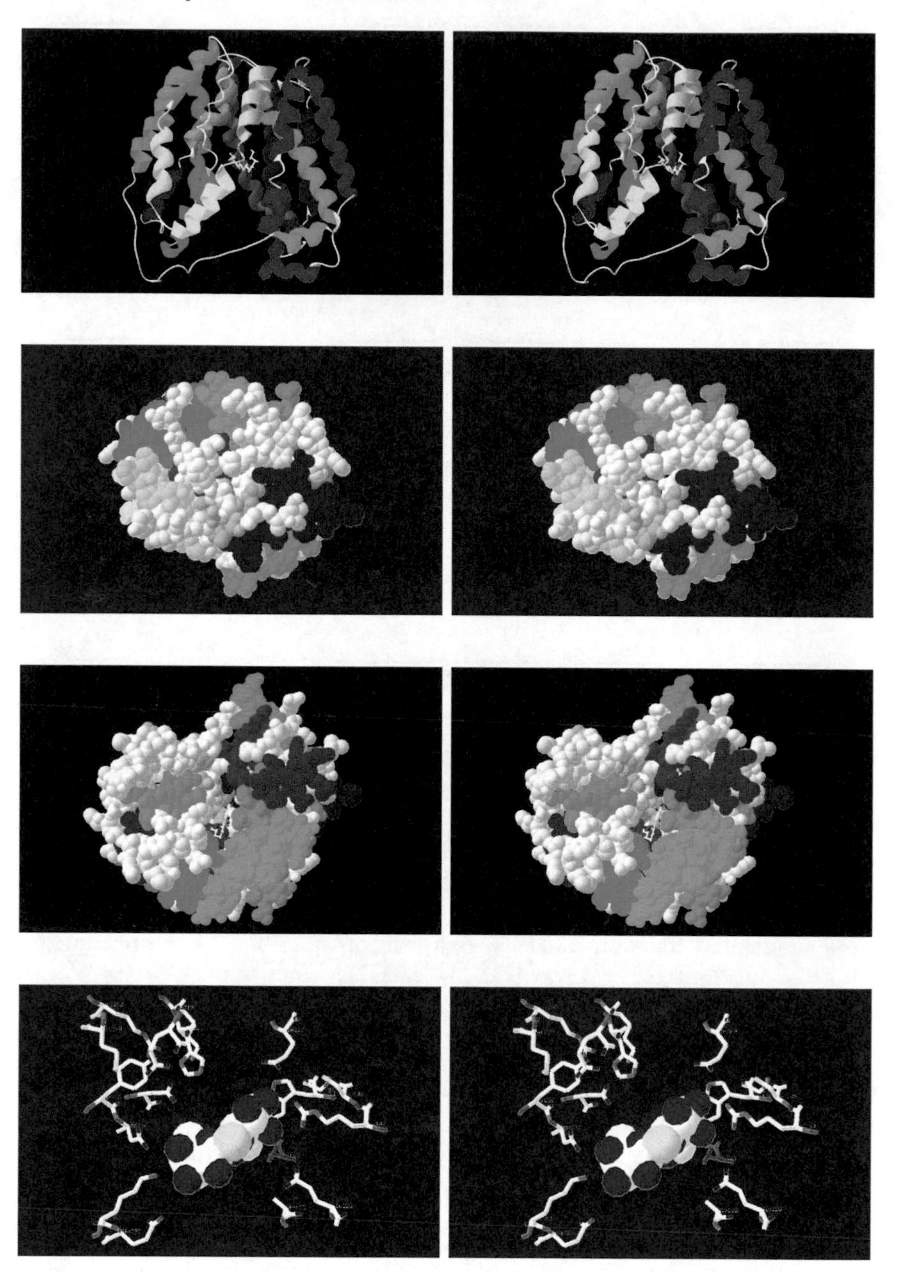

Figure 17.11. The lactose permease LacY from *E. coli* (PDB-code 1PV7) is currently the only major facilitator whose structure has been resolved (to a resolution of 3.6 Å). β-D-galactopyranosyl-1-thio-β-D-galactopyranoside was used as substrate analogue. **Top**: view from the membrane plane (extracellular pointing up), **second**: view from the top onto the closed extracellular binding site, **third**: view from the bottom onto the open intracellular binding site **bottom**: position of important residues with respect to the substrate. For details see text.

O4' via a water molecule. Glu^{359} may also form such an indirect hydrogen bond.

The proton binding site is formed by Tyr^{236}, Asp^{240}, Arg^{302}, Lys^{319}, His^{322} and Glu^{325}. Interaction between His^{322} and Glu^{269} link sugar and proton binding.

Mutations in Cys^{148} and Ala^{122} are known to interfere with substrate binding, in the crystal structure they are close enough to the sugar that steric hindrance can explain this result.

As far as can be deduced from biochemical and crystallographic data, LacY in the outward-facing conformation rapidly binds a proton to Glu^{269}/His^{322}, followed by sugar binding to Arg^{144}/Glu^{126}. This leads to a conformational change in the protein which closes the extracellular and opens the intracellular binding site. During this transition, a salt bridge is formed between Glu^{269} and Arg^{144}, while the bridge between Glu^{269} and His^{322} is broken, leading to proton transfer to Glu^{235}. Lactose is released to the cytoplasm, the resulting conformational change reduces the pK_a of Glu^{235} by interaction with Arg^{302}, Met^{299} and Tyr^{236}, followed by proton release to the cytoplasm. The cycle is closed by a return of LacY to the outward-facing conformation. According to this interpretation, the reaction mechanism of LacY would be ordered bi-bi.

Some bacteria express MFS-transporters which use external binding proteins to catch substrates, just like ABC-transporters do. These binding proteins bind to a 25 kDa integral membrane protein with 4 transmembrane helices that is closely associated with an MFS transporter. Contrary to normal MFS-transporters, these systems, called **tripartite ATP-independent periplasmic transporters (TRAP-T)**, catalyse an unidirectional transport. In some cases, the two membrane proteins have fused during evolution.

Symporter

Na-linked symporters transport nutrients like amino acids and glucose into many animal cells.

For example the 2 Na^+/1 glucose symporter has two domains, a C-terminal glucose channel with 9 transmembrane α-helices and a N-terminal domain which couples Na and glucose flow.

Such a transporter (located in the apical membrane) can very efficiently scavenge glucose from the intestinal lumen into epithelial cells (see fig. 17.14). The basolateral membrane of these epithelial cell contains Na/K-ATPase to provide the low Na-concentration inside the cell, and a glucose uniporter that allows the glucose molecules to move from the cell into the blood stream by facilitated diffusion (blood glucose concentration is lower than that in the epithelial cell).

Antiporter

The Na/Ca-exchanger

A typical example for an antiporter is the Na/Ca-exchanger in cardiac myocytes, which expels a Ca^{2+} in exchange for 3 Na-ions entering the cell. The energy of the combined downhill flow of 3 Na-ions plus the movement of one net positive charge into the cell (which has a negative potential compared to the extracellular medium) allows the maintenance a steep Ca-concentration gradient between the extra- and intracellular environment (2 mM vs 200 nM). As discussed in the section on Na/K-ATPase, inhibition of the Na-pump with ouabain results in lowered Na/Ca-exchange, a rise in cytosolic $[Ca^{2+}]$ and increased heart pumping activity.

The anion exchanger AE1

When the erythrocyte passes through the capillaries in peripheral tissues, oxygen is released from haemoglobin and diffuses into the tissue (small gas molecules can move across the membrane without transporter!). At the same time, carbon dioxide enters the erythrocyte. The carbonic anhydrase inside the erythrocyte catalyses the reaction

$$H_2O + CO_2 \rightleftharpoons HCO_3^- + H^+ \tag{17.8}$$

The proton binds to a His-residue of haemoglobin, lowering its oxygen affinity and aiding in oxygen release (BOHR-effect, see page 113). The bicarbonate ion is transported out of the erythrocyte in exchange for a chloride ion, thus the concentration of bicarbonate in the erythrocyte is low and the conversion of CO_2 into HCO_3^- and H^+ is almost complete.

In the lungs, the opposite process occurs, bicarbonate enters the erythrocyte through AE1 in exchange for a chloride ion and is converted to CO_2 by carbonic anhydrase. The carbon dioxide diffuses out of the erythrocyte and into the lungs. This process abstracts the bound proton from haemoglobin, increasing oxygen affinity again for complete loading.

About 80 % of all carbon dioxide generated in our body is transported as bicarbonate, and 2/3 of that in the plasma. About 5×10^9 bicarbonate ions pass through AE1 transporters back and forth during each passage of an erythrocyte through the body. AE1 is one of the most abundant proteins in erythrocytes and forms a prominent band if erythrocyte membrane proteins are subjected to electrophoresis. Early workers in the field simply numbered the bands visible after electrophoresis, so AE1 is sometimes called "**band 3 protein**".

17.2.3 Facilitated diffusion

Channels

Potassium leakage channels

We have seen that the Na/K-ATPase in the animal cell membrane removes Na-ions from, and pumps K-ions into the cell. The Na-ions are used for coupled transport, but K^+ flows back through special leakage channels, which are normally open. This flow, driven by the concentration gradient across the membrane, results in the formation of an electrical potential difference across the cell membrane, with the inside of the cell negative. This growing potential difference will resist the outflow of K^+, until the outflow of K^+ powered by the concentration gradient is balanced by the inflow powered by the electrical potential. This steady state is achieved when the voltage across the cell membrane is approximately 70 mV. This is slightly lower than the NERNST-potential of K^+ (−95 mV for 134 mM inside, 4 mM outside), the remainder is contributed by other ions.

Thermogenin in brown adipose tissue

Shortly after birth a proton channel, the 30 kDa protein **thermogenin**, dissipates proton gradients generated by lipid oxidation in the mitochondria of **brown adipose tissue**. The energy stored in fat is converted to heat instead of ATP, this is used to keep the body temperature of the newborn up until the body can produce heat by other means. In some hibernating animals the same mechanism is used, in some cases thermogenin is induced even in muscle cell mitochondria.

Shuttles (uniporters)

Shuttles bind their substrates on one side of the membrane and release it on the other.

Uncoupler

The simplest shuttles are the so called "uncouplers", lipophilic acids that dissolve in the membrane of mitochondria, bind protons on the acidic side and release them on the basic. Thus the proton gradient across the mitochondrial membrane is dissipated (the stored energy is converted into heat) and ATP production ceases.

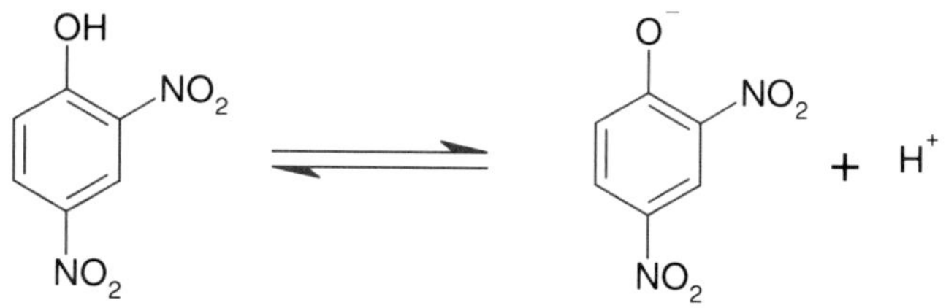

$$\text{2,4-dinitrophenol (OH)} \rightleftharpoons \text{2,4-dinitrophenolate } (O^-) + H^+$$

2,4-dinitrophenol

Figure 17.12. 2,4-Dinitrophenol is a typical uncoupler. The hydrophobic molecule can insert into the mitochondrial membrane and shuttle protons from the acidic to the basic side, short-circuiting the proton gradient.

Uncouplers are used as rat poison and sometimes for sliming. Both the hyperthermia induced and the loss of ATP-synthesis make this highly dangerous, the difference between minimal effective and fatal dose is very small.

The GluT1 glucose transporter

Glucose is one of the most important energy sources for animal cells, some cells can not use other sources like lipids or ketone bodies. Normal blood glucose concentration is about 5 mM, from the blood stream glucose is transported into the cell by the glucose transporter GluT1 (see fig. 17.14), a 45 kDa transmembrane protein. Inside the cell, glucose is immediately phosphorylated to glucose-6-phosphate, thus the glucose concentration inside the cell is virtually zero. GluT1 has a MICHAELIS-constant of 1.5 mM for D-glucose and will be running at 78 % of maximal velocity at normal blood glucose concentrations.

K_m for L-glucose is 3000 mM, 20 mM for D-mannose and 30 mM for D-galactose, GluT1 is very specific for D-glucose and even the change of configuration on a single carbon atom has considerable effects on substrate binding.

Mechanistically, GluT1 has 2 conformations, with a binding site for glucose that can either face to the outside or the inside of the cell (see fig. 17.14). If glucose is bound to the binding site at the outside, a conformational change is triggered, the glucose is transported through the membrane and the glucose

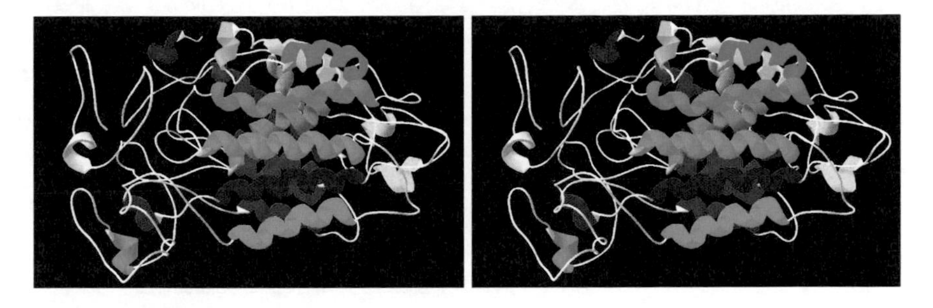

Figure 17.13. Stereo view of the crystal structure of GluT1 (PDB-code 1JA5).

binding site will face the cytosol. Because of the low glucose concentration inside the cell, glucose will dissociate from GluT1. Another conformational change results in the empty binding site facing the outside again. This step is rate determining, if erythrocytes are suspended in a solution of radioactive glucose, the rate of uptake increases with the concentration of non-radioactive glucose inside the erythrocyte, because the conformational change is faster in the glucose-loaded than in the empty transporter.

Water channels: Aquaporins

Water as a polar molecule can not simply cross the biological membrane. Although some water is always taken in the hydration sphere of transported solutes (ions, glucose), most water moves through specialised water pores, aquaporins.

Of the 150 aquaporin types characterised, 10 (Aqp0 to Aqp9) have been found in humans. They occur in a variety of tissues, in highest concentration in the kidney. Aquaporins cycle between intracellular storage vesicles and the cell membrane, controlled by the hormone vasopressin via cAMP. Aquaporins are thus involved in regulating urine volume.

Severity of the consequences of aquaporin mutations depends on the type of aquaporin affected. For example, in COLTON-null disease Aqp1 is missing, but this results only in a slightly reduced capacity to concentrate urine, patients can still have a normal live.

Problems in Aqp2 lead to **diabetes insipidus**, because water can not be re-absorbed in the tubuli. Mutations of the AQP2 gene (on chromosome 12q13) tend to prevent the cycling of Aqp2 between storage vesicles and cell membrane. Although the functional unit of aquaporins is the monomer, the transported unit is the tetramer. Of the several known mutations of Aqp2, some have pseudo-dominant inheritance, because a single mutated molecule can prevent the transport of the entire tetramer. Correct diagnosis of diabetes

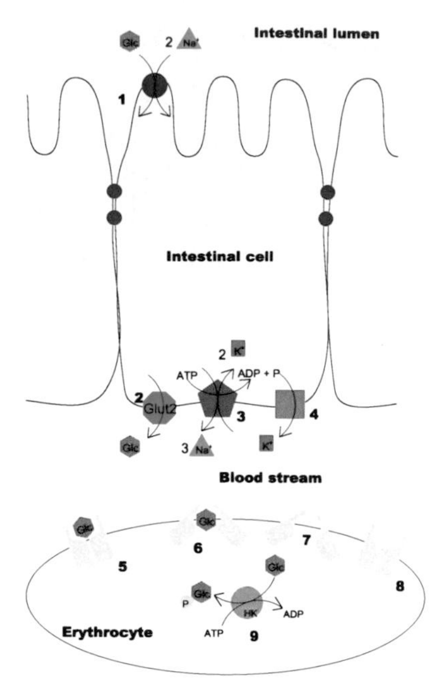

Figure 17.14. Transport of glucose from the intestinal lumen to the consuming cells, here assumed to be an erythrocyte. Some histological features not required for the discussion were left out. **Top**: Cells of the intestinal epithelium have a Glucose/2Na$^+$-symporter (red), that transports glucose from the intestinal lumen into the enterocyte. Coupling with downhill Na-transport results in cellular concentrations of glucose, which can be almost 30 000-fold higher than in the intestinal lumen (**1**, see page 288 for details). Because of the high glucose concentration inside the enterocyte, glucose will move into the blood stream passively through the GluT2 uniporter (green, **2**). In the blood, glucose concentration is maintained at 4–5 mM by the liver. Since GluT2 has a K_m for glucose of 66 mM, blood glucose will not be transported into the enterocyte at an appreciable rate, even if the glucose concentration in the enterocyte drops between meals. The energy for glucose accumulation inside the enterocyte is provided by a sodium gradient, which is produced by Na/K-ATPase (brown, **3**), Na/K-ATPase and the Glc/2Na$^+$-symporter together also ensure efficient uptake of Na$^+$ from the intestine. A potassium leakage channel (orange, **4**) provides a return path for the K$^+$-ions transported by Na/K-ATPase and maintains a stable membrane potential. Direct contact between blood stream and intestinal contend is prevented by tight junctions (magenta). The erythrocyte membrane contains GluT1 (cream, **5**–**8**), which transports glucose into the cell with a K_m of 1.5 mM, thus it works at 3/4 of its maximal activity at normal blood glucose concentrations. Inside the cell the glucose is immediately phosphorylated by hexokinase (gold, **9**), which has a K_m for glucose of 0.1 mM and thus prevents the accumulation of appreciable concentrations of glucose inside the erythrocyte. GluT1 works by binding glucose on an external binding site (**5**), then changing its conformation so that the binding site is exposed to the cytosol (**6**). Dissociation of glucose results in an empty transporter (**7**), which changes its conformation again, bringing its binding site to the outside ready for renewed glucose binding (**8**). The last steep is rate limiting. This example shows how pumps, secondary active transporters and uniporters can work together in our bodies.

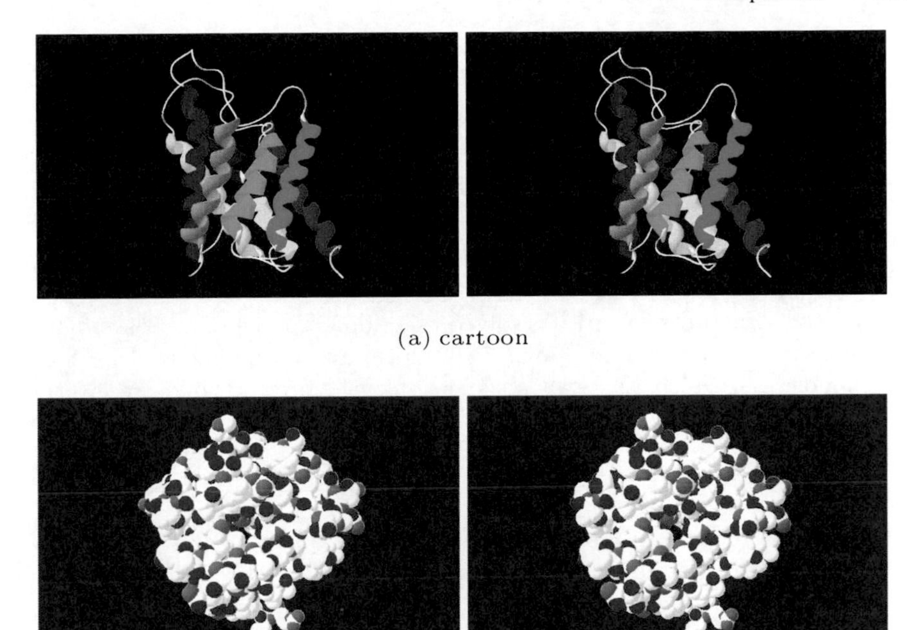

(a) cartoon

(b) space filling

Figure 17.15. A: Stereo cartoon of cattle aquaporin structure at 2.2 Å resolution (PDB-code 1J4N), viewed from the membrane plane. The transmembrane helices are clearly visible. **B**: Stereo view of the crystal structure of aquaporin, viewed from the extracellular side. Note the small channel through the protein, which allows passage of water, but not larger molecules (including ions with their coordinated water).

insipidus in dehydrated patients in an emergency room is essential, as rehydration with i.v. saline leads to hypernatraemia, possibly resulting in severe brain damage. If rehydration can not be performed orally, i.v. glucose may be used.

Cataract and neuronal loss of hearing also may be caused by non-functional aquaporins.

After brain injury excess activity of aquaporins leads to brain swelling, which can be fatal. A good understanding of aquaporin structure and function hopefully will result in the development of specific blockers, which can prevent this.

Each aquaporin molecule can pass about 3×10^9 water molecules per second. A $1\,\text{m}^3$ membrane with densely packed aquaporin molecules would need less than 1 s to desalt 1 l of water by ultrafiltration. Thus aquaporins are also of technical interest.

Water molecules have a size of about 2.8 Å, and the channel of aquaporins is at the narrowest part about 3 Å in diameter. This is however not sufficient to ensure the selectivity of water transport, because an unbroken chain of water molecules, with hydrogen bonding between them, could lead to proton transport by hydrogen bond exchange (GROTTHUSS-mechanism: a proton being bound at one end of the chain and a different hydrogen being released as proton on the other end). This "proton wire" would be physiologically dangerous and is not observed experimentally.

The crystal structure of Aqp1 has solved this riddle: At the narrowest point of the channel two Asn-residues (76 and 192) form hydrogen bonds with a single water molecule, which thus can no longer form hydrogen bonds with other water molecules. This interruption of the water chain in the channel prevents proton leakage.

Advanced material: Computer simulation of enzyme structure and reaction mechanism

The phenomenal increase in computer power over the last 30 years has made it possible to simulate the behaviour of molecules numerically. For this purpose atoms are placed in virtual space, guided for example by the data from X-ray crystallography or NMR-studies. Then there exist forces between these atoms, for example from stretching, compression or torsion of covalent bonds, electrostatic interaction between charged or polarised groups, VAN DER WAALS-attraction and so on. PAULI-repulsion prevents atoms from getting to close to each other. These forces can be modelled as mechanical springs between the particles, the resulting movements can then be calculated by numerical integration of NEWTONs laws of motion (see page 125 for a discussion of numerical integration). This is called **molecular dynamics**.

This way it is possible to refine models of protein structure, if their resolution is not high enough. The virtual molecule is exposed to a high temperature, with rapid movement of the atoms. As the temperature is reduced, movement becomes slower and atoms take the position of minimal energy (**simulated annealing**).

It is also possible to simulate the behaviour of molecules during a reaction, for example the behaviour of water molecules during their passage through aquaporin molecules [28]. The aquaporin molecule was embedded into a lipid membrane, which was immersed in water. The model totalled more than 100 000 atoms. Simulation of this system took several months of calculation time on a supercomputer, resulting in a film that showed the passage of water molecules through the aquaporin. Speed of passage agreed well with measured values, thus the model is probably correct.

The "dance of water molecules" through the aquaporin channel is made possible by the fact that whenever hydrogen bonds need to be broken to allow passage through the narrow channel they are replaced by other hydrogen bonds. Thus the activation energy for water transport is low and the speed can be high. In addition, the aquaporin molecule contains two potential barriers which prevent the passage of both positively and negatively charged ions.

Most residues in the channel wall are hydrophobic to allow water to move through the channel quickly, however, there are 4 water binding pockets with hydrophilic residues as well. These lower the energy required for breaking the hydrogen bonds with other water molecules, which is necessary to transport water molecules in single file.

Exercises

17.1. Uncouplers

a) decrease oxygen consumption.

b) increase heat production.

c) can be safely used for weight reduction.

d) inhibit proton transport across the mitochondrial membrane.

e) increase mitochondrial ATP-formation from ADP.

17.2. The mitochondrial proton-driven ATPase

1) has an F_o-part, which is an integral membrane protein.

2) has an F_o-part, which is a proton channel.

3) has an F_o-part that can be inhibited by oligomycin.

4) has a subunit which rotates driven by a proton gradient.

17.3. Na/K-ATPase

1) is electrogenic.

2) occurs in the mitochondrial membrane.

3) is inhibited by digitalis glycosides.

4) is an ABC-type ATPase.

17.4. Connect each physiological transport reaction to the type of transporter involved:

1) Removal of hydrophobic drugs from the cells of the blood brain barrier	A) P-type ATPase
2) Ca^{2+}-transport into the sarcoplasmic reticulum	B) ABC-type ATPase
3) transport of N-retinylidene-PE from the retinal rod outer segments	C) Antiporter
4) dissipation of proton gradients in brown adipose tissue mitochondria	D) Symporter
	E) Channel

17.5. D-glucose is transported "uphill" against a concentration gradient

1) from the intestinal lumen across the brush border membrane of enterocytes.
2) from the enterocyte into the blood stream.
3) from primary urine into proximal tubulus cells of the kidney.
4) from serum into the erythrocyte.

17.6. Which of the following statements is/are *correct* about solute transport across a membrane?

1) Under conditions of (near) substrate saturation catalysed transport is a reaction of zeroth order.
2) Catalysed transport is always active transport.
3) Active transport is always catalysed.
4) During uncatalysed transport all substrate molecules flow "downhill".

A

Appendix

A.1 Solutions for the exercise questions

We assume that a standard 5-options (A–E) "bubble sheet" is used, these are now quite common since they can be evaluated by scanner/computer. Several types of questions are in common use:

Single selection: There is exactly one correct answer out of a selection of 5. Sometimes, you may also be asked to identify the one wrong answer, then all the others will be correct (the reason is that it may be difficult to find 4 wrong answers which sound credible to a student who has not mastered the subject).

Selection of a combination: Similar to the single selection, but several answers may be correct. Usually 4 possible answers are given, a key is used to map combinations to the possible answers A–E:

if	choose
1, 2 and 3 are correct.	A
1 and 3 are correct.	B
2 and 4 are correct.	C
4 is correct.	D
all 4 are correct.	E

Logical relationship: You will be presented with 2 statements connected by words like "because", "while" or the like. Either statement may be correct or wrong, in addition the connection may be logical or not.

Statement 1	Statement 2	Connection	Choose
correct	correct	correct	A
correct	correct	false	B
correct	false	—	C
false	correct	—	D
false	false	—	E

Connected items: You are presented with two lists of items, and you have to find the items in list 2, which fit to those in list 1. Note that an item in List 2 may belong to several items in list 1.

In addition, some essay type questions and some questions were a short answer needs to be filled in were also presented. Such questions are used to test understanding and problem solving, rather than memorysation. Thus for merely repeating the material given in the lecture you will not receive full marks.

Some hints about exams

Be on time, and bring all required materials with you. It should not be necessary to say that, but experience shows that it is. You will be told in advance which materials are required/allowed in the exam, make sure you completely understand that.

In many exams, you will be supplied with enough paper to write your answers and to make any intermediate calculations. In such cases, you must not use any other paper than that supplied, and you must return all sheets provided, even if they are empty. Failure to obey either rule will count as cheating.

Read the exam completely before you start. Make sure that you understand exactly what is asked of you. Do you have to answer all questions, or do you have to choose some? In the latter case, you will not be given extra points for answering more questions than you were asked to do, so don't waste your time.

Read each question carefully, including all possible answers before choosing one of them. Remember also that in a biochemistry exam you will not be expected to do complicated maths. Numbers will usually be chosen so that calculations — if required at all — can be done in the head, without calculator. So if a calculation gets complicated, you are probably on the wrong track.

The problem with multiple choice exams (from the point of view of the examiner) is that students have a chance of 20 % to find the correct answer by guessing. Thus in some exams negative points are given for wrong answers. If negative points are *not* given, it is to your advantage to mark any answer if you do not know the correct one (since, statistically, you will get 0.2 points for it). However, if negative points are given, guessing will on average lead to a loss of points (4 possibilities for −1 and one possibility for +1 point, so the average will be −0.6 points). Under these conditions it is best to leave the answer open (0 points), unless you have a strong guess.

Even in those places where wrong and missing answers are both given 0 points, examiners have developed ways to guide non-knowing students to the wrong answers. Experience shows for example that guessing students will preferentially choose between two mutually exclusive options, even if both are wrong. There will be some examples for this type of questions below.

Problems of Chapter 1

1.1 C

1) No, Glycine is not. The α-carbon bears only 3 different substituents (1 carboxy and 1 amino group in addition to 2 hydrogens).
2) True. The α-carbon is chiral, in addition also the β-carbon bearing the hydroxy-group.
3) False. Bacterial cell walls also contain D-amino acids.
4) Since all amino acids contain at least one basic and one acidic group, they all have a pI-value.

1.2 The isoelectric point of a compound is the pH, at which it has an equal number of positive and negative charges.

1.3 For the reason given (correctly in the second statement), Gly is *not* chiral (Statement 1 is false). D

1.4 Val (V), Trp (W), Asp (D), Tyr (Y), X is not assigned to any amino acid. 1B, 2C, 3A, 4E

1.5 Trp is aromatic and hydrophobic, Lys positively charged (ϵ-amino group), Gly small (only hydrogen as side chain), Ser is polar (OH-group). Gln does not fit into any of these categories. 1A, 2C, 3D, 4B, 5A

1.6 Below pH 2.2 Lys has 2 positive charges from the fully protonated α- and ϵ-amino groups, and no negative charge because the carboxy 9.74

group will be mostly protonated as well. Above pH 2.2, the carboxy-group will loose its proton, resulting in 1 negative and 2 positive charges, and a net charge of +1. Beyond pH 8.95, the α-amino-group will loose its proton and there will be one negative and positive charge resulting in a net charge of ± 0. Beyond pH 10.53, the proton on the ϵ-amino group will be lost as well, resulting in one negative and no positive charges. Thus the pI is calculated as $1/2 * (8.95 + 10.53) = 9.74$.

1.7 The compound (cystine) does contain two covalently linked amino acids (cysteine), but they are linked by a disulphide- rather than a peptide bond. D

Problems of Chapter 2

2.1 Cys is involved in the formation of disulphide bonds, Ser can be phosphorylated and can carry O-linked sugar trees. Tyr can also be phosphorylated and Asn carries N-linked sugar trees. However, Ala has a rather unreactive methyl-group as side chain, which does not participate in posttranslational modifications. D

2.2 Because the isoelectric point of amino acids were measured in aqueous solutions. Inside a protein interactions between the ionisable groups of the amino acids can modulate their $p\mathrm{K_a}$-values and hence the tabulated pI-values do not apply.

2.3 Rotation around the peptide bond is *not* possible, since it forms a mesomeric system with the carboxy-group, resulting in a bond which has some properties of a double bond. The other statements are correct. A

2.4 A

1) Burying of hydrophobic side chains inside the protein reduces their contact with water, and thus thermodynamically stabilises protein structure.
2) Hydrogen bonds between amino- and carboxy-groups stabilise β-sheets. Note that these bonds are formed between strands, unlike α-helices, where the bonds are formed within a helix.
3) Proper folding of proteins depends on water bridges between (partially) charged groups, these indirect hydrogen bonds are destroyed if salts or water miscible organic solvents compete for the water molecules.

4) False, amino acids in loops are well defined, and coils have important functions in proteins. Expressions like "random coil" or "unordered structure" sometimes found in the literature are totaly misleading.

2.5 D

a) True, beside the α-amino and final carboxy-group, the last amino acid (K = Lys) has an ϵ-amino-group, which bears the second positive charge.

b) True, since the second amino acid is cysteine (C).

c) True, the third amino acid is Asn (N), and it occurs in the Asn-X-Ser/Thr consensus sequence that characterises glycosylation sites.

d) False. Neither Phe (F), Tyr (Y) nor Trp (W) are present.

e) True, there are 3 charged groups and 3 polar (acid amide from Asn, $-$SH from C and $-$OH from Ser) groups present. Ala and Gly are slightly hydrophobic, but with their short chains that effect is not particularly strong.

2.6 Leu, Ile and Val have hydrophobic side chains which can enter the hydrophobic environment of the membrane. Glu is a charged amino acids and is thus hydrophilic. It is rare in transmembrane segments and can occur only at the interface of several such segments. A

Problems of Chapter 3

3.1 C

a) True, the STOKES-radius depends on both size and shape of proteins and determines whether or not they fit into the pores of the gel.

b) True, separation is probably by a mixture of ion exchange and affinity chromatography.

c) False, SDS-PAGE separates by molecular weight. Binding of SDS gives proteins a high negative charge, irrespective of their pI.

d) True, substrate binding reduces conformational freedom and thus stabilises the protein against denaturation.

e) True, as this would result in protein inactivation.

Problems of Chapter 5

5.1

1B, 2D, 3C

1) The removal of hydrogen from ethanol is an oxidation.
2) In this reaction a phosphate group is transferred from the donor ATP to the acceptor glucose.
3) In this reaction functional groups within a molecule change place, creating an isomer.

Problems of Chapter 6

6.1

B

a) Formation of the enzyme-substrate complex is assumed to be a fast, diffusion controlled process, the speed of the reaction is controlled by the turnover of the ES-complex, which is normally slower.
b) V_{max} is determined by the enzyme concentration times the turnover number of the enzyme.
c) The MICHAELIS-constant K_M is equal to the concentration at which the reaction velocity is 1/2 V_{max}, the dissociation constant K_d is the concentration at which half-saturation occurs. Both can, but need not be similar.
d) K_M depends on the reaction conditions (pH, temperature...).
e) At 2 times K_M the reaction velocity is only 2/3 of V_{max}.

6.2 If the enzyme is saturated with substrate, all enzyme molecules occur as ES-complex and the rate of product formation becomes the product of turnover number times enzyme concentration. The HMM-equation is a special case of the law of mass action, derived under simplifying assumptions.

D

6.3 Since the substrate concentration is higher than K_M the reaction velocity needs to be larger than 0.5 * V_{max}, so answers a–c have to be incorrect. On the other hand, V_{max} is reached only at [S] $= \infty$, thus answer e) is incorrect too. That leaves d) as the only possible answer, which can be confirmed by solving the HMM-equation: $v = \frac{3*K_M}{K_M+3*K_M} * V_{max} = 3/4 * V_{max}$.

D

6.4 **A**: In curve 2 the maximal rate is reduced, but the K_M is unchanged. Such behaviour is obtained by reducing the number of enzyme molecules.

B

6.5 If 50 000 g are equivalent to 1 mol of enzyme, then 10 µg are equivalent to 0.2 nmol. Thus 1 molecule of enzyme will turn over 10 000 nmol·min^{-1}/0.2 nmol = 50 000 molecules of substrate per min. Note that a correct answer must also contain the unit!

50 000 min^{-1}

6.6 Higher activity (faster turnover) and lower K_M-value (higher affinity) both are important. Molecular weight and pI have little to do with it. Specificity would influence only the competition of two substrates for the same enzyme.

B

6.7 Isoenzymes are proteins with different primary structure that perform the same reaction on the same substrate.

E

6.8 For a (non-exclusive) list see section 6.2.1

6.9

B

a) Correct. Irreversible destruction of enzyme activity is called inactivation.

b) False. Aspirin® interaction with cyclooxygenase (COX) is a typical example for enzyme inactivation, not inhibition.

c) Correct. The common point is on the x-axis, if $K_i = K_{ii}$, it is above or below the x-axis if the inhibition constants are not equal.

d) Correct.

e) Correct.

6.10

D

1) False. Once the binding/dissociation equilibrium has been established, time does no longer influence the degree of inhibition (unlike inactivation, which is time-dependent).

2) False. Such a reaction would change the apparent substrate concentration and not result in a competitive inhibition pattern.

3) False. Competitive inhibition results when substrate and inhibitor can not bind to the enzyme at the same time. This may, but need not, be the result of inhibitor binding to the substrate site.

4) Correct. In competitive inhibition only the apparent K_M, but not V_{max} change.

Problems of Chapter 7

7.1

D

a) Correct.

b) Correct.

c) Correct.

d) False. The HILL-coefficient may be lower, but not higher than the number of binding sites.

e) Correct.

7.2

B

a) Binding site is between the β-subunits of desoxy-haemoglobin.

b) False. Binding is mainly by ionic interactions.

c) The γ-subunits of fetal haemoglobin bind 2,3-BPG less strongly than the β-chains in adult, resulting in higher oxygen affinity of fetal haemoglobin.

d) This is a regulatory mechanism to increase oxygen transport, for example in people living in high altitudes.

e) Correct.

7.3

A

a) Myoglobin accepts oxygen from haemoglobin and thus needs to have a *higher* oxygen affinity.

b) Myoglobin is a monomer, haemoglobin is a diprotomer.

c) True. Since it has only one oxygen binding site it can not show co-operativity.

d/e) False. Central atom is Fe^{2+}. Fe^{3+} is found in methaemoglobin, which is unable to bind oxygen. Cu^{2+} is found in haemocyanin, the oxygen transport protein of certain invertebrates.

7.4

B

a) Inability to synthesise β-subunit is the cause for β-thalassaemia.

b) Correct, Glu-6 $\rightarrow$ Val.

c) Haem is a prosthetic group, there is no binding equilibrium. Incorporation of haem into haemoglobin is not affected in sickle cell anaemia.

d) The risk is *decreased*. That's why the disease gets enriched in certain populations.

e) Aggregation of HbS is increased in the deoxy-form.

7.5 Sickle cell anaemia is caused by the replacement of the hydrophilic amino acid Glu-6 by the hydrophobic Val. The resulting hydrophobic patch leads to haemoglobin aggregation in erythrocytes, reducing their life expectancy (statement 2 is correct). Malaria parasites spend part of their life cycle inside the erythrocyte, feeding on haemoglobin. If the life span of the erythrocyte is reduced, the parasite can not complete its life cycle, resulting in resistance (statement 1 and connection are correct). A

Problems of Chapter 9

9.1 According to the best available evidence viruses or other nucleic acid-containing entities are not involved in prion diseases, in fact the very name *prion* comes from proteinaceous infectious agent. The other statements are correct. C

Problems of Chapter 10

10.1 E

1) Correct. The ubiquitin marks proteins no longer needed.
2) Correct. This allows the immune system to check which proteins are manufactured inside a cell, and to destroy virus infected and cancer cells.
3) Correct. The proteasome contains an ADP/ATP-exchange stimulating factor Bag1 to aid in protein release from Hsp70/Hsc70.
4) Correct. Hsp70 and its cognate deliver the peptides to the Tap1/Tap2-transporter in the ER.

10.2 See pages 143 and 150, and table 10.2

10.3 Complement is activated by IgG and IgM, not IgD. The other statements are true. B

10.4 C

1/2) IgE binds to the plasma membrane of mast cells via their F_c-ends, cross-linking of two such membrane-bound antibodies releases histamine (from the mast cells, not from macrophages).

3) Transplant rejection is T-cell-, not antibody-dependent.

4) True. This is supposed to be its main function.

10.5 B

1) True, since papain cuts IgG between the C_H1 and the hinge domain.

2) False. IgG is N-glycosylated and the glycosylation site is in the second constant region of the heavy chain.

3) True. That's why it has to be variable.

4) False. The hinge domain is located between C_H1 and C_H2.

10.6 A

1–3) True.

4) There is no bond between the two light chains.

10.7 B

1) True.

2) False. Our body can make about 1×10^{15} different antibodies, but contains only a total of about 3×10^4 different genes, making it impossible to have a separate gene for each antibody. Besides, immunoglobulin building blocks are encoded on chromosomes 2, 14 and 22.

3) True. Each host is exposed to many different antigens, which in turn may contain many different epitopes each. This limits the use of antisera for the diagnosis and treatment of diseases and was the reason for development of monoclonal antibody technology.

4) False, the variable parts of heavy and light chains are encoded by different V-genes, resulting in different primary structures.

10.8 Statement 1 is correct, small molecules like penicillin can not induce an immune response unless they are bound to a carrier (like serum albumin). However, statement 2 is incorrect, haptens can bind to antibodies once those are produced. Haptens are too small to C crosslink 2 membrane-bound antibodies on the B-lymphocyte, which is required for its conversion into an antibody-producing plasma cells.

10.9 B

a) False. Antigenic peptides are much shorter than 50 aa.

b) True. Note however that MHC-I presents proteins produced by the APC itself, while MHC-II presents the foreign antigens collected.

c) False. Binding of antigens to MHC-molecules occurs with low specificity. T-cells then recognise the presented antigens with high specificity.

d) False. They occur mainly in the peripheral tissue (to collect antigens) and in the lymphatic tissues (to present them to T-cells), and only transiently in blood.

e) False. Phosphorylation occurs in regulatory pathways, but has nothing to do with antigen recognition.

10.10 B

1) True.

2) False, loading with antigenic peptides occurs in vesicles.

3) True in most populations. In isolated populations enrichment of specific MHC molecules can lead to the susceptibility to certain diseases.

4) False. CD8 occurs on T_K-cells and is involved in the recognition of MHC-I bound self antigens. MHC-II bound foreign antigens are the responsibility of the CD4 coreceptor on T_H-cells.

10.11 B

1) True.

2) False. T-cell receptor variability is caused by recombination only, somatic hypermutation occurs in B-, not in T-cells.

d) True. Toxic shock syndrome is a clinically important example.

c) False. TCR recognises antigen-MHC-complexes.

10.12 Proteins of the complement system are synthesised in the liver, B-cells are responsible for the synthesis of antibodies. Aggregation of antibodies, specifically IgM and IgG, for example in immune complexes or on the membranes of invaders, lead to the activation of the E complement system, forming anaphylatoxins (which increase vascular permeability) and membrane pores in the attacker.

10.13 Inflammation is the result of complement activation, in the process anaphylatoxins are produced which lead to increased vascular permeability, thus fluid and white blood cells will assemble in the affected area. All 4 symptoms are a direct result of this effect. E

Problems of Chapter 11

11.1

A

1) True. This reaction is controlled by Cdk2.
2) True. One example would be Na/K-ATPase.
3) True. The lining inside the nuclear membrane is formed by lamin A, -B and -C, which are type V intermediate filament proteins.
4) False. Intermediate filament proteins have no known transport function.

11.2 Both statements and their connection are correct. The method is important to determine the origin, and hence proper treatment, of intestinal and breast cancers. A

11.3

A

a) False. Most G-actin molecules contain MgATP, and most F-actin MgADP (rather than the other way round), but ATP hydrolysis is not required for actin polymerisation and ADP/ATP exchange not for depolymerisation.
b) True. Cross-linking is via the protein dystrophin.
c) True. The resulting vesicles are distributed stochastically to the daughter cells during mitosis.
d) True, hair, claws and nails are made from keratin, an IF-protein.
e) True, although colchicine acts by dissociating microtubules, this in term affects the IF.

Problems of Chapter 12

12.1 In skeletal muscle phosphorylation of myosin is not required for contraction, only in smooth muscle is the light chain phosphorylated (hence the slower contraction of smooth compared to skeletal C muscle). Contraction is started by binding of Ca^{2+} to troponin C and myosin light chains. Polymerisation of actin is required for sarcomere formation, and hence only indirectly for contraction.

12.2 D

1) False, release of myosin heads from actin filaments requires ATP, hence lack of ATP leads to *rigor mortis*.
2) False. Tropomyosin and Cap Z cap the actin filament on either end, not the thick filaments.
3) False. Binding is not covalent.
4) True, sarcoplasmic Ca-ATPase is a P-type ATPase.

12.3 Actin filaments contain bound ADP, not GTP. B

12.4 Since the minus-ends of microtubules are capped by the microtubule organising centre (the hub in a)), they have to grow and shrink from the plus-end. The described behaviour is that of actin filaments. The other statements are correct. B

Problems of Chapter 13

13.1 Hydroxylation of Lys and Pro (in collagen), and the formation of desmosin from Lys (in elastin) are important for the stability of the extracellular matrix. Gly is an important component of collagen, but not modified. Tyr phosphorylation is important for the regulation of protein activity, but not in the extracellular matrix. B

13.2 Vitamin C is required to regenerate the enzyme prolyl-hydroxylase, which can become inactivated by oxidation of its catalytic iron centre to Fe^{3+}. If this is not corrected, prolyl-groups of pre-pro-collagen will not become hydroxylated in the ER, and the resulting collagen triple helix will be unstable at physiological temperature. Thus the connective tissue will be weak, leading to the typical symptoms of scurvy like loss of teeth.

13.3 A

a) Only glycine is small enough to fit between the 3 collagen molecules of a triple helix. Mutation of a single glycine can lead to serious inherited defects of connective tissue formation.
c) Collagen 1 is formed from 2 molecules of $\alpha 1(1)$ and 1 molecule of $\alpha 2(1)$. If one copy of the $\alpha 1(1)$ is mutated, only 25 % of all

collagen molecules formed will be functional, resulting in pseudo-dominant inheritance.

d) Since each triple helix contains only 1 molecule of $\alpha2(1)$, mutation in one copy of the gene will result in the production of 50 % functional collagen molecules, thus the mutation is recessive.

e) Helicases are involved in DNA metabolism, formation of the collagen triple helix is not enzyme catalysed.

13.4 Adherens junctions cross-link the actin skeletons of cells, keeping the cells of an epithelium together. C

Problems for Chapter 14

14.1 All proteins of the Hsp70 chaperone family bind to unfolded proteins and prevent the formation of miss-folded states. They can not unfold proteins, this is the job of chaperonins like GroEL/GroES. C

14.2 GroEL/GroES can tear apart mis-folded proteins (in the *trans*-ring) and also form a protected space (ANFINSEN-cage, in the *cis*-ring) in which the protein can fold. However, it can not catalyse folding in any way. A

Problems for Chapter 16

16.1 The amino acid cysteine contains a thiol (thioalcohol, mercaptan, −SH) group, which can enter into disulphide bond formation: $-SH + HS- \rightleftharpoons -S-S- + 2\,[H]$. Removal of hydrogen is equivalent to oxidation. The formation of disulphide bonds takes place in the oxidising environment of the ER, the specificity is determined by disulphide isomerases. Disulphide bonds not only hold light and heavy chain of the immunoglobulins together, but are also important for their correct tertiary structure. The T-form of haemoglobin (with low affinity to oxygen) however is stabilised by additional ionic bonds between α- and β-subunits, which are not present in the R-state. E

16.2 O-linked carbohydrates occur on Ser and Thr. Tyr does contain an OH-group, but is not involved in glycosylation. Asn bears N-, not O-linked sugar trees. B

16.3 **A**: The peptide is glutathione (GSH). Two molecules of glutathione can form a disulphide bond (GSH + GSH $\rightleftharpoons$ GSSG + 2 A

[H]). The hydrogens are available to reduce other compounds, making GSH a reducing agent. Thus both statements are correct, as is their connection.

16.4 All 4 statements are correct. E

16.5 E

a) True. Amino acids C-terminal of the stop-transfer signal do not enter the ER.

b) True. The cytosol has a negative potential with respect to the ER lumen and the extracellular space. Thus positively charge amino acids keep a sequence in the cytosol and determine whether a hydrophobic sequence acts as stop-transfer or signal-anchor sequence (positive inside rule).

c) True, GPI linked to the C-terminus results in type 6 membrane proteins.

d) True, type 5 membrane proteins.

e) False, SRP-interaction is required only for the first transmembrane domain, the remaining interact with the translocon directly, without SRP.

16.6 Cardiolipin occurs in the mitochondrial membrane, specifically thc inner leaflet of the inner membrane. Since heart muscle cells contain many mitochondria cardiolipin was first isolated from heart tissue, hence its name. B

16.7 PI and PE occur in the inner leaflet of the plasma membrane. Cardiolipin occurs in the mitochondrial membrane, triacylglycerol (energy store) is not a component of bilayer membranes. D

16.8 Transmembrane domains in proteins are either α-helices or β-barrels. Lipid anchors can attach a protein to membranes (peripheral membrane proteins). In either case hydrophobic bonds are responsible for the interaction. Although membrane proteins frequently carry sugar trees (either in N- or O-glycosidic bond), these are not involved in membrane-protein-binding. C

16.9 C

a) The different composition of both leaflets originates from lipid synthesis at the ER and is maintained by primary active transport (flippases).

b) Tyrosine autophosphorylation signals ligand binding for example in cytokine receptors (mitogen activated protein kinase pathway).

c) Sugar side chains of glycoproteins always face the extracellular medium.

d) Cholesterol content of the outer leaflet of the plasma membrane can reach 40 %, while for example ER and mitochondrial membranes do not contain significant amounts.

e) True. This disequilibrium is maintained by Na/K-ATPase. Note that bacterial, plant and fungal cells usually have proton gradients instead.

Problems to Chapter 17

17.1

B

a) False. Since the proton-gradient across the mitochondria is short-circuited by uncouplers, ATP-generation in mitochondria stops. The body will increase oxygen consumption in an attempt to generate more ATP.

b) Correct. Since the metabolic energy is no longer channelled into ATP-production it has to go into heat production. This can be physiologically intended (brown adipose tissue in newborn or hibernating animals).

c) False, a dose slightly too high can result in fatal overheating. This procedure is anything but safe.

d) False. They act as transporters of protons, short-circuiting the proton gradient across the mitochondrial membrane, which is required for ATP-synthesis.

e) False. They decrease it.

17.2 All of the above statements are correct.

E

17.3

B

1) Since it pumps 3 Na^+ out of, but only 2 K^+ into the cell per hydrolysed molecule of ATP, each cycle results in the net transport of one positive charge out of the cell, making the cytosol negative with respect to the exterior.

2) False, it occurs in the plasma membrane.

3) True. Although the physiological effects of digitalis are not yet completely understood.

4) False. Na/K-ATPase belongs into the class of P-type (or E_1/E_2-type) ATPases.

17.4 Removal of hydrophobic xenobiotics is the job of Mdr1, an ABC-type ATPase. Ca-transport into the sarcoplasmic reticulum is achieved by Ca-ATPase, a P-type ATPase. The transport of N-retinylidene-PE (product of the conversion of visual purple into visual yellow) is the responsibility of ABCA4p. In brown adipose tissue the proton gradient across the mitochondrial membrane is dissipated by thermogenin, a channel.

1B, 2A, 3B, 4E

17.5 Uptake of glucose from the intestinal lumen into the enterocyte and from primary urine into kidney proximal tubulus cells is by Na/glucose cotransport and occurs against the concentration gradient. Transport from the enterocyte into the blood, and from the blood into other cells like erythrocytes is by facilitated diffusion and follows the concentration gradient.

B

17.6

B

1) True. Once a transporter is saturated, a change of substrate concentration does not lead to a change of transport rate.

2) False. Carrier and channels allow the substrate to flow only "downhill" of the gradient, but catalyse this process.

3) True. Active (uphill) transport is always catalysed, the catalyst couples an energy producing reaction with energy consuming transport.

4) False. Substrate molecules move in both directions, but the rate for downhill transport is higher, thus net transport occurs downhill.

A.2 Short biographies of scientists mentioned in this book

ALPER, TIKVAH South Africa, 1909–1995. Born as fourth daughter of an immigrant from Russia she obtained a scholarship to study Physics at Capetown University, where she received her M.A. at the age of 20. She then moved to Germany, where she worked on the δ-rays from α-particles with L. MEITNER but failed to obtain a PhD because growing antisemitism in Germany forced her to return to South Africa (she was awarded a D.Sc. by London University in 1969). She married the bacteriologist MAX STERN in 1932, their son JONATHAN was born deaf, so she moved to St. Louis (MO) to receive training as teacher for the deaf, in which capacity she worked until she became a lecturer at the University of Witwatersrand and later head of the Biophysics unit at the National Physics Laboratory. In 1951 she signed a circular against growing apartheid and was forced to emigrate to England, where she obtained a grant at the MRC Experimental Radiopathology Research Unit at Hammersmith Hospital. In 1953 she became an official staff member, in 1962 director of this institution, from which she retired in 1973. She was the first to demonstrate that the causative agent for prion diseases does not contain DNA.

ANFINSEN, CHRISTIAN BOEHMER USA, 1916–1995, B.A. 1937 (Swathmore College), M.S. in Organic Chemistry 1939 (University of Pennsylvania), Ph.D. 1943 Harvard Medical School. Served in various positions at Harvard Medical School, John-Hopkins University and the National Institutes of Health. He received the Nobel Price in Chemistry in 1972 for his discovery that proteins fold spontaneously (at least in principle) and that their entire three-dimensional structure is encoded in their amino acid sequence.

BEHRING, EMIL ADOLF VON German, 1854–1917. As the eldest of 13 children of a schoolmaster he had no funds to study, thus he entered the Army Medical College in Berlin in 1874, obtained his medical degree in 1878 and passed his licensing exam in 1880. He was then send to Posen (now in Poland), where he worked as army physician and at the same time continued scientific studies at the Chemical Department of the Experimental Station, working on septic diseases. He found that iodoform (CHI_3) could neutralise bacterial toxins (published 1882). Behring was then send to C. BINZ (Bonn) for further training in pharmacology. In 1888 he was ordered back to Berlin, where he became an assistant to R. KOCH at the Institute for Hygiene and later the Institute for Infectious Diseases (now known as the ROBERT-KOCH-Institute), working in a team that included amongst others P. EHRLICH. In 1894 BEHRING became Professor of Hygiene at the University of Halle, moving to Marburg the following year. He is remembered mostly for his work on diphteria, tuberculosis and tetanus. He found (together with S. KITASATO) that the supernatants of cultures of diphteria and tetanus bacteria, when injected into animals in carefully determined doses, led to the production of *antitoxins* (which now we call antibodies) in their blood, which were able to neutralise the deadly toxins made by these bacteria. The antitoxins produced in one animal could be used to save other animals infected with the bacteria (passive immunisation). BEHRING and F. WERNICKE demonstrated that toxin-antitoxin complexes could also be used for active immunisation of animals, at much lower risk to their well-being than the pure toxins. Following a suggestion of T. SMITH such complexes were used for immunisation campaigns in man. Production of these complexes was done in the **Behringwerke**, a world-renowned company specialising in vaccines (founded in 1914). BEHRING became *Geheimer Medizinalrat* in 1895, and was raised to nobility in 1901, the year he received the NOBEL-price for Physiology and Medicine for his ground-breaking and life-saving work.

BERZELIUS, JÖNS JAKOB Swedish, 1779–1848, studied Medicine at the University of Uppsala and wrote a thesis on the effect of electricity on humans. He became professor of medicine and pharmacy at the Medical College of Stockholm in 1807. He is considered one of the greatest chemists of all times, discovered the law of constant proportions, the elements Ce (1804), Se (1817) and Th (1828) and was involved in the discovery of Li, V and several rare earth metals. He invented chemical formulas (1814). He was the first to understand ionic compounds and coined the name "amino acid".

BLOBEL, GÜNTHER German, born 1936. He fled the Soviet Occupation Zone of Germany in 1954 and studied Medicine at the University of Frankfurt/Main, Kiel, Munich and Tübingen, graduating in 1960. In 1962 he moved to Montreal, where he obtained a Ph.D. in 1967 under V.R. POTTER. Postdoctoral studies under G. PALADE at Rockefeller University, becoming Assistant Professor in 1969, Associate Professor 1973 and Full Professor in 1976. Nobel Price in Physiology and Medicine 1999.

BOHR, CHRISTIAN Danish, 1855–1911, studied Medicine in Copenhagen, M.D. in 1886. Discovered the effect of pH-reduction and carbon dioxide binding on the oxygen affinity of haemoglobin. Father of the physicist NIELS and the mathematician HARALD BOHR.

BOYER, PAUL D. USA, born 1918. B.A. in Chemistry from Bigham Young University, Ph.D. in Biochemistry from the University of Wisconsin in 1943. Postdoctoral training on Stanford University. Assistant Professor of Chemistry at the University of Minnesota 1945, 1963 at UCLA. In 1971 he discovered that the energy of the proton flux is not used to make ATP, but to release it from the ATP-synthase. In the following years he found the co-operation between the 3 catalytic binding sites of ATP-synthase and the concept of rotational synthesis. Nobel price 1997 for his studies on ATP synthase, together with J.E. WALKER & J.C. SKOU.

BUCHNER, EDUARD German, 1860–1917, studied Chemistry and Botany in Munich and Erlangen, Ph.D. in 1888, Professor of Chemistry 1895. Received the Nobel price in Chemistry in 1907 for his discovery (made together with his brother HANS, a well known bacteriologist) of non-cellular fermentation in yeast extracts.

BUCHNER, HANS German, 1850–1902. M.D. 1874, professor of hygiene 1892. Recognised the importance of serum proteins for immune defence and invented techniques for studying anaerobic bacteria. Together with his brother EDUARD he discovered non-cellular fermentation by yeast extracts. He died before his brother received the NOBEL-Price for this discovery.

BURK, DEAN USA, 1904–1988

CECH, THOMAS ROBERT USA, born 1947. Studied at the University of California in Berkely (Doctorate 1975), then went to MIT (1975–1977) and the University of Colorado (1978, since 1983 as professor of Chemistry and Biochemistry as well as for Cell and Developmental Biology). Since 1987 Research Professor at the American Cancer Society. Discovered that the splicing of mRNA is done by the mRNA itself, in other words that some RNA molecules can act as enzymes. For this discovery he received the NOBEL Price for Chemistry in 1989.

CHANCE, BRITTON USA, born 1913. PhD 1940 from the University of Pennsylvania, became professor of biophysics there in 1949. Invented methods to investigate the initial (pre-steady-state) phase of enzyme reactions. Proved the existence of the ES-complex by measuring the absorption spectrum of the haeme group in peroxidase upon binding of H_2O_2.

CLAUDE, ALBERT Belgian, 1899–1983. The early death of his mother from breast cancer left a lasting impression on the 7 year old boy, even though he later claimed that it did not cause him to study Medicine. He obtained a medical degree from the Université de Liège in 1928, then went to the *Institut für Krebsforschung* and the *Kaiser-Wilhelm-Institute Dalem* in Berlin. In 1929 he moved to the Rockefeller Institute, where he spend the rest of his career, although he held additional posts at the University of Brussels and the Catholic University of Louvain. One of the fathers of cell biology, started to use the electron microscope to study biological questions. Identified the mitochondrium as site of oxydation. In 1974 he received the Nobel Price for Physiology and Medicine, together with C. DE DUVE & G.E. PALADE.

CLELAND, WILLIAM WALACE USA. A.B. (Oberlin College), Ph.D. (University of Wisconsin-Madison). M. J. Johnson Professor of Biochemistry at UW-Madison in 1978, member of the National Academy of Science since 1985. Significant contributions to the study of enzyme reaction mechanism by kinetic methods, pioneered the use of isotope effects on enzyme activity.

CORNISH-BOWDEN, ATHELSTAN JOHN English, born 1943, B.A. 1965, M.A. 1967, Ph.D. 1967

CREUTZFELDT, HANS GERHARD German, 1885–1964. Neuropathologist.

DAVY, HUMPHRY English, 1778-1829. Chemist. Showed in 1816 that platinum wire could catalyse the reaction of alcohol vapour with air oxygen, without being used up.

DONNAN, FREDERICK GEORGE British, 1870–1956. He was born in Colombo (Ceylon, now Sri Lanka) as son of Northern Irish parents. At age 3 he returned to Ireland. He studied science at Queens College, Belfast (BSc 1893, BA 1894), Leipzig (PhD 1896), Berlin, London, and the Royal University of Ireland in Dublin (MA 1897), working amongst others with VAN'T HOFF, OSWALD and RAMSAY. He taught Chemistry in various positions in London, Dublin and Liverpool until he retired in 1937. He helped emigrant scientist from Germany during the Nazi rule and contributed to the production of ammonia and nitric acid during second World War as a consultant to the Ministry of Munitions. His most significant scientific contribution was the theory of diffusion equilibria where some ionic species can cross a semipermeable membrane while others can't (1911). The resulting electrochemical potential (found across the membrane of every living cell) bears his name. He is also considered one of the fathers of colloid chemistry and made significant contributions to our understanding of detergent action.

DUCLAUX, PIERRE ÉMILE French, 1840-1904. Director of the Pasteur Institute 1895 to 1904.

DUCHENNE DE BOULOGNE, GUILLAUME BENJAMIN AMAND French, 1806–1875. Studied Medicine in Paris before becoming a practitioner in his hometown of Boulogne (1831). Two years later he invented electro-therapy and electro-diagnostics, which he continued to elaborate for the rest of his life. He described several diseases: DUCHENNE muscular dystrophy, *tabes dorsalis*, tabetic locomotor ataxia, progressive bulbar paralysis and also worked on the consequences of lead poisoning and acute poliomyelitis.

DE DUVE, CHRISTIAN Belgian, born 1917. Studied Medicine at the Catholic University of Louvain from 1934, M.D. 1941. Became army doctor, was captured and fled from the German prisoners camp, returning to Louvain where he worked as an intern and at the same time studied Chemistry since the war prevented him from continuing his research on insulin action. Further studies 1946-1947 at the Medical Nobel Institute in Stockholm with H. THEORELL, Washington University with the CORIs and St. Louis with E. SUTHERLAND. Returned to Louvain in 1947 as Teacher of Physiological Chemistry, becoming Full Professor in 1951. During the following years he developed techniques for the isolation of cell organelles, leading to the discovery of the lysosome and the peroxisome. These techniques are still in routine use today. He was awarded the NOBEL Price for Physiology and Medicine in 1974, together with A. CLAUDE & G.E. PALADE.

EDMAN, PEHR VICTOR Swedish, 1914–1977. Studied Medicine at the Karolinska-Institute (M.D 1947 after an interruption by World War II, which he spend in the Medical Corps). Postdoctoral training at the Rockefeller Institute of Medical Research in Princeton, then Assistant Professor at the University of Lund, where he published his first paper on the determination of protein sequences (now called EDMAN-degradation) in 1950. He became Direktor at the St. Vincent´s School of Medical Research in Australia in 1957, where he developed the first automated protein sequencer in 1967 (with G. BEGG). From 1972 EDMAN continued his work at the MAX PLANCK-Institut for Biochemistry in Martinsried (near Munich, Germany), getting ever longer sequences from ever smaller samples.

EULER, LEONHARD Swiss, 1707–1783, moved to St. Petersburg and Berlin as Professor of Mathematics. Discovered fundamental rules of calculus, and was one of the first to apply maths to physical, economic and musical problems.

FARADAY, MICHAEL British, 1791–1867, started his professional life at age 14 as bookbinder apprentice, becoming an assistant to HUMPHREY DAVY later. In 1831 he started with a series of experiments on electromagnetism, that let him to the invention of dynamo and electric motor. He discovered benzene, the laws of electrochemistry and invented oxydation numbers and the FARADAY-cage. In 1845 he discovered that magnetic fields could turn the direction of polarised light (FARADAY-effect).

FENN, JOHN BENNETT USA, born 1917. Trained at Berea College and Allied Schools in Berea, Kentucky (high school diploma 1932, graduated in chemistry in 1938). Then he moved to Yale, working as an assistant while studying physical chemistry, obtaining a Ph.D. in 1941. After several industrial positions he became an academic staff member at Princeton University working on combustion and molecular beams.

FICK, ADOLF EUGEN German, 1829–1901, discovered the diffusion law named after him (in 1855), described the FICK-principle (1870, cardiac output = oxygen consumption / arteriovenous oxygen difference) and invented the contact lens (1887).

FISCHER, HERMANN EMIL German, 1852–1919, studied in Bonn and Straßburg (now Strasbourg), Ph.D. 1874, habilitation 1878, Professor of Chemistry in Munich 1879, Erlangen 1881 and Würzburg 1888. Worked on dyes and hydrazines, described the structures and metabolism of purines. Between 1882 and 1906 he worked on the stereochemistry of sugars and glucosides. He separated and identified the amino acids in proteins (1899–1908) and discovered the peptide bond. In 1902 he was awarded the Nobel Price in Chemistry. The German Chemical Society has named its medal after him.

GIBBS, JOSIAH WILLARD USA, 1839–1903. Studied in Yale (Ph.D. 1863) and became professor of theoretical physics there in 1871. His work on the driving force behind chemical reactions received little attention at his time because he published it in obscure places.

GOLGI, CAMILLO Italian, 1843–1926, studied Medicine at the University of Pavia, graduation 1865. He continued to work at the Hospital St. Matteo in Pavia, specialising in neurological diseases. In 1872 he became Chief Medical Officer in the Hospital for the Chronically Sick in Abbiategrasso, where he converted a kitchen into a laboratory. He became Extraordinary Professor for Histology at the University of Pavia and later Chair of General Pathology (1881). He identified the three different forms of malaria and their pathogens, and in 1890 succeeded in obtaining photographs of them. He discovered silver staining of nerve cells, the tendon sensory organ, and the GOLGI apparatus. Nobel Price for Physiology and Medicine 1906 together with SANTIAGO RAMÓN Y CAJAL for studies on the nervous system.

GROTTHUS, CHRISTIAN THEODOR VON German, 1791–1822. Published the first theory of electrolysis in 1805.

HALDANE, JOHN BURDON SANDERSON Scottish, 1892–1964, educated in Eton and Oxford. Member of the communist party, which he left after the LYSENKO-disaster. However, he was never happy with the politics of his native country and sought refuge in India 1957, where he also died. Founder of population genetics, worked on applying mathematical principles in biology, thus laying the foundations for our modern understanding of evolution. Coined the word "clone" and contributed to enzyme kinetics.

HENRI, VICTOR French. 1872–1940. Studied psychology at the Sorbonne and in Göttingen, Germany (there Dr. phil. 1897). Worked first at the Sorbonne, then went to Russia (home country of his mother) in 1915, where he worked in the chemical industry with a grant from the French government (Russia and France were allies during WW-I). 1917–18 Prof. of Physiology at the University of Moscow, 1920-30 Prof. of Physical Chemistry at the University of Zurich, Switzerland. Worked as director of the Berre refinery in Marseille 1930–31, before he became Prof. of Physical Chemistry at the University of Liège (Belgium) 1931–40. Worked on enzyme kinetics (first formulation of the HMM-equation 1902), photochemistry and spectroscopy.

HILL, Sir ARCHIBALD VIVIAN English, 1886–1977. Supported by several fellowships he studied Maths and Science at Trinity College, Cambridge (1905–1909). In 1909 he started experimental work in the lab of J.N. LANGLEY. His principal interest was muscle contraction and exercise physiology, a field in which he continued to work throughout his life, using calorimetric methods which he learned during his stay with BURKER and

PASCHEN in Tübingen, Germany (1911). He improved these so much that in 1937 he was able to measure temperature differences of 0.000 15 °C. In addition he worked on nerve physiology, haemoglobin (resulting in the famous HILL-equation), anti-aircraft munitions development (during World War I). He became Lecturer for Physical Chemistry at Cambridge 1914, Professor of Physiology in Manchester 1920, followed by various positions in London. During the war years 1940–45 he served as member of parliament (Independent Conservative for Cambridge) and as member of the War Cabinet Scientific Advisory Committee. Like DONNAN he was a member of the Academic Assistance Council which helped refugees from Nazi Germany to relocate in the West. He received the NOBEL-Price for Physiology and Medicine in 1922 together with his collaborator, the German biochemist O. MEYERHOFF for the discovery that muscle contraction can be fueled by both aerobic and anaerobic metabolism.

HOPPE-SEYLER, ERNST FELIX IMMANUEL German, 1825–1895. Born as E.F.I. HOPPE he became orphan early and was adopted by his brother in law, the priest Dr. SEYLER. Studied Medicine at the Universities of Halle (1846–1847), Leipzig (1847–1850) and Berlin (1850), where he wrote his thesis on cartilage structure. After short stays in Prague and Vienna he started as general practitioner in Berlin 1852, then became research assistant at the University of Greifswald 1854 where he was habilitated in 1855. Prosector and head of the Chemical Laboratories at the Pathological Institute at the Charité Clinic in Berlin 1856 (under VIRCHOW), professor of Applied Chemistry at the University of Tübingen 1860–1871, then professor of Physiological Chemistry at the University of Straßburg (now Strasbourg) from 1872 until his death. He became rector in 1873, founded the first journal for physiological chemistry (*Zeitschrift für Physiologische Chemie*, now *Biological Chemistry*) 1877, and inaugurated the first independent Institute for Physiological Chemistry in Germany 1884. His research work included fermentation, lipid and bile metabolism, urine composition, and the quantification and classification of proteins. Together with his students FRIEDRICH MIESCHER and ALBRECHT KOSSEL he discovered and characterised DNA. He demonstrated reversible oxygen binding to haemoglobin (a word he coined) by spectroscopic methods and found the pathomechanism of carbon monoxide and hydrogen sulphide poisoning. He was the first to crystallise a protein (haemoglobin) in pure form.

JACOB, FRANÇOIS French, born 1920. Started studying Medicine at the Faculty of Paris in 1939 but his education was interrupted by the second world war, during which he served in the Free French Forces in North Africa and in Normandy, where he was severely wounded. He continued his studies (M.D. 1947, M.Sc. 1951, Ph.D. 1954). In 1950 he joined A. LWOFF at the Institute Pasteur, becomming laboratory director in 1956 and head of the Department of Cell Genetics in 1960. In 1964 he became chair of the department for Cell Genetics at the Collège de France. His scientific work centred on the genetics of bacteria and prophages, he discovered bacterial conjugation (with E. WOLLMANN), the circularity of the bacterial chromosome, mRNA and the concepts of operons and allostery (with J. MONOD). Besides he has worked on cell division in cultured mammalian cells (with S. BRENNER). He received the NOBEL-price for Physiology and Medicine in 1965 together with A. LWOFF & J. MONOD.

JAKOB, ALFONS MARIA German, 1884–1931. Neurologist, was the first to describe ALPER's disease and CREUTZFELDT-JAKOB-disease.

JERNE, NIELS KAY English, 1911–1994. Became professor of Biophysics in Geneva 1960, 1962 Head of microbiology at the University of Pittsburgh, 1966 Director of the Paul-Ehrlich-Institute in Frankfurt/M and was director of the Institute for Immunology in Basel 1969–1980. Most well known for his work on the time course of the antibody response (published in 1955), for which he received the NOBEL-price for Physiology and Medicine in 1984, together with C. MILSTEIN & G.J.F. KÖHLER.

KENDREW, JOHN COWDERY English, 1917–1997. Studied chemistry in Clifton College (Bristol) and Trinity College (Cambridge), B.A. 1939. During the war he did research on radar. Returned to Cambridge 1946, Ph.D. 1949, Sc.D. 1962. He established the 3D-structure of myoglobin at 6 Å resolution in 1957. Nobel Price in Chemistry 1962 together with M. PERUTZ.

KIRCHHOFF, GOTTLIEB SIGISMUND CONSTANTIN German, 1764-1833. Learned Pharmacy from his father, who ran a pharmacy in Teterow. Then moved to St. Peterburg (Russia) where he became an employee, later the director of the royal pharmacy. In addition to his duties there he was an important technical chemist, working on ceramics, fire proofing of wood and technical uses of potatoes. In the course of the latter studies he found 1812 that dilute, boiling sulphuric acid could turn insoluble potatoe starch into soluble glucose, without being used up in the process. This was one of the first demonstrations of catalysis. In 1815 he reported that wheat protein had the same property.

KÖHLER, GEORGES JEAN FRANZ German, 1945–1995. Studied Biology at the University of Freiburg, receiving his degree in 1971 and his Ph.D. in 1974. His thesis work on β-galactosidase he performed under F. MELCHERS in Basel. For postdoctoral studies he went to C. MILSTEIN in Cambridge, were both developed a method for obtaining monoclonal antibodies which would earn them the NOBEL-price in Physiology and Medicine together with N.K. JERNE in 1984. In 1975 KÖHLER returned to Basel, where he worked at the Institute for Immunology until his untimely death in 1995.

KOSHLAND, DANIEL. E., JR. USA. born 1919. B.S. in chemistry from the University of California, Berkeley (1941) and Ph.D. from the University of

Chicago (1949). Professor of Biochemistry at the University of California, Berkeley (1965). He is considered one of the most influential scientists of the USA, editor of *Science* since 1985 and of several other journals, member of the Council of the National Academy of Sciences and chairman of the editorial board of its *Proceedings*. Worked on the chemistry of plutonium during the Manhatten Project (under SEABORG), then switched to biochemistry, working on the control of enzymatic reaction (induced fit hypothesis of enzyme-substrate interaction and sequential model of co-operativity), and on the biochemical foundations of memory.

KREBS, SIR HANS ADOLF German, 1900-1981. Studied Medicine in Göttingen, Freiburg/Breisgau, Berlin and Hamburg (M.D. 1925), completing his training with one year of Chemistry in Berlin. Became assistant to O. WARBURG in Berlin-Dahlem (1926–1930), then returned to clinical work. In 1933 he fled from Nazi-Germany to England where he became demonstrator of Biochemistry at the University of Cambridge in 1934. In 1935 he moved to Sheffild as Lecturer, becoming Full Professor in 1945. 1954 he moved to Oxford. His main work was on energy tranformations in living cells, discovering what is now called the KREBS-cycle. He also discovered the urea cycle and worked on membrane transport and the synthesis of uric acid and purines. He was awarded the NOBEL Price for Physiology and Medicine 1953 and a knighthood in 1958.

KÜHNE, WILLY German, 1837–1900. Professor of physiology in Heidelberg, isolated trypsin in 1876. Following the vitalistic view of his time he postulated that biosynthesis can be performed only by the organised ferments inside cells, and suggested the name "enzyme" for the purified preparations in a test tube.

KUPFFER, KARL WILHELM VON German, 1829–1902. Obtained a degree from the medical school at Dorpatt (now Tartu), then worked as general practitioner until he started further training in physiology in Viena, Berlin and Göttingen (1856–1857). He became Professor of Anatomy in Kiel (1866, there he discovered the star-cells in the liver, which bear his name), Königsberg (1875) and Munich (1880, where he retired in 1901).

LINEWEAVER, HANS USA, born 1908, reinvented the double-reciprocal plot (originally proposed by W. WOOLF) while in the lab of DEAN BURK.

LWOFF, ANDRÉ MICHEL French, 1902–1994. Joined the Institute Pasteur at the age of 19, while studying Medicine. M.D. 1927, Ph.D. 1932 with work on development cycle and nutrition of ciliates. Spend one year (1932–33) in Heidelberg with OTTO MEYERHOFF, then 7 months in Cambridge. He became Departmental Head at the Institute Pasteur in 1938 and Professor of Microbiology at the Science Faculty in Paris in 1959 were he studied lysogenic bacteria and polio virus. For his work he received the NOBEL price in Physiology and Medicine in 1965, together with his collaborators F. JACOB & J. MONOD.

MENTEN, MAUDE LEONORA Canadian, 1879–1960. One of the first women to have a scientific career. Studied medicine at the University of Toronto (B.A. 1904, M.B. 1907, M.D. 1911). For her thesis work she had to go to the University of Chicago, as at that time woman were not allowed to do research in Canada. 1912 she moved to Berlin where she worked with L. MICHAELIS, obtaining a Ph.D. in 1916. Worked as Pathologist at the University of Pittsburgh (1923–1950, rising to the rank of full professor) and as Research Fellow at the British Columbia Medical Research Institute (1951–1953). Apart from her work on enzyme kinetics together with L. MICHAELIS based on earlier findings of V. HENRI (resulting in the famous HENRI-MICHAELIS-MENTEN-equation 1913) she characterised bacterial toxins (*B. paratyphosus, S. scarlatinae* and *Salmonella ssp.* 1924 with HELEN MANNING), invented the azo-dye coupling reaction for alkaline phosphatase (1944 with JUNGE & GREEN, still used in histochemistry), and (in 1944 with ANDERSCH & WILSON) found by electrophoresis and ultracentrifugation that adult and fetal haemoglobin are different molecules. She also worked on regulation of blood sugar level and kidney function. Despite suffering from arthritis she was an accomplished musician and painter, her paintings were exhibited in several galleries.

MERRIFIELD, BRUCE USA, 1921–1984. Graduated in Chemistry from UCLA in 1942, Ph.D. 1949. Moved to the Rockefeller Institute for Medical Research in New York where he developed the solid phase synthesis of peptides in 1959, for this discovery he received the Nobel Price in Chemistry in 1984.

MICHAELIS, LEONOR German, 1875–1947. Studied Medicine in Freiburg, graduating in 1897, then moved to Berlin, where he received his doctorate the same year. Worked as assistant to Paul Ehrlich (1898–1899), Moritz Litten (1899-1902) and Ernst Victor von Leyden (1902-1906). In 1906 he started as director of the bacteriology lab in Berlins Charitè clinic, becoming Professor extraordinary at Berlin University in 1908. In 1922 he moved to the Medical School of the University of Nagoya (Japan) as Professor of Biochemistry, 1926 to John Hopkins University in Baltimore (USA) as resident lecturer in medical research and 1929 to the Rockefeller Institute of Medical Research in New York, where he retired 1941. Besides his role in the formulation of the Henri-Michaelis-Menten law (1913) he discovered Janus Green as a supravital stain for mitochondria and the Michaelis-Gutman body in urinary tract infections (1902) and found that thioglycolic acid could dissolve keratin which made him the father of the permanent wave.

MILSTEIN, CÉSAR Argentinian, 1927–2002. He studied Biochemistry in Buenos Aires, obtaining his degree in 1952 and his Ph.D. in Chemistry in 1957 with a thesis on the kinetics of aldehyde dehydrogenase. He then went on a Fellowship to M. DIXON in Cambridge (1958–61), where he worked with F. SANGER on protein translation and obtained another Ph.D. After returning to Buenos Aires he worked on the mechanism of phosphoglyceromutase and alkaline phosphatase, but had to leave the country during the political unrest in 1963. He went to Cambridge again, where he worked at the MRC laboratory of Molecular Biology under F. SANGER, on whose suggestion he shifted the focus of his research from biochemistry to immunology. In 1983 he became the head of its Protein and Nucleic Acid Chemistry Division. He is best known for the development of monoclonal antibodies (together with his postdoc, G.J.F. KÖHLER). In 1984 he was awarded the NOBEL-price for Physiology and Medicine together with N.K. JERNE & G.J.F. KÖHLER.

MITCHELL, PETER DENIS British, 1920–1992, studied in Cambridge, B.A. in 1942 (with very mixed results), Ph.D. 1951 with a thesis on the mode of action of penicillin. Worked as demonstrator in Cambridge, in 1955 he moved to Edinburgh University (Senior Lecturer in 1961, Reader 1962). From 1965 on he used his own house in Cornwall as research laboratory, embarking on a research programm that led him to formulate his chemosmotic hypothesis of oxidative phosphorylation. The scientific debate that followed is known as "the oxphos-war" and became so heated that major journals refused to publish any more papers on the subject. This animosity was aggravated both by the initially scant evidence for the chemosmotic hypothesis and by the difficult personality of P. MITCHELL. However, this discussion was also considered so productive that P. MITCHELL was awarded the Nobel Price in Chemistry in 1978, long before his hypothesis became widely accepted.

MONOD, JAQUES LUCIEN French, 1910–1976. He started studying science in Paris in 1928, obtained his degree in 1931 and his Ph.D. in 1934. After a year at the California Institute of Technology in 1936 and the war he joined the Institute Pasteur as laboratory director under A. LWOFF. In 1954 he became director of the Department of Cell Biochemistry, in 1959 Professor of Metabolic Chemistry at the Sorbonne, in 1967 Professor at the College de France and in 1971 Director of the Institute Pasteur. He was very interested in quantitative description of biological processes and is most famous for his work on the regulation of gene expression and on co-operativity in enzymes. He received the NOBEL-price in Physiology and Medicine in 1965, together with his collaborators A. LWOFF & F. JACOB.

NORTHROP, JOHN HOWARD USA, 1891–1987. Studied Chemistry and Zoology at Columbia University from 1908, B.A. 1912, M.A. 1913, Ph.D. 1915. Started working at the Rockefeller Institute in 1917, Professor of Bacteriology and Biophysics at the University of California 1949. Isolated and crystallised pepsin in 1929, followed by trypsin, chymotrypsin, carboxypeptidase and pepsinogen. Later he worked on bacteria and viruses (managing to crystallise the first virus in 1938), enzyme kinetics and during World War II on the detection of nerve gases. He received the Nobel Price in Chemistry for protein crystallisation in 1946, together with W.M. STANLEY & J.B. SUMNER.

PALADE, GEORGE E. Romanian, born 1912. Studied Medicine at the University of Bucharest (M.D. 1940). Served as physician in the Romanian army during WW II and moved to New York University in 1946 and to the Rockefeller Institute the following year. In 1973 he moved to Yale. Worked on cell fractionisation and electron microscopy, initially in the group of A. CLAUDE. He worked on the fine structure of mitochondria, ER and synapses, discovered the ribosome and characterised the secretory pathway. He can thus be regarded as one of the fathers of modern cell biology. NOBEL-Price for Physiology and Medicine 1974, together with A. CLAUDE & C. DE DUVE.

PAULING, LINUS USA, 1901–1994. Studied Science at the Oregon Agricultural College (now Oregon State University, B.Sc. in Chemical Engineering 1922) and California Institute of Technology (Ph.D. in Chemistry 1925, minors in physics and maths), where he continued as an academic staff member (assistant 1927, associate 1929 and full professor 1931). His field of study was the application of physical principles to chemical problems. In our context relevant is his work on the secondary structure of proteins, on haemoglobin and on the antigen-antibody reaction. He won the NOBEL Price for Chemistry 1954 for his studies on *The Nature of the Chemical Bond* (so the title of his most important book, published 1939) and the NOBEL-Price for Peace 1962 for his stand against nuclear weapons, which makes him the only person ever to win 2 unshared NOBEL-prices. Both the personality of PAULING and the problems of his time shine through his following narration: "A couple of days after my talk, there was a man in my office from the FBI saying 'Who told you how much plutonium there is in an atomic bomb?' And I said 'Nobody told me, I figured it out'." PAULING will be remembered not only for his scientific achievements, but also for his effort in health education (anti-smoking, correct nutrition). To the lay public he became known as an advocate for high-dose vitamin C supplementation, taking up to 18 g per day (the recommended dose is 60 mg).

PAYEN, ANSELME French, 1795–1871. Studied at the École Polytechnique, became director of a chemical factory at the age of 20 and professor of technical and agricultural chemistry in 1835. He invented a synthesis of borax from boric acid and soda, breaking a dutch monopoly on borax imported from East India. He also invented the use of activated charcoal to decolourise sugar, methods to produce starch and alcohol from potatoes and an analytical procedure for nitrogen determination. From malt solution he, together with J.F. PERSOZ, isolated the first enzyme, diastase, by alcohol precipitation in 1833, showing it to be heat labile. In 1834 he discovered cellulose.

PERSOZ, JEAN-FRANÇOIS French, 1805–1869. Started his career as an apothecary's apprentice, then became an assistant to J.L. THENARD at the Collège de France in Paris. Professor of chemistry at the Faculty of Science at Strasbourg 1833. Later moved to Paris as professor of technical chemistry. Best known for his book "Traité théorique et pratique de l'impression des tissus" (1846). Together with A. PAYEN he isolated the first enzyme, diastase from barley malt (1833) and demonstrated its presence in saliva.

PERUTZ, MAX FERDINAND Austrian, 1914–2002. Started studying Chemistry at Vienna University in 1932, moved to Cambridge in 1936 where he stayed for the rest of his career. Became research assistant to Sir LAWRENCE BRAGG in 1939 and head of the Medical research Council Unit of Molecular Biology in 1947, with J.C. KENDREW as his only staff. His work on the crystal structure of haemoglobin started in 1937 and continued until 1959. During that time he solved the phasing problem by isomorphous replacement with heavy atoms (1953). He received the Nobel Price in Chemistry in 1962.

PRUSINER, STANLEY B. USA, born 1942. Studied Chemistry and Medicine at the University of Pennsylvania (A.B. 1964, M.D. 1968). After Internship at the UCSF he went to the National Institutes of Health for three years and returned as Resident in Neurology to UCSF. He became Assistant Professor in 1974, Lecturer 1976, Associate Professor 1980 and Full Professor in 1984. In 1972 he encountered his first patient with CJD, which turned his interest to the spongiform encephalopathies and their causative agents, then believed to be slow viruses. Like T. ALPER before him he found that preparations of the causative agents appeared to consist only of protein, not of nucleic acids. This led to his publication of the prion-hypothesis in 1982. He and his collaborators went a long way to characterise this protein. He received the NOBEL-price for Physiology and Medicine in 1997.

RAMACHANDRAN, GOPALASAMUDRAM NARAYANA IYER Indian, 1922–2001. Studied physics at Madras University (Master 1942, D.Sc.), then Cavendish Laboratory in Cambridge (Ph.D. 1949). Invented an X-ray mirror and an X-ray microscope and made important contributions to the field of crystallography. After returning to Madras University in 1952 his interest shifted to macromolecular structure, discovering the collagen triple helix (1954 with G. KARTHA) and inventing the RAMACHANDRAN map in 1962. He is considered one of the fathers of the field of molecular biophysics.

RÉAUMUR, RENÉ ANTOINE-FERCHAULT DE French, 1683–1757. Moved to Paris in 1703, where he became member of the Academy of Science in 1708. One of the great universal scholars in the history of science, worked on palaeontology, zoology and chemical technology. Invented the alcohol/water-filled thermometer, a process for steel production, the incubator and recognised corals as animals rather than plants. Fed food bits packed in small metal cages to sea gulls, when the birds regurgitated the cages, the food inside was partly digested. If the cages contained little sponges, a juice pressed from them could hydrolyse meat. Thus RÉAUMUR recognised digestion as chemical process.

SANGER, FREDERICK British, born 1918, B.A. 1939, Ph.D. 1943, both in Cambridge. Introduced 2,4-dinitrofluorbenzene (SANGERs reagent) into protein chemistry in 1945. Was the first to determine the primary structure of a protein (insulin) after years of work, Nobel Price in Chemistry 1958. He later switched to nucleic acid sequencing, developing the chain termination method (1975) for which he received a second NOBEL-price in 1980.

SCHWANN, THEODOR German, 1810–1882. Studied medicine in Berlin (M.D. 1834), where in 1836 he prepared the first enzyme from animal sources, pepsin by adding mercury chloride to stomach juice. Became professor at the University of Louvain in 1838 and Liège in 1848. With M.J. SCHLEIDEN he formulated the cell theory, introduced the term "metabolism", and worked on muscle and nerve cell excitation (discovering the myelin sheath around nerve axons named after him) and fermentation. He is considered the founder of embryology.

SKOU, JENS C. Danish, born 1918. Studied medicine at the University of Copenhagen 1937–1944. Then started Ph.D. work in Aarhus on anaesthetics in 1947 (finished in 1954) which led to the discovery of Na/K-ATPase. Skou stayed in Aarhus, became professor and chairman of the Department of Physiology in 1963. Received a Nobel price in Physiology and Medicine for his discovery of Na/K-ATPase in 1997, together with P.D. BOYER & J.E. WALKER.

SØRENSEN, SØREN PEDER LAURITZ Danish, 1868–1939. PhD in chemistry 1899 in Copenhagen. Head of Carlsberg Laboratory, Copenhagen. Recognised the importance of hydrogen ion concentration for chemical and in particular enzymatic reactions and introduced the logarithmic *p*H-scale (in 1909).

SPALLANZANI, LAZZARO Italian, 1729–1799. Studied law and theology at the University of Bologna, but turned to science, becoming professor of logic, metaphysics, and Greek at the University of Reggio aged 25. In 1768 he proved that cells could not be spontaneously generated but came from air and could be killed by boiling. However, since he used closed vessels to keep his preparations germ-free his results were not generally accepted until L. PASTEUR repeated the experiment in the open swan-neck bottle.

He described the role of sperm and ovum in mammalian reproduction and performed the first artificial insemination (in a dog). He studied the regeneration of body parts in lizards and he proved that the active ingredient in stomach juice (pepsin) lost its activity upon storage. Investigating bat orientation he found that the animals could still fly with their eye sight blocked, but lost their bearing when their ears were covered. The reason of course eluded him and became clear only after the discovery of ultrasound.

STANLEY, WENDELL MEREDITH USA, 1904–1971. B.A. at Earlham College 1926, M.A. 1927 and Ph.D. in Chemistry 1929, both at the University of Illinois. Postdoc in Berlin, assistant at the Rockefeller Institute 1931, Professor of Biochemistry at the University of California 1948. Worked on lepracidal compounds, sterols, tobacco mosaic and influenza virus. Nobel Price in Chemistry 1946 together with J.B. SUMNER & J.H. NORTHROP.

STOKES, GEORGE GABRIEL Irish, 1819–1903. Studied at the University of Cambridge where he also became Professor of Mathematics in 1849. He worked on pure maths and on mathematical and experimental physics, contributing greatly to hydrodynamics, optics and the theory of wave propagation.

SUMNER, JAMES BATCHELLER USA, 1887–1955. Lost his left arm in a hunting accident aged 17, studied Chemistry at Harvard, graduating in 1910. After brief spells at various positions he continued his studies in Biochemistry at Harvard, obtaining a Ph.D. in 1914. Assistant professor at Cornell 1914, full professor 1929. He started to work on the purification of jackbean urease in 1921, succeeding in 1926. Urease was the first ever enzyme to be purified enough to be crystallised, proving that proteins were defined compounds. Nobel Price in Chemistry 1946, together with J.H. NORTHROP & W.M. STANLEY.

SVEDBERG, THEODOR (THE) Swedish, 1884–1971. Studied at Uppsala University, B.A. 1905, M.A. 1907, Ph.D. 1908. His thesis, "Studien zur Lehre von den kolloiden Lösungen" is still a classic in his field, he tested EINSTEINs theories on BROWNian motion and thereby proved the existence of molecules and founded the field of molecular hydrodynamics. Svedberg became Lecturer of physical chemistry in 1909, professor in 1912. He constructed the first ultracentrifuge (1923) which for the first time allowed the determination of the molecular weight of proteins. Together with TISELIUS he invented free-flow electrophoresis of proteins. Worked on nuclear chemistry, radiation biology, photochemistry and polymer-chemistry. He was awarded the Nobel Price in Chemistry in 1926 for his work on colloids.

TANAKA, KOICHI Japanese, born 1959. Started studying electrical engineering at Tohoku University in 1978 and joined the central research laboratory of Shimadzu Corporation in 1983. In 1984 he started working there on mass spectrometry, by chance inventing matrix assisted laser desorbtion/ionisation (MALDI, with glycerol as matrix), one of two methods used to ionise protein molecules for mass spectrometric analysis. For this important contribution to protein science he received the nobel Price in Chemistry for the year 2002, together with J.B. FENN and K. WÜTHRICH.

THEORELL, AXEL HUGO THEODOR Swedish, 1903–1982. Studied Medicine at the Karolinska Institute from 1921, Bachelor of Medicine 1924, associate assistant 1924, temporary associate Professor 1928, M.D. 1930 with a dissertation on blood plasma lipids. Moved to Uppsala to work with T. SVEDBERG in 1931 and became Associate Professor in Medical and Physiological Chemistry at Uppsala University in 1932. From 1933–1935 he worked with O. WARBURG in Berlin-Dalem, discovering FMN. In 1936 he was appointed Director of the Biochemistry Department of the newly founded NOBEL Medical Institute, which opened in 1937. Here he continued his work on the reaction mechanism of redox-enzymes, developing new methods of rapid enzyme kinetics, for which he received the NOBEL-price for Physiology and Medicine in 1955.

TRAUBE, MORITZ German, 1826–1894. Wine merchant and private scholar. Worked on biological oxidation and osmosis. Invented an apparatus for measuring the osmotic pressure of a solution in 1867 (incorrectly known as PFEFFER's cell, PFEFFER build a similar device 10 years later). He was the first to recognise that enzymes were proteins.

TSVETT, MIKHAIL Russian botanist, 1872–1920. Invented chromatography to separate leaf pigments.

WALKER, JOHN E. British, born 1941. B.A. in Chemistry from St. Catherines College, Oxford in 1964, D.phil. 1969. Postdoctoral training 1969–1974 in various position, then joined the Medical Research Councils Laboratory of Molecular Biology. Started work on membrane proteins in 1978, in particular ATP-synthase. Discovered the ATPase-motive named after him and crystallised the F-domain of ATP-synthase. Nobel price in Chemistry 1997 together with P.D. BOYER & J.C. SKOU for his studies on the reaction mechanism of ATP-synthase.

WARBURG, OTTO HEINRICH German, 1883–1970. Studied Chemistry in Berlin under E. FISCHER, obtaining a doctorate in 1906. He then went to Heidelberg and obtained an M.D. in 1911. During WW I he served in the Prussian Horse Guards, in 1918 he was appointed Professor at the Kaiser-Wilhelm-Institute of Biology in Berlin-Dahlem, in 1931 he became Director of the Kaiser-Wilhelm-Institute for Cell Physiology. He worked on polypeptides, biological oxidation, carbon fixation in plants, radiation biology and tumor-metabolism. He discovered the roles of FMN and $NAD(P)^+$ (together with A.H.T. THEORELL, NOBEL-Price 1955), and of iron-containing enzymes. Although he held the rank of Professor he was never a teacher and spend his time entirely on research. For his research on respiratory enzymes he was awarded the NOBEL-Price in Physiology and Medicine in 1931.

WILLEBRAND, ERIK ADOLF VON Finnish, 1870-1949. Internist, published his study on "hereditary pseudohaemophilia" in a family of 66 members from the Åland-islands in 1926. Unlike haemophilia (lack of clotting factor VIII) females too may show this disease. It took 60 years until the defective gene for what is now known as VON WILLEBRAND-factor could be identified.

WILLSTÄTTER, RICHARD German, 1872–1942. Studied Science at the University of Munich and the Technical School in Nürnberg, PhD 1894 as student of BAEYER. Lecturer 1896 and Professor 1902. Moved to ETH Zürich in 1905, Kaiser-Wilhelm-Institute in Berlin 1912 and University of Munich 1916. In 1925 he resigned his post as professor in protest against growing antisemitism and fled to Switzerland in 1939. Nobel Price in Chemistry 1915 for his work on natural dyes, in particular chlorophyll and haemoglobin, whose structure he solved. His enzymological studies laid the ground for chromatographic purification of proteins, even though he himself never believed enzymes to be proteins.

WÜTHRICH, KURT Swiss, born 1938. Studied science and sports 1957–1962 at the University of Berne, then from 1962–1964 at the University of Basel, where he obtained a Ph.D. in chemistry. His project on the catalytic activity of copper compounds required the use of electron paramagnetic resonance spectroscopy (EPR). For postdoctoral studies he went to Berkeley (1965–1967), where he started using nuclear magnetic resonance (NMR). This work he continued at Bell Laboratories (1967–1969), where he was in charge of the first NMR-spectrometer operating with supercooled coils, working at 220 MHz. Here he started his NMR-work on proteins, initially with haemoglobin. In 1969 he joined the ETH at Zürich where he remained for the rest of his career. There he developed new methods of multi-dimensional NMR. This allowed the de-convolution of the signals of the many different atom-atom interactions in a protein. For this work he received the Nobel Price in Chemistry in 2002, together with J.B. FENN and K. TANAKA.

A.3 List of symbols

c	concentration (mol/l)
D	diffusion coefficient (cm^2/s)
e	elementary charge (1.6022×10^{-19} C/mol)
$\mathcal{E}$	absorbance (pure number)
E	energy (J)
	potential difference (V)
$\mathfrak{E}$	electrical field strength (V/m)
$\mathfrak{F}$	force (N)
f	friction coefficient (kg/s)
$\mathfrak{g}$	gravitational acceleration ($9.8067\,m\,s^{-2}$)
$\Delta G'^0$	GIBBS free energy (under standard biological conditions, J/mol)
h	PLANCK's quantum (6.6262×10^{-34} J/s)
k	Boltzmann constant (1.3807×10^{-23} J/K)
	reaction velocity constant (unit depends on order of reaction)
k_{cat}	turnover number (s^{-1})
K_a	association constant (M^{-1})
K_d	dissociation constant (M)
K_m	MICHAELIS-constant (M)
K_p	partitioning coefficient (pure number)
l	length (m)
m	mass (kg)
M	molecular mass (pure number, but Da is often used)
N_A	AVOGADRO's number ($6.022 \times 10^{23}\,mol^{-1}$)
n_H	HILL-coefficient (pure number)
n	number
pH	hydrogen ion tension (pure number)
pI	isoelectric point (pure number)
pK$_a$	strength of an acid (pure number)
Q	electrical charge (C)
r	radius (m)
R	universal gas constant ($8.3143\,J\,mol^{-1}\,K^{-1}$)
t	time (s)
$t_{1/2}$	half life period (s)
T	absolute (thermodynamic) temperature (K)
v	reaction velocity (mol/s)
V_{max}	maximal reaction velocity (kat = mol/s)
V	volume (l)
z	number of elementary charges transferred in a reaction

θ molar fraction of enzyme with bound substrate (pure number)
λ wavelength (nm)
τ relaxation time (s)
ϕ dihedral angle around the N-C$^{\alpha}$-bond (°)
ψ dihedral angle around the C$^{\alpha}$-C'-bond (°)

A.4 Greek alphabet

alpha	α	A
beta	β	B
gamma	γ	Γ
delta	δ	Δ
epsilon	ϵ, ε	E
zeta	ζ	Z
eta	η	H
theta	θ, ϑ	Θ
iota	ι	I
kappa	κ	K
lambda	λ	Λ
mu	μ	M
nu	ν	N
xi	ξ	Ξ
o	o	O
pi	π, ϖ	Π
rho	ρ, ϱ	R
sigma	σ, ς	Σ
tau	τ	T
upsilon	υ	Υ
phi	ϕ, φ	Φ
chi	χ	X
psi	ψ	Ψ
omega	ω	Ω

A.5 The genetic code

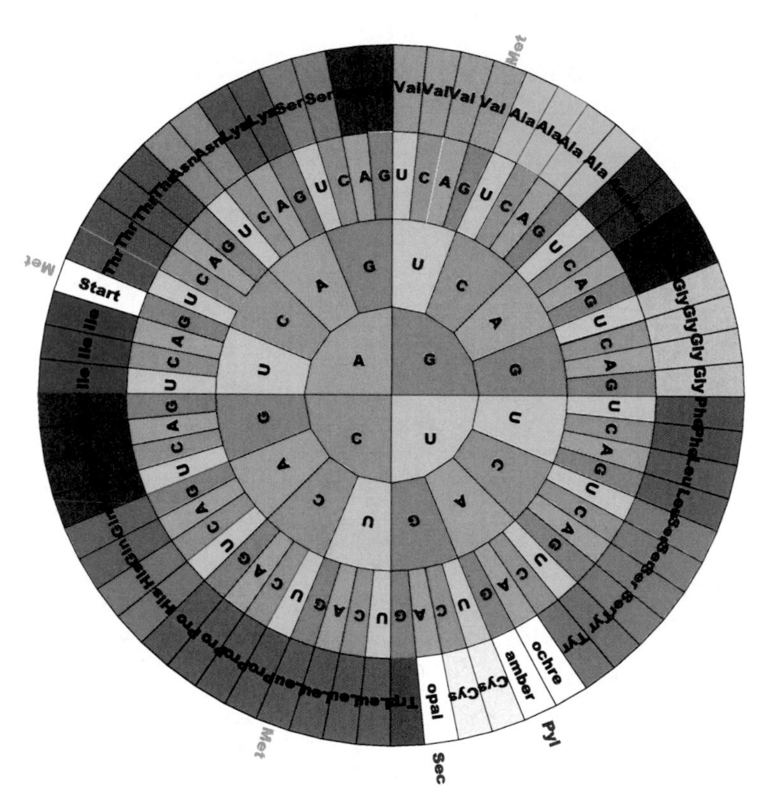

Figure A.1. mRNA-codons and the corresponding amino acids. Alternative uses of codons are marked on the exterior. The colours used to symbolise different compounds are known as "Shapely colour set", a quasi-standard in molecular modelling.

A.6 Acronyms

ABC ATP binding cassette, major family of primary active transporters

ADP adenosine 5'-diphosphate

AE1 anion exchanger 1, also called band-3 protein, equilibrates chloride and bicarbonate across the erythrocyte membrane

AIDS acquired immune deficiency syndrome, life-threatening sexually transmitted disease caused by HIV

AMP adenosine 5'-monophosphate

AMP-PCP adenosine 5´[β,γ-methylene]triphosphate, non-hydrolysable ATP-derivate, the oxygen between the β- and γ-phosphate is replaced by an methylene (-CH_2-) group.

AMP-PNP adenosine 5´[β,γ-imino]triphosphate, non-hydrolysable ATP-derivate, the oxygen between the β- and γ-phosphate is replaced by an imino (-NH-) group

AP adaptor protein, occur in CCV

apoA-I apolipoprotein A-I

Aqp aquaporin, water conducting pore in the cell membrane

ARF ADP-ribosylating factor, small G-protein involved in COP-vesicle formation

ATP adenosine 5'-triphosphate, energy carrier for metabolic reactions

ATPase adenosine triphosphatase, enzymes that hydrolyse ATP to ADP and P_i

BiP Ig heavy chain binding protein, synonym for Grp78

2,3-BPG 2,3-Bisphosphoglycerate, compound used to modulate the oxygen affinity of haemoglobin

BSE bovine spongiforme encephalopathy, prion-disease in cattle, also known as "mad cow disease"

Btu B twelf uptake, ABC-transporter required for the uptake of vitamin B_{12} in bacteria

C1Inh C1-inhibitor, inactivates complement. Lack of this protein causes hereditary angioneurotic oedema

C4BP C4-binding protein, protein which inactivates complement

CAM cell adhesion molecule, proteins required for cell-cell-interactions

cAMP cyclic adenosine 5'-monophosphate, an important second messenger

CCV clathrin coated vesicles, required for endocytosis, transcytosis and GOLGI-endosome transport

Cdk cyclin dependent kinase, has 2 isoforms (Cdk1 and Cdk2)

CFTR cystic fibrosis transconductance regulator, a chloride channel

CH calponin homology, domain in some actin-binding proteins

CIITA Class-II transactivator, regulates MHC-gene expression in an interferon dependent manner. Lack of this protein causes bare lymphocyte syndrome, an inherited immunodeficiency

CJD CREUTZFELDT-JAKOB-disease, a naturally occurring prion disease in humans

CK creatine kinase catalyses the reaction Phosphocreatine + ADP $\rightleftharpoons$ creatine + ATP. Phosphocreatine is a storage form of high energy phosphate especially in muscle, but also in other cells.

CLIP Class II associated Invariant chain Peptide, stabilises MHC-II until antigenic peptide is bound

cMOAT canalicular multiple organic anion transporter

CMP cytosine monophosphate

COP β-coat protein, in vesicles for ER/GOLGI transport and for transport between GOLGI-stacks

CTAB cetyltrimethyl ammonium bromide, positively charged detergent which can solubilise membrane proteins without denaturation

DAF decay-accelerating factor, protein which inactivates complement. Inability to synthesise its glycolipid tail leads to paroxysmal nocturnal haemoglobinuria

DNA deoxyribonucleic acid, carrier of genetic information

DnaJ ADP/ATP exchange stimulating factor of DnaK

DnaK chaperone in prokaryotes and organelles derived from them

DNP 2,4-dinitrophenol, uncoupler

DTT dithiotreitol, used in the lab as an antioxidant to protect SH-groups in proteins (CLELANDs reagent)

EC Enzyme Commission, body responsible for enzyme nomenclature within IUBMB

ECM extracellular matrix, "glue" between cells

EDTA ethylenediamino-N,N,N',N'-tetraacetic acid, reagent that forms strong complexes with some metal ions

EGF epidermal growth factor

ELISA enzyme linked immunosorbent assay, very sensitive method to detect and quantify either antibodies or their antigen

Elk-1 Eph-like kinase 1

EM electron microscopy

Eph ephrin receptor tyrosine kinase, receptor for ephrins, a group of small proteins required for morphogenesis of the nervous system

ER endoplasmic reticulum, membrane system inside a eukaryotic cell

ERAD ER associated protein degradation, mechanism to destroy mis-folded proteins in the ER

Erk extracellular signal regulated kinase, protein in the MAP-kinase pathway

EU European Union, economic and political association of (at the time of writing) 25 European countries

F_{ab} immunoglobulin molecule whose F_c-end has been removed enzymatically

F_c carboxy-terminal end of a immunoglobulin molecule, interacts with effector-cells

$FADH_2$ flavine adenine dinucleotide, reduced

FFI fatal familial insomnia, naturally occurring prion-disease in humans

FMN flavine mononucleotide, prosthetic group of some flavoproteins

Fuc fucose, a sugar used in glycoproteins and glycolipids

GABA γ-aminobutyric acid, a neurotransmitter

Gal galactose

GalNAc N-acetylgalactosamine

GDP guanosine diphosphate

Glc glucose

GlcNAc N-acetyl-glucosamine

GluT glucose transporter

GPI glycosyl phosphatidylinositol, C-terminal anchor of type-6 membrane proteins

Grb2 growth factor receptor bound protein 2, protein in MAPK-pathway

Grp78 78 kDa glucose regulated protein, chaperone of the ER

GrpE ATP-hydrolysis stimulating factor of DnaK

GSH glutathione, reduced

GSS GERSTMANN-STRÄUSSLER-SCHEINKER-disease, a natural prion disease in humans

GSSG glutathione, oxidised

GTP guanosine triphosphate

HbA normal adult haemoglobin (α_2/β_2)

HbA$_{1c}$ glycated HbA

HbC haemoglobin C, Glu-6 $\rightarrow$ Lys

HbF fetal haemoglobin, ($\alpha_2\gamma_2$)

HbH haemoglobin H, tetrameres of β-subunits, formed in embryos with α^0-thalassaemia

HbS sickle cell haemoglobin, Glu-6 $\rightarrow$ Val

HDL high density lipoprotein, VLDL from which cholesterol has been removed

HIV human immunodeficiency virus, causative agent of acquired immune deficiency syndrome (AIDS)

HLA Haupt-Lymphocyten Antigen, (Ger.), synonym for MHC, today mainly used in tissue typing for transplantation

HMM HENRI-MICHAELIS-MENTEN, biochemists who described the relationship between substrate concentration and enzyme reaction velocity

Hsc10 10 kDa heat shock cognate, GroEL

Hsc70 70 kDa heat shock cognate, constitutively expressed cytosolic isoform of Hsp70

Hsc60 60 kDa heat shock cognate, GroES

Hsp40 40 kDa heat shock protein, eukaryotic homologue of DnaJ

Hsp70 70 kDa heat shock protein, a molecular chaperone

Hsp90 90 kDa heat shock protein, a molecular chaperone

HSV herpes simplex virus

IEF isoelectric focussing, electrophoretic method to separate proteins by their isoelectric point.

IF intermediate filament

IFAP intermediate filament associated proteins

IFN interferon, a family of cytokines

Ig immunoglobulin

IgA immunoglobulin A

IgD immunoglobulin D

IgE immunoglobulin E

IgG immunoglobulin G

IgM immunoglobulin M

IL interleukin, a family of cytokines

IMAC immobilised ion affinity chromatography, method to separate proteins by the number of exposed His-groups

IUBMB International Union for Biochemistry and Molecular Biology, an international body which, amongst other things, is responsible for standardisation of the nomenclature of molecules (in part together with IUPAC)

IUPAC International Union for Pure and Applied Chemistry

Kir6. together with Sur part of the K_{ATP} channel that regulates insulin secretion.

LCFA long chain fatty acid, 14–24 C-atoms

Lck tyrosine kinase of the immune system

LDL low-density lipoprotein

M6P mannose-6-phosphate, residue required for protein sorting into the lysosome

mAb monoclonal antibody, obtained from fusing antibody-producing B-cells with immortal hybridoma cells

Man mannose

Mant-ATP Methylantroyl-ATP, ATP labelled with a fluorescent reporter group

MAP mitogen activated protein, proteins phosphorylated in response to mitogen stimulation of a cell. Stimulation occurs via a cascade of MAP-kinases (MAPK), MAP-kinase-kinases (MAPKK) and MAP-kinase-kinase-kinases (MAPKKK). Also used for microtubule associated protein

MASP MBL associated serine protease, protein in the complement pathway

MBL mannan binding lectin, protein of the complement pathway that recognises bacterial cell walls

MBP microtubule binding protein

Mdr multiple drug resistance transporter, group of ABC-type transmembrane ATPases

MEK MAP/Erk kinease, protein in MAPK-pathway

MFS major facilitator superfamily, secondary active transporters

MHC major histocompatibility complex, antigen-presenting protein on the cell surface

MΦ macrophage, amoeboid cells of the innate immune system

MMP: matrix specific metalloprotease, break integrin/ECM interactions

MPF maturation promoting factor, Cdk1 associated with Cyclin B

Mpp matrix processing protease, clips of the signal sequence of nuclear encoded proteins after their import into the mitochondrium

mRNA messenger ribonucleic acid, working copy of genetic information

Mrp Multidrug resistance related protein

mtDNA mitochondrial DNA

MTOC microtubule organising centre, central hub where the minus-ends of microtubules meet

MW molecular weight

NAD^+ nicotine adenine dinucleotide, oxidised, acceptor of activated hydrogen, usually in catabolic reactions

$NADH + H^+$ nicotine adenine dinucleotide, reduced, carrier of activated hydrogen

$NADP^+$ nicotine adenine dinucleotide phosphate, oxidised, acceptor of activated hydrogen

$NADPH + H^+$ nicotine adenine dinucleotide phosphate, reduced, donor of activated hydrogen, usually in anabolic reactions

$NAD(P)^+$ either NAD^+ or $NADP^+$

NAD(P)H either $NADH + H^+$ or $NADPH + H^+$

NANA N-acetylneuraminic acid, sialic acid

NBD nucleotide binding domain

NEM N-ethylmaleimide, reactive compound used to label SH-groups

NKCC Na/K/2Cl Cotransporter

NMR nuclear magnetic resonance, a method to determine the distance of atoms in a molecule

NSF NEM-sensitive factor, protein required for vesicle fusion

OmpA outer membrane protein A from *E. coli*

PAF platelet activating factor

PC phosphatidyl choline

PCR polymerase chain reaction, highly sensitive method for the detection and quantitation of nucleic acids

PDI protein disulphide isomerase, acts on disulphide bridges in proteins

PE phosphatidyl ethanolamine

PEG polyethylene glycol, compound used in protein purification and crystallisation

PEP phospho*enol*pyruvate, super energy rich phosphoester formed during glycolysis

Perk regulator of the unfolded protein response

PFIC progressive familial intrahepatic cholestasis, group of inherited diseases of the liver

Pgp permeability glycoprotein, synonym for Mdr1

PHHI persistent hyperinsulinaemic hypoglycaemia of infancy

P_i inorganic phosphate

PIP_2 phosphatidylinositol bisphosphate

PITC phenyl isothiocyanate, reagent used in EDMAN-sequencing of proteins

PK pyruvate kinase, enzyme that catalyses the reaction phospho*enol*pyruvate + ADP → pyruvate + ATP

PPI peptidyl prolyl *cis/trans*-isomerase, catalyses the conversion between the *cis*- and *trans*-conformation of Pro in protein chains

prion proteinaceous infectious agent

PrP prion protein, comes in two variants, cellular (normal) PrP^{C} and scrapy (abnormal) PrP^{SC}

PS phosphatidyl serine

Rab group of small G-proteins involved in vesicle fusion

Raf receptor associated factor, protein in the tyrosine kinase receptor signaling cascade

Rag recombination activating gene, encodes for proteins that are required for recombination of DNA

Ras rat sarcoma viral oncogene, protein in MAPK-pathway, constitutively active homologs of this protein are encoded by some tumor viruses.

RNA ribonucleic acid

RNase ribonuclease, enzyme that digests RNA

ROS reactive oxygen species, molecules resulting from incomplete reduction of oxygen, like $O_2^{-\cdot}$, $HO_2^{\cdot}$, H_2O_2 and $HO^{\cdot}$

rRNA ribosomal RNA

SDS sodium dodecylsulfate, anionic detergent used to destroy the secondary structure of proteins

SEC size exclusion chromatography, synonym for gel filtration

Sec61 61 kDa secretory protein, transports membrane proteins into the ER

SHR steroid hormone receptor

sHsp small heat shock proteins

SIR-2 silent information regulator 2, class III protein deacetylase

SIV simian immunodeficiency virus, causative agent of an AIDS-like disease in monkeys

SLE systemic *lupus erythematosus*, debilitating autoimmune disease caused by auto-antibodies against nuclear antigens

SNAP soluble NSF-attachment protein

SNARE SNAP receptor

SOS son of sevenless, protein in MAPK-pathway

SPGP Sister of Pgp, ABC-type transporter of bile salts

SR sarcoplasmic reticulum, specialised ER in muscle cells

SRP signal recognition particle, recognises membrane proteins during their synthesis and directs them to the ER

Sur sulphonyl urea receptor

Tap transporter associated with antigen processing, ABC-type ATPase that transports antigenic peptides from the cytosol into the ER

TCR T-cell receptor, protein on the surface of T-lymphocytes which binds to the MHC/antigen-complex

TdT Deoxynucleotidyl-transferase

TGN *trans*-GOLGI network

T_h helper T-cell

TIM transporter of the inner membrane, required for import of nuclear encoded proteins into mitochondria

T_k killer T-cell

TMD trans-membrane domain

TNF tumour necrosis factor, a family of cytokines

TOM transporter of the outer membrane, required for import of nuclear encoded proteins into mitochondria

TRAM translocating chain associated membrane protein

TRAP-T tripartite ATP-independent periplasmic transporters, secondary active transporters with extracellular binding proteins

TRiC **T**ailless complex polypeptide 1 **Ri**ng **C**omplex, eukaryotic homologue of GroEL/GroES

γTU-RC γ-tubulin ring complex, structure in the MTOC

UbA E1-ubiquitin ligase

UbC E2-ubiquitin ligase

UbL ubiquitin-like modifiers, involved in many regulatory processes

UDP uridine 5'-diphosphate

UTP uridine 5'-triphosphate

UV ultraviolet, light with wavelengths shorter than 400 nm

UVIS visible and ultraviolet

vCJD new variant CREUTZFELDT-JAKOB-disease, human prion-disease caused by eating meat from animals suffering from bovine spongiforme encephalopathy

VLCFA very long chain fatty acid, > 24 C-atoms

VLDL very low-density lipoprotein, transport cholesterol in blood

VSV vesicular stomatitis virus, laboratory model for protein synthesis and -transport studies

WASP Wiscott-Aldrich syndrome protein, involved in the regulation of actin filament branching

XLSA/A X-linked sideroblastic anaemia/cerebrellar ataxia, disturbance of iron metabolism due to mutation in ABCB7

References

[1] J. Abramson, I. Smirnova, V. Kasho, G. Verner, H.R. Kaback, and S. Iwata. Structure and mechanism of the lactose permease of *Escherichia coli*. *Science*, 301:610–615, 2003.

[2] T. Alper. The exceptionally small size of the Scrapie agent. *Biochem. Biophys. Res. Commun.*, 22:278–284, 1966.

[3] T. Alper, W.A. Cramp, D.A. Haig, and M.C. Clarke. Does the agent of Scrapie replicate without nucleic acid? *Nature*, 214:764–766, 1967.

[4] C. Anfinsen. Principles that govern the folding of protein chains. *Science*, 181:223–230, 1973.

[5] H. Bisswanger. *Enzyme Kinetics, Theories and Methods.* VCH, Weinheim, third edition, 2002.

[6] N. Boggetto and M. Reboud-Ravaux. Dimerization inhibitors of HIV-1 protease. *Biol. Chem.*, 383:1321–1324, 2002.

[7] G.E. Briggs and J.B.S. Haldane. A note on the kinetics of enzyme action. *Biochem. J.*, 19:338–339, 1925.

[8] E. Buxbaum. Cationic electrophoresis and electrotransfer of membrane glycoproteins. *Anal. Biochem.*, 314:70–76, 2003.

[9] C.M. Cabral, Y. Liu, and R.N. Sifers. Dissecting glycoprotein quality control in the secretory pathway. *Trends Bichem. Sci.*, 26:619–624, 2001.

[10] B. Chance. The kinetics of the enzyme-substrate compound of peroxidase. 1943. *Adv. Enzymol. Relat. Areas. Mol. Biol.*, 73:3–23, 1999.

[11] S. Chowdhury, C. Ragaz, E. Kreuger, and F. Narberhaus. Temperature-controlled structural alterations of an RNA thermometer. *J. Biol. Chem.*, 278:47915–47921, 2003.

[12] S. Clarke. Protein methylation. *Curr. Opin. Cell Biol.*, 5:977–983, 1993.

[13] W.W. Cleland. The kinetics of enzyme-catalyzed reactions with two or more substrates or products 1. Nomenclature and rate equations. *Biochim. Biophys. Acta*, 67:104–137, 1963.

[14] W.W. Cleland. The kinetics of enzyme-catalyzed reactions with two or more substrates or products 2. Inhibition: Nomenclature and theory. *Biochim. Biophys. Acta*, 67:173–187, 1963.

[15] W.W. Cleland. The kinetics of enzyme-catalyzed reactions with two or more substrates or products 3. Prediction of initial velocity and inhibition patterns by inspection. *Biochim. Biophys. Acta*, 67:188–196, 1963.

[16] International Human Genome Sequencing Consortium. Initial sequencing and analysis of the human genome. *Nature*, 409:860–921, 2001.

[17] International Human Genome Sequencing Consortium. Finishing the euchromatic sequence of the human genome. *Nature*, 431:931–945, 2004.

[18] R.A. Copeland. *Enzymes. A practical Introduction to structure, mechanism and data analysis.* Wiley-VCH, New York, second edition, 2000.

[19] D.M. Cyr, J. Höhfeld, and C. Patterson. Protein quality control: U-box containing E3 ubiquitin ligases join the fold. *Trends Biochem. Sci.*, 27: 368–375, 2002.

[20] W. Darge. *The Ribosome.* Ribosomen-Verlag, Krefeld, 2004.

[21] M.O. Dayhoff. Computer analysis of protein evolution. *Scientific American*, 221(1):86–95, 1969.

[22] J.M. Denu. Linking chromatin function with metabolic networks. Sir-2 family of NAD^+-dependent deacetylases. *TIBS*, 28:41–48, 2003.

[23] R. Eisenthal and A. Cornish-Bowden. The direct linear plot. A new graphical procedure for estimating enzyme kinetic parameters. *Biochem. J.*, 139(3):715–720, 1974.

[24] D.E. Feldman and J. Frydman. Protein folding *in vivo*: the importance of molecular chaperones. *Curr. Opin. Struct. Biol.*, 10:26–33, 2000.

[25] J.E. Ferrell, Jr. Tripping the switch fantastic: how a protein kinase cascade can convert graded inputs into switch-like outputs. *Trends Biochem. Sci.*, 21(12):460–465, 1996.

[26] S. Fleischer and M. Kervina. Subcellular fractionation of rat liver. *Meth. Enzymol.*, 31:6–41, 1974.

[27] Y. Fujiki, S. Fowler, H. Shiu, A.L. Hubbard, and P.B. Lazarow. Isolation of intracellular membranes by means of sodium carbonate treatment: Application to endoplasmic reticulum. *J. Cell Biol.*, 93:97–102, 1982.

[28] Y. Fujiyoshi, K. Mitsoka, B.L. de Groot, A. Philippsen, H. Grubmüller, P.Agre, and A. Engel. Structure and function of water channels. *Curr. Opin. Struct. Biol.*, 12:509–515, 2002.

[29] Gavin et al. Functional organisation of the yeast proteome by systematic analysis of protein complexes. *Nature*, 415:141–147, 2002.

[30] B. Gutte and R.B. Merrifield. The total synthesis of an enzyme with ribonuclease A activity. *J. Am. Chem. Soc.*, 91:501–502, 1969.

[31] J.B.S. Haldane. Graphical methods in enzyme chemistry. *Nature*, 179: 832, 1957.

[32] J.B.S. Haldane and K.G. Stern. *Allgemeine Chemie der Enzyme.* Steinkopf, Dresden, 1932.

[33] K.A. Hasselbalch. Die Berechnung der Wasserstoffzahl des Blutes aus der freien und gebundenen Kohlensäure desselben und die Sauerstoffbindung des Blutes als Funktion der Wasserstoffzahl. *Biochem. Z.*, 78:112–144, 1916.

[34] G. Heinzel. Beliebig genau: Moderne Runge-Kutta Verfahren zur Lösung von Differentialgleichungen. *c't*, 8(8):172–185, Aug 1992.

[35] V. Henri. Theorie generale de l'action de quelques diastases. *Comptes rendues l'Academie Des Sci.*, 135:916–919, 1902.

[36] A.V. Hill. The possible effects of the aggregation of the molecules of haemoglobin on its dissociation curves. *J. Physiol.*, 40:190, 1910.

[37] B.K. Ho, A. Thomas, and R. Brasseur. Revisiting the Ramachandran plot: Hard-sphere repulsion, electrostatics and H-bonding in the α-helix. *Protein Sci.*, 12:2508–2522, 2003.

[38] J. Horwitz. Alpha-crystallin. *Exp. Eye Res.*, 76:145–153, 2003.

[39] Howitz et al. Small molecule activators of sirtuins extend saccharomyces cerevisiae lifespan. *Nature*, 425:191–196, 2003.

[40] B. Hörnlimann, D. Riesner, and H. Kretzschmar. *Prionen und Prionkrankheiten.* de Gruyter, Berlin, New York, 2001.

[41] F. Jacob and J.Monod. Genetic regulatory mechanisms in the synthesis of proteins. *J. Mol. Biol.*, 3:318–356, 1961.

[42] C.A. Janeway et al. *Immunobiology: The Immune System in Health and Disease.* Garland Publishing, New York, NY, 2001.

[43] D.E. Koshland. Enzyme flexibility and enzyme action. *J. Cell. Comp. Physiol.*, 54(Suppl. 1):245, 1959.

[44] D.E. Koshland, G. Nemethy, and D. Filmer. Comparison of experimental binding data and theoretical models in proteins containing subunits. *Biochemistry*, 4:365–385, 1966.

[45] S.A. Kuby. *A Study on Enzymes.* CRC Press, Boca Raton, 1991.

[46] U.K. Laemli. Cleavage of structural proteins during the assembly of the head of bacteriophage T4. *Nature*, 227:680–685, 1970.

[47] H.J. Lee and I.B. Wilson. Enzymic parameters: Measurement of V and K_m. *Biochim. Biophys. Acta*, 242(3):519–522, 1971.

[48] H. Lodish et al. *Molecular Cell Biology.* W.H. Freeman and Company, New York, fourth edition, 2000.

[49] B.W. Matthews. Solvent content of protein crystals. *J. Mol. Biol*, 33: 491–497, 1968.

[50] A. McPherson. *Crystallization of Biological Macromolecules.* Cold Spring Harbor Laboratory Press, Cold Spring Harbor, NY, 1999.

[51] P.S. McPherson et al. A presynaptic inositol-5-phosphatase. *Nature*, 379: 353–357, 1996.

[52] O. Medalia, I. Weber, A.S. Frangakis, D. Nicastro, G. Gerisch, and W. Baumeister. Macromolecular architecture in eukariotic cells visualized by cryoelectron tomography. *Science*, 298:1209–1213, 2002.

[53] L. Michaelis and M.L. Menten. Die Kinetik der Invertin-Wirkung. *Biochem. Z.*, 49:333–369, 1913.

[54] J. Monod, J. Wyman, and J.P. Changeux. On the nature of allosteric transitions: a plausible model. *J. Mol. Biol.*, 12:88–118, 1965.

[55] P. Nash et al. Multisite phosphorylation of a CDK inhibitor sets a threshold for the onset of DNA replication. *Nature*, 414:514–521, 2001.

[56] D.L. Nelson and M.M. Cox. *Lehninger Principles of Biochemistry.* Worth Publishers, New York, third edition, 2000.

[57] J.E. Nielsen and J.A. McCammon. Calculating pKa values in enzyme active sites. *Protein Sci.*, 12:1894–1901, 2003.

[58] C.J. Noren, J. Wang, and F.B. Perler. Disecting the chemistry of protein splicing and its application. *Angew. Chem. Int. Ed. Engl.*, 39:450–466, 2000.

[59] L. Ornstein. Disk electrophoresis: I. Background and theory. *Ann. N. Y. Acad. Sci.*, 121:321–351, 1962.

[60] A.J. Parodi. Protein glycosylation and its role in protein folding. *Annu. Rev. Biochem.*, 69:69–63, 2000.

[61] H. Paulus. Protein splicing and related forms of protein autoprocessing. *Annu. Rev. Biochem.*, 69:447–496, 2000.

[62] M.F. Princiotta et al. Quantitating protein synthesis, degradation, and endogenous antigen processing. *Immunity*, 18:343–354, 2003.

[63] S.B. Prusiner. Prions. *Proc. Natl. Acad. Sci. USA*, 95:13363–13383, 1998.

[64] G.N. Ramachandran and V. Sasisekharan. Conformation of polypeptides and proteins. *Adv. Protein Chem.*, 23:283–437, 1968.

[65] H. Rattle. *An NMR Primer for Life Scientists.* Partnership Press, Fareham (GB), 1995.

[66] J.S. Richardson. The anatomy and taxonomy of protein structures. *Adv. Protein Chem.*, 34:167–339, 1981.

[67] J.L. Rigaud. Membrane proteins: functional and structural studies using reconstituted proteoliposomes and 2-D crystals. *Braz. J. Med. Biol. Res.*, 35:753–766, 2002.

[68] D.J. Roberts et al. Rapid switching to multiple antigenic and adhesive phenotypes in malaria. *Nature*, 352:595–600, 1992.

[69] M. Salemi and A.M. Vandamme. *The Phylogenetics Handbook: A Practical Approach to DNA and Protein Phylogeny.* Cambridge University Press, Cambridge, 2003.

[70] I.H. Segel. *Enzyme Kinetics.* Wiley, New York, 1975, reprinted 1993.

[71] P.H.A. Sneath and R.R. Sokal. *The principles and practice of numerical taxonomy.* Freeman, San Francisco, 1973.

[72] S.P.L. Sørensen. Enzymstudien II. Über die Messung und Bedeutung der Wasserstoffionenkonzentration bei enzymatischen Prozessen. *Biochem. Z.*, 21:131–304, 1909.

[73] J. Sumner. The isolation and crystallization of the enzyme urease. *J. Biol. Chem.*, 69:435–441, 1926.

[74] L. Tagliavacca, T. Anelli, C. Fagioli, A. Mezghrani, E. Ruffato, and R. Sitia. The making of a professional secretory cell: Architectural and

functional changes in ER during B lymphocyte plasma cell differentiation. *Biol. Chem.*, 384:1273–1277, 2003.

[75] F. Takagi, N. Koga, and S. Takada. How protein thermodynamics and folding mechanism are altered by the chaperonin cage: Molecular simulations. *Proc. Natl. Acad. Sci. USA*, 100:11367–11372, 2003.

[76] H.A. Tissenbaum and L. Guarente. Increased dosage of sir-2 gene extends lifespan in *Caenorabditis elegans*. *Nature*, 410:227–230, 2001.

[77] J.J. Tyson et al. Chemical kinetic theory: understanding cell-cycle regulation. *Trends Biochem. Sci.*, 21:89–96, 1996.

[78] J.C. Venter et al. The sequence of the human genome. *Science*, 291: 1304–1351, 2001.

[79] S. Walter and J. Buchner. Molecular chaperones — cellular machines for protein folding. *Angew. Chemie Int. Ed.*, 41:1098–1113, 2002.

[80] O. Warburg and W. Christian. Pyridin, der wasserstoffübertragende Bestandteil von Gährungsfermenten. *Biochem. Z.*, 287:291–328, 1936.

[81] P. Wentworth, Jr. et al. Evidence for antibody-catalyzed ozone formation in bacterial killing and inflammation. *Science*, 298:2195–2199, 2002.

[82] W.C. Winkler, A. Nahvi, A. Roth, J.A. Collins, and R.H. Breaker. Control of gene expression by a natural metabolite-responsive ribozyme. *Nature*, 428:281–286, 2004.

[83] G. Zandomeneghi, M.R.H. Krebs, M.G. McCammon, and M. Fändrich. FTIR reveals structural differences between native β-sheet proteins and amyloid fibrils. *Protein Science*, 13:3314–3321, 2004.

[84] K.L. Murray. Vertical transmission of variant CJD. *J. Neurology, Neurosurgery and Psychiatry*, 76:1318, 2005.

Index

Printed by Publishers' Graphics LLC
MO20120814